Two Paths Toward Sustainable Forests

Public Values in Canada and the United States

Two Paths Toward Sustainable Forests

Public Values in Canada and the United States

Edited by

Bruce A. Shindler
Thomas M. Beckley
Mary Carmel Finley

Oregon State University Press
Corvallis

The paper in this book meets the guidelines for permanence and durability of the Committee on Production Guidelines for Book Longevity of the Council on Library Resources and the minimum requirements of the American National Standard for Permanence of Paper for Printed Library Materials Z39.48-1984.

Library of Congress Cataloging-in-Publication Data
Two paths toward sustainable forests : public values in Canada and the United States / edited by Bruce A. Shindler, Thomas M. Beckley, Mary Carmel Finley.
p. cm.
Includes bibliographical references (p.).
ISBN 0-87071-561-5 (alk. paper)
1. Sustainable forestry—Social aspects—Canada. 2. Sustainable forestry—Social aspects—United States. 3. Sustainable forestry—Economic aspects—Canada. 4. Sustainable forestry—Economic aspects—United States. 5. Forest policy—Canada. 6. Forest policy—United States. I. Shindler, Bruce A. II. Beckley, Thomas M. III. Finley, Mary Carmel. IV. Title.
SD145.T86 2003
333.75'097—dc22

2003017128

Printed in the United States of America

Oregon State University Press
101 Waldo Hall
Corvallis OR 97331-6407
541-737-3166 • fax 541-737-3170
http://oregonstate.edu/dept/press

Contents

Acknowledgements

As with any book, there are a number of people whom we need to thank and without whom this volume may never have come to fruition. Some of these people were "just doing their job," and may feel that no thanks are required (though we disagree). Others deserve our thanks for more personal support and sacrifices.

Initially we wish to thank the Canadian Embassy for its seed investment in the concept of a book extending across our two borders. Embassy administrators in Washington, D.C. and the Canadian Consulate in Seattle were particularly supportive of our efforts. Several granting agencies also contributed by sponsoring the research that resulted in many of the book chapters and generally tolerated being put under the scrutiny of social sciences. Among these are the Model Forest Network and the Sustainable Forest Management Network in Canada and the USDA Forest Service and USDI Bureau of Land Management in the United States. As well, our respective departments, faculties, and universities offered significant logistical support.

Don Field and Rabel Burdge deserve recognition for having the vision to organize the International Symposium for Society and Resource Management. That forum has consistently brought together researchers interested in resource management from across a wide array of social science disciplines. The idea for this volume was launched at the 2000 ISSRM meeting in Bellingham, Washington, and the diversity of its contributors is a testament to the broad and inclusive research community that the ISSRM has spawned.

At the outset we were three university academics new to the editorial game. As a result, we relied heavily on the wisdom, expertise, and professionalism of many staff and affiliated employees at the OSU Press. Among these were Mary Braun, Jo Alexander, Paul Merchant, and Tom Booth. Their collective guidance enlightened us and made our task much easier. We also would like to thank the contributors to this volume as well as those whose chapters were cut for lack of space. Our colleagues demonstrated great patience and understanding throughout the publishing process. Finally, the close scrutiny and keen insights of our anonymous reviewers have significantly raised the quality of this work. We are grateful to you all.

Bruce Shindler
Thomas Beckley
Mary Carmel Finley

Part One

Setting the Stage

In 1850, British captain Richard Hinderwell stopped at Olympic Peninsula's Discovery Bay to cut trees for masts and spars. One of the biggest problems confronting his ship's carpenter, William Bolton, was in finding Douglas-fir small enough to fit the specifications of two-foot-thick butts. It took the ship's crew and a band of hired Clallam Indians four months to cut seventeen spars. The trees were so dense it was difficult to find room to let them fall over.[1]

It was easy for Captain Hinderwell to define his forest values. He needed masts and spars so his ship could return to England. But for North Americans at the beginning of the twenty-first century, defining our collective forest values is considerably more difficult. Forests are not only the source for the equivalents of today's masts and spars, but they provide many other things. Forests are home to wildlife and legally protected endangered species. They supply critical environmental services of cleaning our water and our air. They are important places for recreation and spiritual places of solitude and beauty. Over time, we have steadily expanded the different commodities and amenities we want forests to provide. Economists now calculate forest carbon sequestration as a means to reduce greenhouse gas concentrations and to slow global warming. Researchers look for sources of new and important drugs. As we continue to harvest, replant, and conserve the forests of North America, we find we are engaged in a fundamental reconceptualization of what we want them to be. This book is about the changing forest values in North America and the challenges we face in developing strategies that incorporate these evolving values into management, practice, and policy.

Canada and the United States share many things. They share similar ecological regions, a common border, similar history, and a political commitment to democratic institutions and values. They also share comparable conflicts over how to manage public forests. Despite differences in institutions and political governance, both countries are seeking a path toward goals that are principally the same—sustainability of their forests and the communities that depend upon them.

We are in the midst of a shift away from the long-held, traditional forms of managing forests to one that recognizes the ecological and social complexity of these systems. As more people believe forests should be managed for a range of values, federal and provincial forest agencies are moving toward more holistic approaches. The Canadian form is called "sustainable forest management" (SFM) while the United States is following a model called "ecosystem management" (EM). These terms have been used somewhat interchangeably across our borders, largely because they reach for similar goals: sustaining healthy forests, providing for some appropriate level of timber extraction, attention to the contextual characteristics of landscapes, recognition of local communities (ecological and human), and an expanded role for citizens in decision-making. Both SFM and EM embrace the philosophy that natural resources should be managed on the basis of science as well as social and political factors. They are grounded in the premise that resource management is a largely political process, and that solutions must reach for ecological and socioeconomic sustainability.

One problem often obscured in discussions of sustainability is clearly identifying what forest components we might choose to sustain.[2] An equally important obstacle is reaching agreement about these choices. There has always been a large gap between what we know about the biological components of forests and how we understand the sociopolitical interactions that determine how we will manage them. Certainly economists have provided us with methods to quantify timber values and tell us at what point harvesting is most efficient. But fiber is no longer the only forest value that matters. Indeed, in many places it is no longer even the most important. And while assessment tools for valuing attributes such as aesthetics and recreational uses are advancing, we are still at an early developmental stage in our ability to quantify other forest values such as spiritual benefits, biodiversity, carbon storage, or contributions to climate stability. Despite the difficulty in measuring less conventional values, they play a central role in the political arena because they represent a large share of what people are attempting to sustain in forests.[3] We have arrived at a point where the politics of forestry has grown particularly contentious in both Canada and the United States, fueled by public distrust in our government institutions and an increasing demand that citizens have an expanded role in decision-making.

The need to sort and make sense of public expectations and institutional responses is one of the challenges of sustaining forest ecosystems. *Two Paths Toward Sustainable Forests* examines the social dynamics of ecosystem-based forest management in Canada and the United States. The team of authors includes sociologists, foresters, economists, political scientists, and geographers, as well as scholars in recreation and tourism. Many have lived and worked in both countries, and this volume represents their collective observations and assessments. Our approach is to use survey evidence from forest communities and the general public in both nations to examine management strategies, analyze trends, expose trouble spots, review policies, evaluate innovation, and to explore a course for the future. Just where these two paths through the forest will finally emerge is uncertain. But as anyone who has explored a forest path knows, it is the journey that is most interesting.

1

Seeking Sustainable Forests in North America

Bruce A. Shindler, Thomas M. Beckley, and Carmel Finley

Introduction

The primary purpose of this book is to examine societal influences on U.S. and Canadian policies as both countries seek to manage public forests on a sustainable basis. We examine in detail public values for forests and public opinion regarding how forests ought to be managed. Other recent works have documented changes in policy outcomes,[4] or offered specific advice to resource management agencies.[5] Our contribution focuses more on the values underlying policy change, but also on alternative pathways to reform forest policies in contrast to a path of incremental change in regulatory systems. What we ultimately argue is that more fundamental institutional change is required to address public demands for management that provides a broader spectrum of forest values.

Our common interest in this theme brought many of the authors represented here to Bellingham, Washington, in 2000. We met as part of a larger conference on society and natural resource management to discuss the interesting parallels in how our two countries were approaching sustainable forestry.[6] Most of us had been investigating the questions surrounding sustainability and ecosystem management throughout the 1990s. We quickly agreed on the need for a book that examined the challenges of sustaining forests, public expectations about their use, and how management institutions respond to these shifting societal demands. Such scrutiny is both important and timely. Although considerable effort has been made to define the biological functions of forests, as well as to measure their value to society as commodities, relatively little investment has been made in understanding the public's views of sustainability and the conditions under which people might come to support ecosystem-based strategies. In the following chapters we examine citizens' judgments about forest practices and management institutions. Collectively, our objective is to penetrate the vagueness of the sustainability concept in ways that reflect its potential and limitations.

The second purpose of this book is to benefit from the decade-long U.S. and Canadian experiment with policies and institutional innovation in pursuit of sustainable forestry. Authors take advantage of these parallel management approaches as a useful method for examining experiences. However, this volume is not a comparative analysis of the two management models. There simply has not been enough sociopolitical study to compare U.S. and Canadian attempts at sustainable forest management. Although the concept of sustainable forestry has been with us for years, its specific application as a more holistic ecosystem-based approach is little more than a decade old. Under the best of circumstances, this time frame would barely be sufficient to evaluate the extent to which innovations have been "successful" and if people have adopted this type of policy experiment.[7] For example, we are just beginning to find out how (or how well) people weigh the costs and benefits of specific forest management problems and evaluate the potential solutions. Up to now, social scientists commonly have used opinion polls to ask people about their preferences for managing resources, and from these have characterized public attitudes and values. But this is little more than asking people what they want, not if they understand the issues or if they recognize the trade-offs involved. Of course, one reason research has been slow to develop in both the United States and Canada is the lack of places where citizens can see ecosystem management practices playing out on forest landscapes. This in turn translates to a lack of opportunities for citizens and resource professionals to examine together cause and effect relationships.

Notwithstanding the shortage of experimentation and experiences, we are two nations with a common problem. Our forest management institutions are attempting to develop long-term policies that are both sustainable and supported by citizens. This volume is a response by social scientists who have observed and evaluated the social context within which forestry professionals operate today. These assessments have been organized into four parts, reflecting a logical progression from problem identification to future perspectives. The first section frames the sustainability challenge and provides context and background for the reader on forest history and recent developments in forest policy. Chapters in the second section demonstrate new societal demands on forest ecosystems and management agencies by using case studies of forest communities. These studies include both communities of place (i.e., forest-dependent communities) and communities of interest (stakeholder groups with shared values).

The third section examines agency responses to public demands for a range of objectives from forests and describes the key challenges that lie ahead. Chapters in the final section offer in-depth analysis of some of the most pressing problems for the near future. Authors critically examine the certification of forest products, globalization and trade, and our need for more inspiring and effective management institutions. On any long journey, one inevitably gets asked, "Are we there yet?" In the closing chapter we summarize our contributors' collective judgment on this question, and conclude that we have not yet arrived at sustainable ecosystem-based management on a broad scale, despite contrary claims by the Canadian federal government.[8] However, we review recent evidence that suggests our two countries are headed in the right direction and we describe what more is needed from our respective citizenry and research and management institutions.

Many of these chapters are based on empirical research such as the U.S. national opinion study detailed by Steel and Weber and a provincial perspective of public values in Alberta by McFarlane and colleagues. These large-scale analyses help us to see patterns among populations, anticipate shifts in resource use, and identify general support for sustainable forestry and ecosystem management. Other chapters have a more specific focus to help us learn from localized experimentation. As is often the case with local studies, we are able to see common themes emerge among communities and across forest settings.

During the creation of this volume, one colleague asked if we were describing a crisis in North American forestry and, if so, whether it is a crisis of science, of bureaucracy, or of democracy. Our response was "yes, yes, and yes." The relevant crisis in science is the growing lack of consensus among scientists and forestry professionals that has been widely publicized in the media. This is not limited to forest science; it includes medical science (stem cell research), agricultural science (genetically modified foods), and other disciplines. The forest management bureaucracy attempts to base its management upon sound science, but when there is no consensus from the scientific community about what to do, this can create a crisis in bureaucracy. The relationship between science-based decision-making and democratic decision-making is tenuous at best. As society increasingly recognizes that our decisions about forest use are value decisions (the scientific data do not tell us what is the *right* thing to do) we verge on a crisis in democracy.

One result is that our institutional framework does a poor job of interpreting public values and incorporating them into forest management. More traditional foresters tend to say that we just need to do a better job of "educating the public."[9] But policy-makers do not need to manage the public better; rather they need to allow their actions and decisions to be grounded more by public values. And the public increasingly says that the most important role for forests is in maintaining environmental quality, biodiversity, and ecological integrity, not as a driver of the economy. It is apparent to us that institutional change is lagging behind the recognized need for change, and this only exacerbates the growing degree of citizen distrust of public agencies. A variety of perspectives are represented in this volume, offering different ideas about the degree of salience of these challenges and proposed solutions.

Two Paths: Canada and the United States

Forests have been one of the defining elements of North America. Trade in forest products links Canada and the United States to one another and to the world. Canada is also the world's largest exporter of forest products. The United States is not only the world's largest producer of wood products; it is the largest consumer as well. However, it is for the non-timber benefits of forests that the world is increasingly becoming concerned. Canada contains 9 percent of the Earth's usable water but only .005 percent of its human population.[10] One-quarter of the world's remaining, large, intact frontier forests are in Canada,[11] a fact that has global significance for regulating climate, air and water quality, as well as providing habitat for animal, plant, bird, and fish populations. The U.S. system of national parks, wildlife preserves, and wilderness areas has been a model for other nations. Forestry has historically been a cornerstone of the economic engine in both Canada and the United States. At the same time British Columbia and the U.S. Pacific Northwest are renowned for their temperate rainforests, which many translate to mean "really big trees" and the ecological conditions they associate with old-growth forests. Each different set of values can evoke deep responses. Over the past decade the debate over how forests should be managed has become a noisy one, driven by our dependence on various wood products and forest amenities and expressed in the loud, powerful language of ideological opinions. In 1993 the largest demonstration of civil disobedience in Canadian history was over the forests of Clayoquot Sound, where 840 people were arrested for standing in the way of logging crews. That same year U.S. President Clinton personally convened a Forest Conference in Portland, Oregon, to break political tension and gridlock over management of federal forestlands. He challenged scientists, managers, and citizens to work together to find solutions that were both ecologically sound and addressed the interests of a range of constituencies. At the heart of this similar debate in our two countries are conflicting values and priorities about what is "right" for the environment and the degree to which we can, or should, manipulate forest systems.

At this stage it is not clear if the paths toward sustainable forestry taken by Canada and the United States will cross and merge or remain distinctly separate. Although the two countries share a common cultural heritage and democratic tradition in the British Empire and the American Revolution, historical commentators frequently point out differences. Canada is often seen as having a deferential culture and a tendency to opt for collective policies, such as a more comprehensive social safety net. In contrast, the United States is viewed as being individualistic and entrepreneurial.[12] The "similar but different" theme is important especially in how the two countries perceive each other's progress and the likelihood that one system may influence the other. For example, Lipset[13] provided the metaphor of two trains moving down parallel tracks, always remaining different even though they are moving in the same direction. Alternatively, others who fear Canada's submersion into a monolithic American culture call for maintaining Canada's cultural and political independence from the "Goliath to the South."[14] The perceived danger, of course, is that encroaching social values from the United States may lead to disintegration of Canadian distinctiveness. Or that due to imbalance in political power, the United States may simply exert its will and values on Canada, as it has attempted to do for over two decades regarding tenure reform through the softwood lumber dispute.[15]

Differences

Irrespective of these sentiments, there are a number of important structural and institutional differences that influence forest management in both countries. The most obvious is population and forest area. The

United States has 285 million people and approximately 750 million acres of forestland. Canada's population is only 31 million with over one billion acres of forest area. As the world's largest exporter of forest products Canada is more sensitive to the needs and demands of the international market. Although the eastern United States has relied on Canadian forests for a good share of its newsprint, America has traditionally been more sheltered from international pressure because it could otherwise supply its own market. But this situation is changing. For example, environmental regulation and forest health concerns (particularly in the west) have curtailed harvesting on U.S. federal forests and resulted in the need to import more forest products—often cedar, fir, and spruce from Canada—to accommodate American appetites for wood building materials.

Another key structural difference between the two nations is that 95 percent of Canada's forest is public land (referred to as Crown land), while only 27 percent of U.S. forestland is under government tenure. Both nations have a greater preponderance of private land in the east, where European settlement occurred first. Institutional jurisdictions are different as well. Despite the birth of the United States as an alliance of independent states and Canada's origins as a centralized administrative unit, Crown land is now administered by the provinces, while the vast majority of public land in the United States is under federal management. Differences are not limited to formal management institutions; policy styles are also distinct. For example, the legal system plays a much greater role for forestry issues in the United States, in contrast to the informal policy network in Canada that links government with forest industry.[16] In evidence, U.S. courts have held a prominent position in deciding environmental disputes for at least two decades.

The different styles are also reflected in how each country historically has dealt with ownership and use of aboriginal lands. In the United States, formal treaties were made between Native American tribes and the federal government. Largely irrespective of the centuries of mistreatment heaped upon Native Americans, these treaties articulated legal arrangements for ownership of tribal lands and became law. In Canada, early treaties with First Nations people were "peace and friendship treaties" that did not cede land or rights. Over time, Canada has clarified issues by further negotiating title to lands and rights for the use of natural resources; however, this has been slow to occur in the western provinces (essentially British Columbia and Alberta). Recent court decisions, most notably a case known as Delgamuukw, confirm that aboriginal rights continue to exist and are protected under Canada's constitution (still, this legislation did not fully define the scope of title for any particular First Nation in British Columbia).[17] Over all, these agreements specified smaller reserves than in the United States, but also left Canada with an uncertain legacy regarding aboriginal rights off reserve.

The fundamental issue in each country involves legal process, not only who has right to which property, but also how far rights should extend to the use of natural resources. Through Delgamuukw and similar cases, courts in each country have slowly spelled out that federal and provincial governments have an obligation to ensure that resource management decisions do not infringe upon historical tribal rights for the use of their traditional resource base. Now with the move toward sustainable forestry and ecosystem management, governments are recognizing opportunities to work more closely with First Nations and Native Americans (as several authors describe in later chapters). But while there is evidence that tribal and government forest managers are working together—most notably in fire suppression activities in the United States and planning for model forests in Canada—prevailing laws and forest boundaries suggest that separatist policies will continue to be the norm.

Similarities

Despite the differences between our two nations, there are many similarities and thus reasons why Canadians and Americans would be wise to pay attention to forestry change and innovation in their respective "backyards." First, there may be no two nations on the planet more culturally similar, yet more distinct from the rest of the world. Western expansion, struggles with Native rights, social movements for women's suffrage, the environment, the experience of the Great Depression, and other historic events affected both nations on similar timelines. In his popular *Nine Nations of North America*, Joel Garreau[18] described important congruities within regions that span Canada and the United States. When he redrew the map of North America to reflect economic linkages and cultural similarities he arrived at the nine nations of his title. No fewer than five of these transcend the current Canada—U.S. border. In many respects we concur with this assessment. In our experience, for example, Maine and New Brunswick are more similar to one another in terms of their forest endowment, historical use of the resource, land tenure, and values than they are to Washington state and British Columbia respectively. Garreau's version of North America is shaped in no small degree by the regional differences in ecology that transcend national boundaries. Eastern Canada's boreal forests are much different from British Columbia's more temperate forests, just as eastern U.S. forests are different from those in the South or the Pacific Northwest.

There are also few examples of two nations more economically intertwined. Extremely strong links between the two nations have been forged over time through the exchange of large numbers of people, trade, and shared political alliances. Tens of thousands of Loyalists from all thirteen Colonies headed to what was to become Canada after the American Revolutionary War. The expulsion of Acadians from the Maritimes sent thousands to the United States, while New England's mills drew additional tens of thousands of Quebeckers south in the 1900s. More recently, young Americans headed north to Canada to evade the draft in the 1960s and 1970s. Today, although the terrorist activities of 9-11 and the war with Iraq have contributed to a slowdown in the U.S. economy, the Canadian media continue to report on the "brain drain" of well-educated Canadians to the United States, drawn by more rapid economic growth and the lure of strong American dollars. The emergence of shared postindustrial values in the two countries is apparent in the merging of media markets, increasingly integrated economies, growing alliances in the entertainment field including professional sports, and the commonality of issues ranking high on policy agendas.[19]

Of course, comparisons go beyond ecological and economic factors. Sociopolitical factors, such as the level of activism over forests, are often more intense these days in British Columbia, Oregon, and Washington than in New Brunswick and Alabama. In the west, people tend to get extremely excited about clearcuts and the continent's remaining big trees. However, the current cross-boundary deliberations in forestry are more about issues such as sustainable harvest levels, forest health, or whether forest communities should convert to recreation and tourism as a means of retaining economic stability. Confrontational forest politics are likely to expand to places like New Brunswick and Alabama despite their lack of big trees or their political legacies that strongly support industrial uses of forests. Even the long-held Canadian tradition of demonstrating greater trust toward political and social institutions is beginning to erode and look more like the sentiments of its southern counterpart. Forest advocacy coalitions have developed in Canada resembling those in the United States. These have initiated contentious debates over issues of sustainability and the role of citizens in environmental policy.[20]

This unique combination of a shared history and shared values, but different political institutions and forest management regimes, suggests that Canada and the United States have much to learn from one another. Indeed, former Chief of the U.S. Forest Service, Jack Ward Thomas, recently offered a similar view. He observed that the same patterns of forest management emerge in both countries, but for some unknown reason, they frequently first emerge in the United States.[21] He pointed out that changes in the last two decades have been dramatic—many in the United States have been costly and vigorously disputed—and suggested that Canadian foresters may want to keep a watchful eye on developments to the south. Thus the most applicable cross-fertilization of ideas in forest management is likely to come from right across the border rather than from Sweden, Costa Rica or India. Will our two paths converge? Americans are renowned for their preference for "made at home solutions," and Canada is famous for its love-hate relationship with its southern neighbor. However, the stakes are too high to maintain trivial differences and traditional rivalries. As a result of our shared forest history, we have come to a similar crisis point at roughly the same time.

Challenges of Forest Sustainability

The numerous challenges associated with sustainable forestry typically begin with a list of ecological considerations (covered in depth in other studies), but many social challenges are of equal importance. Notable among them is the point mentioned previously regarding how our bureaucratic institutions do a poor job of tracking public values and incorporating them into decisions about forest management. In this section we highlight several others.

Coming to Terms with Sustainability

One of the most common (and observable) difficulties in both Canada and the United States is the language of sustainability and ecosystem management. Many people use the terms and language of sustainable forestry without clear reference to what it means (to them or others). For example, environmentalists may think about sustainability as an absence of clearcuts, while industry personnel may believe sustainability is sustained-yield fiber farming. Official definitions do not help a great deal except to supply the broadest of contexts. For instance, The Society of American Foresters' *Dictionary of Forestry* tells us sustainable forestry is the "practice of meeting the forest resource needs and values of the present without compromising the similar capability of future generations."[22]

Governments have not helped much either, whether in defining the specific components of sustainable forestry or, more important, in implementing recognizable forms of sustainable management. Agency critics have made both points frequently.[23] In the United States, for example, the Forest Service spent a good part of the 1990s selling ecosystem management to the public as a way to "have it all" from forests; however, it did not spend much time explaining that this did not mean they could "have it everywhere."[24] Certainly one solution is for agencies to define these ideas more clearly, but creating a proper forum for these discussions—where managers and citizens together can reach agreement on common objectives—is no small detail, a problem several authors describe in this volume.

Whatever different parties say they want from forests, very real problems appear as soon as the ideals of sustainability meet the real world of forest resource extraction. In both Canada and the United States, the dominant policy for most of the last 125 years involved maximizing and sustaining a flow of fiber products (e.g., dimensional lumber, pulp, paper, fiber board). Traditionally, this policy direction has been represented by government resource agencies, large forest product companies, professional forestry associations, and policy experts, particularly economists.[25] As other players have joined the debate, discussions of sustainability have shifted in emphasis from sustaining products to sustaining ecological function, as well as a broader range of human benefits. The emerging model is one that sustains forests not just for boards, but also for birds, bugs, and beautiful vistas.

Integrating Values

Numerous scholars argue it is this need to integrate both science and human values that is the central challenge facing today's forestry institutions in the United States and Canada.[26] Their premise is that there is an inherent instability to resource policies—for sustainable forestry or otherwise—that do not adequately integrate the concerns of citizens.[27] One need only look at the daily headlines to appreciate the extent to which adverse public judgments can postpone, modify, or revoke a management program, irrespective of the rigor of the underlying science. Rather than accept unpopular decisions, citizens have access to a variety of options to influence forest policy.[28] They can invoke the courts, lobby legislators, attract media attention for their cause or—as is the case in the United States—organize statewide ballot initiatives to change forest practice laws. When the public seeks forums that better reflect their values, their methods often circumvent the authority of management agencies.

Although there may be instability to resource policy developed in the absence of citizen involvement, an equally compelling counter-argument exists: There can also be instability in conducting forest management based entirely on the demands of current citizens. If social values are changing, then the demands and desires of today's population may not suit the values of future generations. Wildlife management policies provide a case in point. In recent years considerable effort has been directed toward protection of game species. But as hunting declines in both the United States and Canada, long-held strategies to maintain high populations of game species are not likely to serve subsequent generations who may be more concerned about biodiversity as a management goal. This is just one example, but it suggests the challenges of sustainability are long-term. The time it takes managed forests in most of North America to mature can span three or four human generations (assuming a generation every twenty years); in northern Canada rotations are considerably longer. Sustaining these systems over time will not be achieved by public referendum or by scientific juntas. Given these timelines, as well as the uncertainty of future forest values, sustainability may really be about maintaining flexibility in our forest systems. Forest management then becomes a mediated process between what is ecologically possible and what is politically feasible.

Such thinking emphasizes the various roles of stakeholders. The growing expectation among citizens that they will have a legitimate role in forest planning is a primary force in how management agencies now conduct themselves. It has been a rocky road for organizations like the U.S. Forest Service, which, until the 1980s, was viewed as something of a superstar among federal agencies.[29] More recently, the organization has struggled. An article in the *New York Times*[30] underscores the gap between the Forest Service's view of its role as government steward of public forestland and the current attitude of the public. The agency organized a public forum in Seattle to ask citizens how it might better serve them as customers. Their answer? "We are not your customers. . . . We are your owners." The failure to heed sociopolitical concerns was a key factor in the Forest Service's shift to an ecosystem management philosophy.[31]

One recent method for reorganizing management approaches is shared by our two governments. Both Canada and the United States established geographically bounded areas at about the same time to experiment with sustainable management models. In the mid-1990s Canada created ten (now eleven) model forests and the United States set aside ten adaptive management areas (AMAs). These sites were established as places to learn "how to do" sustainable forestry and ecosystem management. Although the Canadian system is truly a national experiment (with sites in almost every province, while the AMAs are only in Washington, Oregon and northern California), each was established so that attention could be focused on local forest conditions and relevant social context. In chapter 11, reviewing both models, Ryan notes that these similar programs with similar goals evolved through very different political systems, and that both have been implemented under different conditions of authority and influence.

One could make the case that model forests generally are more accomplished than AMAs[32]—dedicated funding and a clear organizational priority are major reasons—but it is still too early to judge overall success for either program. Canada's model forest program also appears to be more resilient to political shifts. Model forests have continued to be supported over time, whereas in the United States the Bush administration has viewed the AMAs as part of the failed Clinton Northwest Forest Plan. President Bush has promised to overhaul the plan and is considering, for example, "charter forests" where local concerns (generally considered to be economic) can receive greater emphasis.[33]

Sustaining Communities

These efforts point to another important challenge shared by both countries. This is the growing concern about the ability of forest communities to sustain themselves through fluctuating environmental and political conditions. Many forest communities have been affected by economic changes such as technological innovation, market differentiation, and global restructuring. Frequently the results are mill closures, loss of traditional ways of life, a search for substitutes, and a general shift in values among residents (those who remain and those who move in). As Nadeau and colleagues describe in chapter 4, the relationship between a community and its surrounding natural resources goes far beyond economic dependency. The question then becomes why some communities weather the conflicts better than others.

An upshot of the many challenges of sustainability is a different way of thinking about how to reinvent both government and natural resource management. Many now contend that effective environmental programs require complex, collaborative partnerships among diverse government, civic, and business actors at the state and local levels. The trend has many names, among them community-based management, watershed democracy, collaborative conservation, and the watershed movement. Chapter 10, by Weber and Herzog, describes the trend as grass-roots ecosystem management. They argue that public activism is grounded in the belief that mutually beneficial policies for environmental, economic, and community benefits can be sustained. These days, however, the movement is also powered by a widespread loss of faith in government institutions. Any solution will require a greater role for citizens and involve methods (most often at the local level) that can rebuild the public's trust in our forest management agencies.

Seeking Evidence of Progress

As contributors to this book demonstrate, how we integrate the views of multiple publics and organize ourselves to answer the difficult questions ahead is critically important to citizens in Canada and the United States. Our policies and processes in each country may be somewhat different, but we are seeking the same outcome: to create strategies for managing forests that result in a more sustainable future for all citizens. Social scientists—who study human interactions, public opinion, political processes, and the capacity of our forest agencies and communities to respond to change—are well positioned to offer a critical look at our progress, or lack thereof, toward the goal of sustainable forest management.

A guiding question for this collection of essays is whether all the talk about managing differently is adding up to substantive and tangible change in meaningful places—in on-the-ground management units, in forest communities, in the political decision-making arena, and in the minds and hearts of citizens in both countries. And if the answer is *yes*, or even *it depends*, can we then capture the nature of these changes and learn from them? Still other challenges lie ahead, and this volume will likely raise more questions than it will answer. If existing institutional responses such as model forests and adaptive management areas hold some promise for ecosystem-based practices, how do we expand their geographical context or translate positive experiences more broadly? If certification is an effective market-based incentive for adopting sustainable practices, how do we facilitate the orderly expansion of certification processes to more operators and more forests? We speculate on these questions in our concluding chapter, but more conclusive answers will only come with the passage of time. For the moment, we document the state of the nations with respect to their pursuit of sustainable forestry at the turn of the century and their attempts to incorporate changing public values into management. Just as the previous turn of the century was a pivotal time for shaping the practice of forestry in two emerging industrial nations, the entry into a new century will prove to be just as decisive for the future of forestry in North America, if not the world.

2

Forests, Paradigms, and Policies through Ten Centuries

Thomas M. Beckley

Introduction

Scientists, policy makers, and forest managers are increasingly recognizing the dynamic nature of forests. We now understand forests as complex systems that constantly change, even in the absence of human disturbances. Fire, disease, insect infestation, and wind work alone and in combination and dramatically reconfigure forests over time. In North America, the first human inhabitants of the continent often worked with these cyclical changes and were the source of some ecological disturbance themselves. Native Americans and First Nations members practiced shifting cultivation of agricultural crops within forest patches. They practiced controlled burning of grassland and forestland to provide habitat for desired game species. In addition, they relied on forests for fuel, food, medicine, and tools, and they manipulated the forest environment to provide these goods.

In the five hundred years since European contact, longer-term and more broad-scale shifts have occurred on the forest landscape of North America. Vast tracts of land have been transformed from forests to cities, suburbs, agricultural land, transportation corridors and other uses. Human use of forest resources has had dramatic effects on forest type, wildlife habitat, and susceptibility or resilience of forests to natural disturbances.

The intent of this chapter is to provide background and context regarding North America's forests and how humans have used and altered them during the last millennium. Other authors in this volume take up the recent history of policy activity (Duinker et al., chapter 3) and forest use change. Nadeau et al. (chapter 4) describe the history of forest-dependent communities in North America, and policies related to their support and maintenance. Most of the chapters in this volume review current trends and look to the future. The intent of this chapter is to look backward to review the path taken to this place—this particular constellation of values, policies, and our current endowment of forest. The scope of this chapter is broad, and thus omits many of the fascinating details of forest history on this continent. Many of the texts cited as sources for this chapter offer more depth on how our two forest nations developed. I begin with a description of the major forest types of North America and follow with a broad portrait of forest use history over ten centuries.

An Overview of North America's Forests

The landmass of North America contains more forest than any other cover type. Nearly half of Canada is cloaked in forest (418 million hectares, 1.03 billion acres). The United States contains less forest both in raw numbers and proportion. The 300 million hectares of forest in the United States (741 million acres) represents roughly a third of its territory.[1] Various methods are used to classify ecosystems. In Canada, Rowe[2] continues to be a popular classification scheme. Rowe describes eight different broad forest types in Canada. These are Boreal, Subalpine, Montane, Coast, Columbia, Deciduous, Great Lakes-Saint Lawrence, and Acadian. The territory of Canada is dominated by the boreal forest, which extends very nearly from coast to coast, from the Yukon Territory to Newfoundland. Other forest types such as Montane and Subalpine are found in much less abundance.

In the United States, the Forest Service generally describes twenty zones of forested land. These are divided into ten eastern and ten western forest types. The descriptions are based on one, two, or three dominant tree species, such as Spruce-Fir, Oak-Hickory, or Maple-Beech-Birch in the east and Larch, Hemlock-Sitka spruce, or Fir-Spruce in the west. To an ecologist, these classifications are very broad and are only a starting point for more fine-tuned ecosystem classification (for example, in New Brunswick there are seven ecozones for an area of only 7.3 million hectares). Walker describes twelve forest types across North America and goes into detail on the geography, history and ecology of each type.[3] However, for a simple and general description, the five forest bio-regions described by Ricciuti[4] are adequate. Ricciuti divides North America into Boreal, Eastern Mixed Forest, Pacific Coastal Forest, Mountain Forest, and Southeastern Forest. Four of these five forest types exist in both Canada and the United States, with only the Southeastern Forest being unique to one of the two nations.

The Boreal Forest is predominantly a coniferous forest that extends from 43 degrees north in Wisconsin and Maine to the tree line all across the Sub-Arctic. The boreal forest is the largest in North America, and the least disturbed by human intervention. Most of the boreal is located north of major population centers, even in Canada. This forest is characterized by relatively low species diversity in both its flora and fauna relative to other forest types in North America, and much of the boreal has poor soil. Fire has historically been the major source of disturbance in the boreal forest. Insect outbreaks in the east have also played a major role in forest decline and renewal. The patterns of disturbance and renewal from fire and insects results in large areas of even-aged stands of three types—boreal hardwood (trembling aspen, birch, etc.), softwood (spruce, fir and jack pine), and boreal mixed wood.

Eastern Mixed Forest is the second largest type in terms of area. The range of this forest type extends from Louisiana across the southern U.S. uplands and up the eastern seaboard to Maine and southern Nova Scotia. It also extends across to Michigan, Wisconsin, Minnesota, and in a thin band through Manitoba and Saskatchewan into Alberta.[5] The Eastern Mixed Forest contains a dramatic variety of species diversity. Hardwoods such as various species of poplar, maple, birch, oak, and hickory are found in the Eastern Mixed Forest. Pine, spruce, hemlock, and fir are common coniferous species found throughout this forest. The eastern forests are home to the greatest human population densities in North America, and this region has the longest history of human disturbance in the post-European contact period. Despite massive conversions of land to agriculture and pasture during the colonial period and the ravages of wasteful forestry practices throughout the industrial revolution, this forest persists. In fact, in many parts of the east, this forest type is increasing in area. Areas that were once cleared for pasture have grown up in forest once again. The stone walls and cellar holes that dot the forest land of New England are testament to this dual conversion of forest to agriculture land and back to forest once again.[6]

The Southeastern Forest is perhaps the least well known forest type among the general populace of North America. This forest runs in a thin band from east Texas across the coastal plain of the Gulf of Mexico and up the eastern seaboard to Maryland. The upland regions of the southern Appalachians fall into the Eastern Mixed Forest category. Although the region of the Southeastern Forest is relatively small, it contains a diverse set of ecosystems and species. The sub-tropical conditions support a wide diversity of tree species as well. Loblolly pine and short-leaf pine are common. Pine species regenerate and grow quickly. Over the past few decades the southeastern region of the United States has become dominant in the production of softwood species, resulting primarily from the rapid growth rates and short rotation time of pine species.

Mountain Forests extend from the southern Yukon Territory in Canada through interior British Columbia and into the intermountain west. In the United States, Rocky Mountains host mountain forests, and examples of this forest type also exist in large blocks on the smaller mountain ranges that dot the intermountain west as far south as New Mexico. Mountain forests host a range of microclimates based on orientation and elevation. They are also home to a wide variety of species. Softwoods predominate in the Mountain Forests, but birch and trembling aspen are also found. Softwoods include western red cedar, larch, Douglas fir, lodgepole pine, and hemlock. The forest industry in North America developed first from east to west and later from the west coast inland, so the Mountain Forests were among the last to be exploited for industrial purposes in North America. Mountain Forests exist in some of the least densely populated areas of the continent. This fact has also minimized the human disturbance to these forests, at least relative to most other forest types.

The Pacific Coastal Forest stretches in a thin band from Alaska to central California. Heavy rainfall, mild temperatures, and huge, dominating trees characterize this forest. The Pacific Coastal Forests are famous for their giant trees—sequoias up to one hundred meters tall, ten meters in diameter and thousands of years old. Even some of the less dramatic specimens, such as the Sitka spruce, Douglas fir and western red cedar can grow to hundreds of feet tall and would tower over the trees of any other forest type on the continent. In forests where these species dominate, they create a tall canopy under which an incredibly lush and diverse under story exists. The moist Pacific air protects these forests from fire, and the diversity inherent in these forests protects the entire system from large insect outbreaks. The major natural disturbance regimes for this forest are wind, lightning and old age, all of which leave small gaps in the forest rather than large areas that grow up in even aged stands. Because of their massive trees, and their proximity to the coast, these forests attracted early attention from both loggers and conservationists.[7]

Early History of Forest Use in North America

Some westerners maintain the conceit that forest management began only a century ago in North America. In fact, people have been manipulating forest systems to their benefit for thousands of years on this continent. Exactly when native peoples crossed the Bering land bridge and entered North America from Siberia is still a matter of contention. Recent theories and evidence suggest that Native Peoples may have been here longer than was previously believed. According to their own creation stories, they have been here all along. What is clear is that there is over ten thousand years of human experience living in and with the native forests of North America. There were tribes and bands in every forest type listed above. Some could support larger population densities than others, but all were inhabited, used, manipulated, and so in a manner, managed.

Native North Americans enjoyed the luxury of an abundant land. Scarcity existed, but only locally, and most tribes and bands were mobile, at least periodically. For example, the Hurons built stockades surrounding their villages and spent years in one locale, but they would move the entire location of the village once game and firewood was depleted in the local area, and the soil fertility in the immediate area adjacent to the village was played out. Forests were the first peoples' grocery and pharmacy.[8] Thousands of years of experimentation with species from the local environment produced a vast wealth of traditional knowledge. First Peoples demonstrated a remarkable degree of ingenuity. The production of maple syrup by placing heated rocks in birch bark containers full of collected sap is not an intuitive process. There are legends about how the knowledge to make syrup arose, but it is clear that thousands of years of trial and error and experimentation are behind the vast collective forest knowledge and wisdom of First Peoples on the continent. The first ten or twenty thousand years of history of forest use on the continent was characterized by subsistence use of a vast array of goods from roots to bark, to flowers, to animals, to wood, and all manner of other forest goods to provide a livelihood.[9]

The Colonial Period, 1600-1780

For the first two hundred and fifty years following the arrival of Europeans to North America, the primary items of interest to Europeans were non-timber forest products—namely furs. The courtiers of Europe favored all manner of North American furs. It was a sign of status to wear furs from exotic animals from the New World. Later, furs declined somewhat in popularity, but from 1600 to 1800 the pelt of the beaver became highly prized.[10] The exploration, settlement and development of much of Canada was shaped by the beaver trade.[11] Competition for beaver habitat and trading partners drove the English and the French deeper into the Canadian interior. However, the reason for procuring territory at the time was not for settlement, but rather to control the trade of furs.

In the portion of the Colonies that was to become the United States, religious dissidents, adventurers and profiteers arrived with somewhat different objectives. These settlements were more permanent in the minds of their settlers (as opposed to the Hudson's Bay Company and Northwest Company outposts that dotted Canada's interior). And with this idea of permanent settlement, came the idea of "taming the land." To Europeans, this meant clearing the land and converting it to agricultural use. MacCleery suggests that to the colonists, the forests were a mixed blessing. On the one hand, they provided building materials, fuel, and a food source (primarily wild game). Lumber also became a tradable commodity. But on the other hand, they viewed the forest as an impediment to development.[12] Despite the Natives' reliance on the forest for sustenance for thousands of years, in the European worldview the land needed to be stripped of trees and planted to crops to be considered productive and valuable. And the settlers set to this task with a fervor that is still the stuff of legend.

Over time, during the early Colonial Period, timber became an important commodity for the New World. However, this was slow to develop. The first masts were shipped from New England in 1634, but the dreams of some that the region would become a manufacturing powerhouse were slow to be realized. Labor was in chronic undersupply in the Colonies, and life was hard. Many arrived with dreams of setting up manufacturing enterprises, but they became preoccupied for years or even decades just maintaining a semi-subsistence lifestyle.[13] Eventually trade in timber did develop, but initially focused more on the West Indies than on Europe. Settlers in the colonies relied on West Indies cotton for clothing and their most marketable commodity to trade was timber in the form of planks, clapboards and staves.[14]

Despite this incipient trade, most of the forest harvested in the New World during the Colonial Period went to subsistence needs. Whitney reports that in the late eighteenth century, roughly two thirds of the timber harvested in Canada and the United States was for fuel, and most of this for domestic fuel. With inefficient fireplaces as the primary source of heat in most dwellings, between twenty and forty cords of wood were required for a single heating season.[15] In addition, vast quantities of wood were used for fencing. By 1850, there were an estimated 3.2 million miles of wooden fence in the United States alone.[16]

Throughout the Colonial Period, the low population density of European settlers and their relatively crude technological toolkit meant that land clearing had only a marginal impact on the overall landscape. Even at the beginning of the nineteenth century, the land from the Eastern seaboard to the Great Plains was largely characterized by vast tracts of unbroken forest. The advent of a myriad of technological advances and an exponential increase in the rate of immigration changed this in the nineteenth century.

The Industrial Assault on North America's Forest, 1780-1900

The wars of the late eighteenth and early nineteenth centuries settled the dominion of North America. First, France was virtually expelled from the Eastern Seaboard in the Seven Years' War (French and Indian War to the Americans). Not long after, the Revolutionary War in America expelled the British from what is now the United States. The stalemate of the War of 1812 guaranteed that both British and American interests would remain in the New World. Interestingly, these military disputes were of little interest to the commoners, who simply wanted to get on with their cutting, sawing, milling of wood and their planting and harvesting of agricultural crops. However, these conflicts eventually paved the way for a new era in North America. This era was characterized by the rapid expansion of the industrial capacity in the United States and in British North America.

In the United States the population increased by seven times from 1785 to 1850.[17] Canada grew at a somewhat slower rate, and had only 3 million inhabitants compared to the United States' 23.5 million by 1850. Nonetheless, significant demands were placed on Canada's forest resource as well. Demand for wood increased by six times from 1800 to 1850. At this time the primary use of wood was still for cooking and heating. A huge amount of wood was simply wasted. It was viewed as an impediment to agriculture. Trees were girded, felled, and those not used for cooking and heating were burned in the fields. Today, many North Americans express horror and consternation over the wasteful practices of forest conversion in the Amazon Basin. A scant 150 years ago, it was the Europeans who were aghast at the wasteful and unsightly slash and burn practices that were widespread across North America at the time.[18]

Trade in timber became significant during this period.[19] The trade began in the East, in New Brunswick, Maine and Quebec, and worked its way across the continent in the course of one century. There were always competing claims as to what city was the "timber capital" of North America. However, the legitimate claimants to that title (based on exports) increasingly came from further and further west. After New England, it was New York, then Pennsylvania, the Ottawa Valley, and Michigan. The vast pine resources of the Upper Great Lakes States were depleted in a matter of a few decades. That region also suffered the ravages of many catastrophic fires due to great piles of slash and wasted wood left in the wake of logging. The Midwest, originally mostly forested, was well suited to agriculture. Ohio was 96 percent forested at the beginning of the nineteenth century, and only 25 percent forested by the turn of the twentieth century.[20]

In addition to farmland conversion, industry began to exert a greater demand for wood to fuel the industrial revolution. Great Britain converted to coal early in the century to meet its increasing fuel demand. In North America, people turned to resources that were close at hand, abundant, and available with relatively simple technology—namely wood. Ironworks were responsible for consuming between five and six million acres of forestland to supply charcoal to the nineteenth century. This area was small compared to the wood consumed and wasted in the rush to convert land to agriculture, but it resulted in severe local wood shortages, nonetheless.

Ironworks were small consumers compared to the railroads. Both Canada and the United States were built and bound together through their vast transportation networks. During this century waterways were eventually abandoned in favor of the railroad as the means for transporting commodities and getting goods to markets. Steam powered river craft were major consumers of wood in their own right, but by the end of the nineteenth century, railroads were responsible for 25 percent of all wood used in the United States.[21]

While the quest for large, accessible, high quality lumber took loggers and lumbermen further and further west, in their wake new entrepreneurs began to eye the forests that grew following exploitative timber harvests. In the East, lower value, lower quality wood grew up in the aftermath of the two centuries of logging. Unfortunately, the genetic legacy of two centuries of high grading (taking only the best and leaving lower quality stock) was that less robust trees replaced those that were harvested. However, papermaking technology emerged in the latter part of the nineteenth century that made it feasible to produce paper from wood fiber.[22] A whole new crop of forest-dependent communities grew up in areas of the East where there was abundant wood (of any quality) and the potential for hydropower. Many of these communities still exist and have continued their reliance on local forest resources to the present.[23]

The Forest Management Era, 1900-1980

MacCleery presents a rather stark picture of the condition of forests in America in 1900. At that time, wildfires ravaged between twenty to fifty million acres annually; the volume of wood cut surpassed forest growth, wood was still cheap and vast quantities were wasted through inefficient processes in the woods and at mills; conversion to agricultural land continued apace (13.5 square miles per day); and whitetail deer, wild turkey, pronghorn antelope, black bear, moose, bighorn sheep and bison had been eliminated from vast areas of their former range.[24] The situation in Canada was less bleak, but not as a result of good stewardship. Rather, a harsher climate and lower population density meant that there were more areas of wilderness in Canada, but this was due to climate and historical accident, not enlightened policies or practice.

The troubled condition of North America's forests was not lost on a few visionaries. Concern over the state of the continent's forests began during the Industrial Era. The Thoreau-inspired preservationists like John Muir began lobbying for the protection of the Sierra Nevada in the 1860s. The first American Forestry Congress was held in Montreal in 1882, in large part out of a concern over "cut and run" policies of the timber industry and the degraded landscapes and impoverished communities left in their wake. Toward the end of the eighteenth century, pragmatic and utilitarian conservation advocates such as Bernard Fernow and Gifford Pinchot began to lobby for improved forestry practices and regulations to enforce better management. These two pioneers in forestry imported German-style management to North America and helped to establish the institutions that would implement scientific forest management. These included both university schools of forestry and government agencies charged with stewardship of natural resources. These two streams of forest advocacy—preservationists on the one hand, and conservationists on the other—have continued an uneasy co-existence throughout the twentieth century.[25]

The Forest Management Era can be broken down into several sub-eras. In the early part of the century Conservation was the forestry gospel. During this period the U.S. Forest Service was established (1905) with Pinchot at the helm. In Canada, Elihu Stewart was named Superintendent of Forests in 1899, though the Forestry Branch was not officially created until 1906. Stewart and Pinchot consulted with one another during this time. Pinchot was a strong advocate of Canada maintaining a centralized, federal forestry agency.[26] However, in 1930 the federal government of Canada turned over control of public forests to provincial government. Eventually other provinces gained control of their forests as well, resulting in this major difference between forest tenures in Canada and the United States. During that period, schools or faculties of forestry were also established on both sides of the border—some by the same individuals. Bernard Fernow established the forestry school at Cornell University in New York State before moving to Canada to establish the school at the University of Toronto. Scientific forestry required scientifically trained foresters, and schools sprang up in both countries.

In the 1930s, the concept of sustained-yield gained widespread acceptance among the small but growing cadre of professional foresters. In essence the sustained yield policy codified the principles of German forestry brought to North America by Fernow and Pinchot. Nadeau et al. (Chapter 4) describe the origins of this policy in more detail. Interestingly there was an explicitly social welfare component to the sustained yield policy. Conservation *for use* was Pinchot's mantra, and while the preservationists had a significant following, the Great Depression and World War II ensured the predominance of Pinchot's vision for the middle decades of the twentieth century. During these major historical events, preservation seemed too much of a luxury. During the Depression, people needed work more than they needed wilderness, and during the War, the nations' forests were put into production to support the military effort. Sustained yield policies became embodied in law in both Canada (1940s in various provinces)[27] and the United States (1944).[28]

By the early twentieth century professional foresters were succeeding in convincing the public that forests were not simply obstacles to be overcome in the pursuit of agriculture or industry. The idea that a local economy based on forestry did not have to be a temporary stage in regional development began to take hold. Foresters argued that the forest industry could be an economic mainstay in its own right, and that regional development centered on a strong forest industry was a viable option. Once policy makers were convinced of this, sustained yield policies became codified in law.[29]

The post-WWII era was characterized by tremendous economic growth in both Canada and the United States. The decades of restraint and sacrifice that characterized the Great Depression and war years gave way to unbridled consumerism in the late 1940s and 1950s. Along with just about everything else, demand for wood and wood-based products was high. This was especially true for dimensional lumber for construction. A building boom began in the late 1940s as returning war veterans settled back into civilian life. During this period outdoor recreation became a major pastime, and improved road networks and greater access to automobiles meant that more people (including urban people) were enjoying the woods for camping, fishing, hiking, skiing, and other pursuits. While the sustained yield paradigm and policies still held the day, forest managers began to manage for other values as well, such as wildlife and recreation. There was not so much a paradigm shift as an evolution. When new demands for public forests emerged, managers began to take them into account. This process happened unevenly, of course. In forests accessible to the more densely populated regions of both nations, recreational demands arrived more rapidly. There is no question that these demands were increasing. MacCleery reports that recreational visits to national forests in the United States increased from 18 million in 1946 to 93 million in 1960.[30]

The new management style, managing primarily for timber but attempting to satisfy a myriad of other public demands, was known as Multiple Use Management. In the United States, the Multiple Use-Sustained Yield Act was passed in 1960,[31] essentially codifying incremental changes that had been occurring over the previous decade. In Canada, multiple use policies arrived somewhat later. The British Columbia Ministry of Forests redefined its mandate to be consistent with the principles of multiple use in 1978.[32] In New Brunswick, the Crown Lands and Forests Act, planned and discussed since the mid-1970s, was not enacted till 1982.

In association with the widening mandate associated with multiple use, the established forestry schools began to broaden the scope of their curricula. Government departments responsible for forest management experienced similar growth and diversification of their resources, their mandates, and their personnel. The development of technological advances in the post-war era led to a realization of the limits of the forest resource in North America. Old technologies and practices such as river drives and horse logging persisted, even on an industrial scale, into the 1970s. However, in the post-war era, technological advances in felling, skidding and hauling eventually made even the general public aware that our capacity to harvest the forest had finally outstripped its capacity to replenish itself.

This awareness led to new policies, and the preservationist stream of the Conservation movement gained some prominence once again. In 1964 the Wilderness Act was ratified by the U.S. Congress, setting aside permanent roadless areas across the United States. In Canada, once again, due to lower population density and greater abundance of forested land, there was less pressure to delineate protected areas. For example, large-scale industrial development of the forest did not arrive in the boreal forest of northern Prairie Provinces until the late 1980s. The intolerant hardwoods common to this region had little commercial value until pulping technologies finally made these species usable in the mid-1980s. As such, the forest industry in northern prairies up to that point was small and only operated on scattered portions of the total forestland base. Large-scale industrial development, and opponents to large-scale industrial development (and proponents of protected areas), did not emerge in the region until the late 1980s.[33] The social movement to protect the environment (and particularly forest ecosystems) was larger in the United States, and it occurred earlier there, but as in most things related to forests, forest policy, and forest values, Canada finds itself on a similar trajectory but in a slightly different time frame.

During the 1980s, in both nations, the concept and language of sustainability began to gain favor in the forestry community. For a decade or more, various new titles and labels were attached to the new ways of thinking about forests and forestry. For a time, the term "New Forestry" was used in the United States, but this gave way to "Ecosystem Management" (EM). In Canada, in the 1980s and early 1990s, the term "Integrated Resource Management" gained favor, and then was quickly abandoned for the term "Sustainable Forest Management" (SFM).

Critics argue that sustainable forest management and ecosystem management are nice new names but that they have not really resulted in significantly different practices in the forest. The intent of this book is not to judge whether or not these new management regimes result in more ecologically friendly forest practices. Our concern is primarily with the degree to which these new regimes take a different stance with respect to public opinion and public values. Both SFM and EM claim to do a better job of involving the public in forest management. The chapters that follow present evidence on whether meaningful and lasting change in the way government and industry deal with the public has taken place.

Key Similarities and Differences between Canada and U.S. Forests

The intent of this chapter is to provide readers with an overview of the current condition of our forests and the current array of forest policies, institutions, and management paradigms in both Canada and the United States. As previously mentioned, one of the notable similarities between these countries is in the forests themselves. The forests are different in their distribution (e.g. much more Boreal in Canada, more Eastern Mixed in the United States), but four of the five forest types reviewed in this chapter are common to both countries. This has meant that management strategies and technological and scientific developments that are advanced in one nation are often applicable in various forest types of the other.

As is the case with many other aspects of North American history, there is a tremendous amount of similarity in the trajectory of forestry developments in Canada and the United States. The political distinction between the two nations was only established three centuries after the first efforts of Europeans to colonize North America. And despite Canada's historic ties to Great Britain, the values, practices and orientation toward the forest was and is uniquely North American. However, it is more accurate to say that Canada and the United States have had a shared historical trajectory rather than to say they have a shared history. The similar outcomes in policy and practice have more to do with similar pressures, similar environments, and similar immigrant populations with similar goals and expectations in relation to land, land settlement, and land use, than with shared institutions or explicit collaboration in the development of forest policy. There has been a significant exchange of ideas on an informal level between the two countries. Fernow was the first, but by no means the last, forester to begin his professional career in one country, but to finish it in the other.

As noted several times in this chapter, the social, economic and ecological pressures that led to change in forest practice or forest policy were often felt more acutely first in the United States. This was due to greater population density, more rapid industrial development, and the proportionately greater resource base that Canadians enjoy relative to their southern neighbors. In some ways, this places Canada at a tremendous advantage, as Canadians can watch and learn and anticipate future conditions (both in the forest and in the people that use the forest) by observing developments in the United States. Writers from the United States have warned Canadians not to make the same mistakes that the United States made in the past, and they continue to do so.[34] Sometimes such advice is heeded and sometimes not. Canadians and Americans tend to favor "homegrown" solutions, but the fact remains that Canada enjoys somewhat of a time lag relative to changing forest conditions and issues and in some cases has been able to profit by observing change in the United States and anticipating such change in Canada. For example, some areas in Canada were first exploited for timber decades after the concept of sustained yield was fully realized. Thus they were not working with a degraded forest, but were able to implement sustained yield management in a virgin forest.[35]

While Canada and the United States share forest types, forest values, and similar historical trajectories, there are significant differences in a few critical areas. The first of these is land ownership. Because of the extensive explicit policies to distribute land in the United States based largely on Jeffersonian ideals of agrarian democracy, the United States has much less land in the public trust. The state and federal governments own around one third of all land in the United States, but most of this is located in the West. The federal government controls only 3 to 8 percent of the land east of the Mississippi River, and only a few states in that half of the nation own significant portions of land. In Canada, the pattern is similar but the proportions are different. The eastern, Maritime provinces (also settled in the Colonial Era) of New Brunswick (50 percent) , Prince Edward Island (90 percent) and Nova Scotia (72 percent) are predominantly private land—not unlike the eastern states. In the west, Canadian provinces are dominated by public land. Overall, however, 96 percent of Canada's forests are Crown (e.g., publicly owned) land. This is a major difference relative to land holding in the United States.

A second significant difference is the locus of management for the public lands of both nations. In the United States, a number of federal agencies (USDA Forest Service, Bureau of Land Management, and Department of Defense) have management responsibility for public land. A few states have state-owned land or county land that was obtained through tax delinquency after the forests were cut and the timber barons moved on. However, public land of this sort does not amount to much. In Canada, public land was once controlled by the federal government, but in the 1930s responsibility for Crown lands was transferred to the provinces.[36] The provinces thus enjoy the royalties gathered from stumpage on Crown land. Provincial governments work out the long-term lease arrangement with companies that operate within their boundaries, and provinces regulate the same companies through provincial Acts and laws that govern forest management.

The tenure arrangements on public land in Canada are also quite different from those in the United States. In order to attract investment from the forest sector and to encourage investors (Canadian, U.S. based and international) to operate in remote, northern locations, Canadian provincial governments have typically entered into long-term agreements with corporations sometimes on extremely favorable terms (see Beckley and Bonnell, Chapter 9). Often these involve defined geographical areas. In other instances companies are guaranteed quotas or volumes of wood.[37] In the United States, timber from public land is typically sold at auction. Thus no companies have long-term defined rights to land under the public trust. There are advantages and disadvantages to both systems. However, in recent years the issue of the benefits these different systems confer has come to a head. The different tenure systems in the two countries are the basis for the long-standing softwood lumber dispute. The debate is chronicled in more detail in Alavalapati et al. (chapter 14). The gist of the debate is that U.S. policy-makers argue that Canada provides subsidies to its softwood lumber producers through favorable stumpage rates and the provision of infrastructure. Canadian policy-makers argue that it is their right, as a sovereign nation, to set whatever policies they wish, and furthermore that favorable stumpage rates and infrastructure provisions are critical for attracting and keeping the forest industry in Canada.

The only reason for the United States leverage on this issue illustrates the final significant difference between the forests of both nations. Canada's forest industry and forest output are heavily geared toward exports, and the vast majority of those exports are sent to the United States. In 2001, Canada exported over 44 billion dollars worth of forest products, around 80 percent of which went to the United States.[38] Again, the issues of population density, resource abundance, and proximity lead to this situation. The United States consumes most of its own production. This has led to the current conflict over trade in softwood between the two nations, but it also has other implications. For example, as a result of Canada's dependence upon exports, it is much more sensitive to external social pressures, such as the current trend toward certification.

Conclusion

To extend the metaphor of two paths used throughout this volume, it is clear that Canada and the United States have had distinct but similar forest histories and traditions of forest use, forest policy, and institution building. While Canada lags slightly behind the United States in policy innovation, it does so only because it has also lagged behind in the magnitude and intensity of social problems and pressures related to its forests and forest use. The story of forest use, misuse, conservation and management in the two nations is remarkably similar. In both jurisdictions, the view of forests has completely changed in two centuries. At the turn of the nineteenth century, forests were still largely viewed as an impediment to progress. According to the values of the day, progress was defined as agricultural development, industrial development, and westward expansion. In either case, forests were often viewed as standing in the way, and thus were felled and dispensed with over vast areas of North America, especially in the United States and the more populated regions of Canada (southern Quebec and southern Ontario). Because forests were viewed as an obstacle, the prevailing values of European settlers for forests two hundred years ago were primarily negative. In addition to impeding progress, forests were still home to wild beasts, unpredictable original inhabitants of the continent, and any number of imagined dangers.

Contemporary inhabitants of both nations view forests quite differently from their ancestors. Instead of regarding forests as an impediment to progress, many view them as a critical element in the economy. Furthermore, people hold extremely positive views towards forests, whether they view them as entities of inherent value, or objects of value due to their potential to create wealth, jobs, and income. The idea that forests are inherently valuable in their own right is increasingly gaining credence. We are also discovering new things to value in forests, such as biodiversity and carbon sequestration. Clearly, society is diversifying in the values that it holds for forests. First Nations and Native Americans have always maintained a deep regard for the forests of North America. They recognized that forests sustained them physically and spiritually. European settlers to North America have gone through a progression of values toward forests, beginning with the negative, progressing through the utilitarian, and arriving at a recognition of the intrinsic value of forests. More and more the public desires, and in many cases demands, that this wider set of values be recognized in the way that the public and private forests of North America are managed. The mechanisms for doing this are still being debated. Forest management will likely always react to public opinion, and there may always be a gap between public opinion and existing practice, but with sincere effort and experimentation with new methods of public involvement, we may close that gap. The remainder of this volume explores the nexus between public opinion, forest values, forest policy, and the implications for forest management.

3

Sustainable Forestry in Canada and the United States: Developments and Prospects

Peter N. Duinker, Gary Q. Bull, and Bruce Shindler

Introduction

This chapter aims to give readers a glimpse into the key concepts associated with contemporary sustainable forest management (SFM) in Canada and the largely parallel shift to ecosystem management (EM) in the United States. Indeed, it can only be an overview; the make-up of the forest sector in the two countries is far too complex to describe fully in these few pages. Our purpose is to help provide both background and context for the chapters that follow. In keeping with the intent of this volume, we focus on policies and institutions as opposed to field practices, which of course have experienced their own interesting metamorphoses.

We will discuss the time period most relevant to both SFM and EM, essentially the past twelve years—largely the 1990s—during which the current form of sustainable forestry was born and grew to become the prevailing paradigm of forest management in North America. We will also venture into the future, during which a number of important issues will need to be addressed with enthusiasm and energy. We do not deal in any depth with forest management prior to the 1990s (it has been written about extensively elsewhere and summarized by Beckley in the previous chapter), except where we might need to point out how different things are today by comparison with the late 1980s and earlier. We first describe SFM and EM as coming of age in the 1990s and then discuss key formative concepts from each country. The final section of the paper identifies several themes that we believe will challenge forestry professionals across North America as they attempt to implement SFM and EM in the coming years. Again, our purpose in introducing these themes here is to help set the stage for the in-depth deliberation of these ideas by authors throughout this book, most notably our colleagues who bring the discussion full circle in the final two chapters.

SFM and EM Come of Age in the 1990s

Sustainable development as a concept was discussed even before 1987,[1] but the concept was immensely popularised by the World Commission on Environment and Development,[2] often referred to as the Brundtland Commission and, subsequently, the Brundtland Report. The commission issued a clarion call for development that meets the needs of the present human population without compromising those of future generations. Such development, it was declared, needed to put social and ecological issues at the same level as economic ones. Thus came the conception that sustainable development embodies the best integration possible of three co-dependent pillars—economy, society, and environment.

Canadians heard the call and responded swiftly. The first order of business was to structure the discussion to sort out what sustainable development should mean in a Canadian context. Government at both the federal and provincial level quickly established roundtables to discuss the environment and the economy. In 1990, the federal government concocted and implemented a "Green Plan" to demonstrate its commitment to sustainable development. While that Plan has run its course, other initiatives have risen to prominence. One was the establishment of a Commissioner of the Environment and Sustainable Development as part of the Auditor-General's office. Another is the Cabinet-directed requirement that all federal departments should produce sustainable-development strategies to guide their ongoing activities and directions.[3]

As Beckley has described in chapter 2, the concept of national forest policy in Canada needs to be taken loosely. Most of the forest land—roughly 90 percent nationally, but varying widely among provinces—is publicly owned (Crown land), but governed at the provincial level. The federal government has a small amount of its own forestland, mostly associated with military ranges, national parks, and Indian reservations. It is also more directly responsible for forestland in the Territories, but due to small tree size and poor growing conditions only a fraction of these are managed for fibre. The federal government also retains authority for matters dealing with international trade of forest products and forest related foreign-affairs (e.g., international criteria-and-indicators processes), and takes a leadership role in such matters as forestry research. The federal government also wields its influence on provincial forests through its programs of transfer payments (i.e., tax monies collected federally and doled out to the provinces for their internal development). Related to forests, these funding transfers were substantial through the 1970s and 1980s, but have all but disappeared in the 1990s. To summarize, while the federal government directly controls little of Canada's forestland, it has many other roles in sustaining the national forest-sector.

Perhaps the response to SFM at the provincial level came most quickly in British Columbia. Over the last decade, British Columbia's forests have been at the center of conflict[4] among key policy actors: environmentalists, industry, aboriginals, loggers and others. Calls for reform have been numerous[5] and recent.[6] The provincial government responded with policy innovations in an attempt to create a refreshed economic, ecological, social, and legal foundation for SFM. The policies include a comprehensive protected area program, a different relationship with First Nations, and new forest legislation guiding field practices and relationships with communities. For the sake of brevity, many other important policy developments are not described here, but taken altogether they indicate that British Columbia has been on a path of revolution in the forest sector[7].

Governments and forestry professionals in the United States were much slower to respond to the sustainability challenge issued by the Brundtland Commission. Previous to the commission's report, it was primarily biologists and ecologists who were thinking about how to manage large areas of land to conserve biodiversity. During the 1970s and 1980s much of the concern grew out of the need for a management system that could adequately provide for threatened and endangered species. These ideas, which eventually evolved into EM, emphasized maintaining the ecological functions of the landscape rather than producing forest products.[8] But as we mentioned, adoption of the concept in the United States was slow to arrive. EM was controversial among forest managers who had been trained to emphasize the perpetual production of timber[9], and during the remaining boom years of production forestry on public lands (the 1970s and early 1980s), traditional foresters were still in control of the management agenda.

In time, however, and through an accretion of events—the environmental movement, forestry research stemming largely from universities, and admonishment from these scientists that the intent of the Multiple Use-Sustained Yield Act of 1960 was going unmet,[10] together with Congressional action resulting in a series of laws aimed at greater environmental protection—the federal agencies responsible for the nation's forests all adopted an ecosystem approach to natural resource management. The first to announce this decision formally was the Forest Service. In a 1992 white paper Chief F. Dale Robertson decreed that the agency would follow a policy of ecosystem management. Within a year, the three other primary land management agencies—the Bureau of Land Management, the National Park Service, and the Fish and Wildlife Service—independently announced similar decisions. At about the same time, it also became widely recognized that the country's research institutions (e.g., U.S. Geological Survey, National Biological Survey, Environmental Protection Agency) were engaged in the study of how to manage on an ecosystem basis.[11]

In late 1993, President Clinton reconfirmed these decisions with an executive order establishing ecosystem management policies across the federal government. The order directed that EM be phased in using selected demonstration projects placed strategically around the U.S.[12]; eventually, four projects were funded that featured old-growth forests in the Pacific Northwest, the Everglades in South Florida, the urban watershed of the Anacostia River in Maryland and District of Columbia, and Alaska's Prince William Sound that had been damaged by the oil spill from the Exxon *Valdez.* During that same year, a new Chief of the Forest Service was named—Jack Ward Thomas, a wildlife biologist, the first non-traditional forester to be appointed to the position. During his tenure, he reinforced the shift to EM by emphasizing that certain factors be used in forest planning:[13]

- increase the spatial scale used in assessments
- consider longer time frames and the dynamic nature of ecosystems
- frame management within the confines of the range of natural variability
- think across ownership and political boundaries
- consider human needs and interests

While ecosystem management per se was largely a federal initiative in the United States, similar ideas caught on at the state level in varying degrees, particularly as environmentalist attitudes began to permeate public opinion in forest regions. Regulations began to show up in state forest practices acts (especially in western states like California, Oregon, and Washington) that addressed timber management reform on both public and private lands. Included in these reforms were increased protection of wildlife habitat, broadened definitions of riparian systems, and the creation of scenic buffers along state highways.

During the ensuing years, resource professionals at all levels struggled with not only how to interpret these new guidelines, but also how best to implement them in the face of numerous demands on our forest systems. Recently, the country experienced a change in administration as a result of the 2000 presidential election and it appears that interpretation of laws about managing ecosystems will be relaxed. There has already been an easing of air quality standards for companies involved in energy production as well as much talk about managing federal forests for increased timber production to stimulate western economies. Large-scale wildfires throughout the west in 2000 and 2002 are being presented as evidence that forest systems in the United States are out of balance and greater management intervention is necessary. How the nation views ecosystem-based management practices is likely to undergo new scrutiny in the near term.

National Developments

Although SFM and EM have followed similar trajectories in the two countries, individual differences are notable and have contributed to two separate paths to the development of sustainable forest policies in Canada and the United States. Thus we believe a useful framework is first to outline these developments from each nation's perspective, then to bring the discussion back to a composite view in the second half of this chapter.

Canada

We start the SFM story in early 1990 when the Canadian Council of Forest Ministers (CCFM, an organization formed in 1985 which will feature prominently in this discussion) convened a national forum on "Sustainable Development and Forest Management."[14] Although attendance was limited to a few dozen individuals—mostly forest-sector leaders who had much to say about the prospects of sustainability in Canada's forest management—we believe that the forum was an eye-opener. It succeeded in driving home the message that the forest sector had better (fervently) embrace sustainable developments concepts if it hoped to retain any kind of social license and was to continue to thrive in Canada. From that point on, the sector attempted to make progress with several key nationwide initiatives including the following five SFM-related programs. Developments in certification are relevant here, but they are covered thoroughly by McDermott and Hoberg in chapter 13.

We should mention that forests have never had a stable home in the Canadian government. In earlier times they were covered by a service within a department of government (i.e., the Canadian Forestry Service). A shift from a service to a full-fledged department occurred in 1989 with the establishment of Forestry Canada. In the department's legislation, the Minister was charged with the sustainable development of the nation's forests, something we believe might be the first reference to sustainable development of forests in any senior government's legislation. Alas, the department was short-lived, for in 1993, with a change in governing party following a general election, Forestry Canada was folded into a new department called Natural Resources Canada (which also includes energy and minerals), and returned to a slightly modified old name, the Canadian Forest Service (CFS). We recount this fluid state of affairs to warn readers about potential confusion over institutional names in the paragraphs that follow.

National State-of-the-Forest (SOF) Reporting

The Green Plan's allocation of resources to Forestry Canada was used partly to start up an annual program of reporting to Parliament on the state of Canada's forests. It is interesting that the first such report[15] was entitled "The State of Forestry in Canada: 1990 Report to Parliament", but the apparent mistake in focus was, by 1992, corrected, so that all subsequent reports were entitled "The State of Canada's Forests" (with years). Each report consists of a small dataset about forests, province by province (and territory), taken from the national forest database. Of greater interest to most readers are the substantive articles that, mostly in qualitative terms, relate what kind of progress Canada is making on a variety of SFM fronts. Articles about SFM criteria and indicators have appeared annually in the later reports.

No one was really paying any close attention to the reports' subtitles until the eleventh report recently appeared in our mailboxes. Its subtitle—"Sustainable Forestry: A Reality in Canada"[16]—represented a rather strong claim, especially surprising when other organizations such as Global Forest Watch Canada[17] were simultaneously publishing words to the contrary. What the CFS[18] portrayed in the report was actually a dozen examples of real SFM (read "real SFM progress") from across Canada. As most knowledgeable observers would agree, while many signs were pointing in the right direction, claiming that SFM has already arrived was seriously stretching the truth. However, the federal government can at least take credit for a solid effort at reporting annually to Parliament and, perhaps just as important, to the people of Canada on the state of the nation's forests and the state of SFM developments.

The Model Forest Program

Forestry Canada's Green Plan initiatives also included an aggressive program of experimentation with and demonstration of SFM approaches, epitomized best by the Model Forest Program. Each of Canada's model forests (ten in 1992, now eleven), using substantial funding from the federal government, was expected to find ways to respond locally to the global sustainability challenge set out by the Brundtland Commission. In a word, model forests were to be living partnership-based laboratories "where people with a direct interest in the forest, supported by the most up-to-date science and technology, would participate in decisions about how the forest could be sustainably managed."[19]

The model forests have been interesting ventures because each has a defined territory and a substantial budget but no direct management responsibilities for the land or forest cover. In large measure, the first five years (1992-97) of these partnerships were spent learning how various forest stakeholders should work together in search of SFM, and then conducting research on some of the key questions related to SFM locally. During the second phase of the program (1997-2002), an SFM indicators initiative was undertaken by all model forests[20] that substantially advanced the forest sector's understanding of how to gauge local progress in the pursuit of SFM.

On this latter point, two important questions can now be asked of the model forests: 1) to what degree can their achievements and accomplishments be disseminated for successful application elsewhere in Canada's forests, and 2) to what degree will an infusion of money be necessary for other groups in other areas to arrive at the same state of progress with SFM? There is little doubt that significant strides have been made by the model forests in both the political and scientific dimensions[21] of SFM. During the writing of this chapter, another five-year phase of the Canadian Model Forest Program was funded by the federal government. This extension will help substantially to address these questions.

The Canada Forest Accord and the National Forest Strategy

The Canadian forest sector had produced a rather skeletal National Forest Strategy by 1982, and a second emerged in 1987, following some intense but narrowly restricted consultation.[22] Largely as a result of the sustainable-development movement of the late 1980s, the Canadian Council of Forest Ministers (CCFM) determined that the third national forest strategy must be decidedly more advanced in several respects. First, it should focus on SFM as its key theme (indeed, the title is telling in this regard: "Sustainable Forests: A Canadian Commitment"). Second, it should be developed using a much more open consultation process so that a wider range of people across the country could contribute to its development. Finally, it should represent a set of commitments to SFM, not just a set of recommendations that governments and other sector players could take or leave.

The CCFM also determined that the strategy[23] should be accompanied by a high-order statement or accord that could be signed by all forest-sector stakeholders. Thus the Canada Forest Accord[24] came into being. The 1992 document held twenty-seven signatures; by 1998 the number was forty-two, and in 2001 it climbed to fifty-two. The Accord is a set of general statements including a national forest goal, vision statement, set of beliefs, and commitments to action. It is linked implicitly to the National Forest Strategy by obligating signatories to develop action plans to carry out each commitment statement.

There are opposing views as to the meaning and value of the strategy.[25] On one hand, many see it as merely a confirmation of what Canadians are already doing, or planning to do imminently, as they pursue SFM. On the other hand, it is seen as an exercise in reminding all sector participants of the breadth and depth of what still must be done to claim real achievements in SFM. Probably the strategy is serving both functions, as well as communicating to people outside Canada the attitudes held and actions taken in this country about getting on with SFM. Evaluations of strategy progress[26] thus far point generally to the obvious—on some fronts we seem to be doing well, whereas in others we have made little progress. A final evaluation of accomplishments[27] and a new round of consultations will be used to create the fifth National

Forest Strategy in 2003, just in time for dissemination at the World Forestry Congress in Quebec in September of that year.

Criteria and Indicators of SFM

Canada made a commitment in the 1992 National Forest Strategy[28] to develop a set of national criteria and indicators for SFM by the mid-1990s. Through a consultative process engaging a wide array of participants in the forest sector, the mission was accomplished in 1995. Canada's criteria and indicators for SFM[29] are generally consistent with those of the Montreal and Helsinki Processes, and feature the following six criteria:

1. conservation of biological diversity
2. maintenance and enhancement of forest ecosystem condition and productivity
3. conservation of soil and water resources
4. forest ecosystem contributions to global ecological cycles
5. multiple benefits to society
6. accepting society's responsibility for sustainable development

Under twenty-two sub-criteria (or, as they have become known, elements), there are over eighty indicators. National-scale reporting has occurred both in 1997[30] and again in 2000 as part of the Montreal Process. In addition, annual reporting to a modest extent has occurred in the yearly State-of-the-Forest reports to Parliament.

We can make the following observations about the national criteria and indicators for SFM. One is that four of the six criteria are ecological, criterion five is arguably the economic one, and criterion six is the social one. This imbalance has not gone unnoticed in the forest community. Some say it is about time that ecological factors got their fair share of the attention in sustainability analyses. Others have complained that social and economic indicators have unfairly taken a back seat to ecological ones. Regardless, it is clear that Canada's criteria and indicators for SFM are dominated by ecological concerns.

Another observation is that the indicators seem to have been hastily assembled and of rather uneven quality.[31] As we recall the indicator development process, there was no systematic evaluation of indicators using a set of quality criteria or screens.[32] This has led to innumerable problems in obtaining data for the indicators, as well as great variation in our ability to interpret the data in sustainability terms. Despite their shortcomings, the national criteria and indicators have substantially pushed forward the SFM agenda. They have served as inspiration for a wide range of regional and local indicator exercises, not the least of which are the endeavours of the model forests. They were also built directly into the Canadian Standards Association's (CSA) SFM standard.[33] A process for revisiting and revising the indicator suite commenced in 2002.

The SFM Network of Centres of Excellence

Finally we address the subject of SFM research. In an effort to rejuvenate and focus Canada's federally funded research efforts, the national granting councils developed a program of networks of centers of excellence (NCEs) in the mid-1990s. The idea behind the program was to create collaborations not only among Canadian universities but also between university people and people from business/industry and government to undertake research on issues of strategic national importance. A group coordinated by the University of Alberta applied to the program to establish an NCE on SFM, and won its birth in 1995.

The SFM Network, as it is now called, grew from humble beginnings into an organization that boasts participation by about one hundred researchers at some thirty Canadian universities, along with dozens of partners representing provincial governments, forest-industry companies, First Nations and environmental organizations.[34] In its early days, the SFM Network's research was dominated by ecological work on understanding the natural disturbance mechanisms of Canada's boreal forests. Over the years, Network research has expanded significantly into the management, silvicultural, social, economic, institutional, and policy realms of SFM, and has expanded beyond the boreal forest to embrace all forest types in Canada.

Each NCE is eligible for up to fourteen years of funding under the federal program. The SFM Network recently won approval for its second phase of seven years (2002-2009), after which it must become independently funded. The fact that the Network received all the funding it requested at renewal time seems to be testament to both the national relevance of SFM and the quality of its research. The SFM Network seeks to influence sustainable forest practices and policy by creating and disseminating knowledge of direct utility to forest-sector stakeholders. Direct beneficiaries of the Network's activities include its senior researchers (i.e., professors), its junior researchers (i.e., graduate students), and its research partners (i.e., professionals in government, industry, and other organizations). In its first seven years, the SFM Network managed to build up a strong forest-oriented research enterprise. In its second seven years, it should become a major driving force bringing new ideas to SFM developments in Canada.

United States

Although the move to EM in the United States was formally announced over ten years ago, implementation of these policies has been sporadic. A good amount of time has been spent in determining what EM should consist of or in arguing about whether existing practices were already meeting intended goals. The scientific community took up the call for EM—particularly biologists and ecologists who found renewed relevance for biodiversity research and conservation of habitats in the EM concept—while forest managers were often split among those who welcomed these ideas and those who voiced the opinion that "we've always been doing ecosystem management." Over the last decade, implementation of EM has lurched steadily forward as scientists and managers attempt to sort out key principles and wrestle with how best to apply them on forest landscapes. During this period several important processes have been initiated to help move EM from concept to reality.

Regional Assessments and EM

Public land resource professionals came under increasing pressure to manage forests with sustainable approaches that include adaptive management strategies and that consider large regional ecosystems. In the mid-1990s considerable federal resources were directed at conducting bioregional assessments to determine which essential components ecosystem-based management should consider and which practices are most appropriate for managing at the landscape scale. The impetus for the assessments stemmed largely from policy problems, often accompanied by high levels of public and legislative scrutiny and short time frames.[35] For example, the first major regional analysis was conducted in the Pacific Northwest by the Forest Ecosystem Management Assessment Team (FEMAT) in response to a presidential mandate to break the gridlock over federal forest management in Washington, Oregon, and Northern California. Natural resource agency and university scientists were given sixty days to produce a plan that brought together economic and social considerations with the legal requirements for species and forest protection. Major criticisms of the process included the short time-frame given to the assessment and that it was largely an endeavour by scientists without input from land managers. The FEMAT group produced the 1994 Northwest Forest Plan and federal managers have been attempting to implement it ever since.

A second large-scale analysis was completed in 1996 that involved the Southern Appalachian region. This assessment targeted the social, cultural, and economic forces that were a part of the ecosystems in the Appalachian Mountains in Virginia, North and South Carolina, Tennessee, Georgia, and Alabama.[36] The analysis was accomplished by a research cooperative of sixteen federal and state agencies. It attempted to use the natural boundaries of ecosystems rather than the artificial boundaries of counties, state, or national forests.

A third example was the Interior Columbia Basin Ecosystem Management Project (ICBEMP) that included more than three hundred scientists and agency and private sector personnel. Some regard the range of participants involved as an attempt to avoid the criticism levied at FEMAT. The priorities of the project were protection of ecosystems, restoration of deteriorated ecosystems, and the provision of multiple benefits for people of the region.[37] The ICBEMP covered 144 million acres, portions of seven states, one hundred counties and twenty-two tribal governments, and included some of the most diverse ecosystems in the United States. Although the plan was finally completed in 1999, the four-year process was marred by congressional wrangling (largely from those opposed to the Clinton Administration), the unwieldiness of such large numbers of participants, and the sheer geographic scale the project attempted to examine. The most recent regional assessment to appear is the 2002 Southern Forest Resource Assessment, the first to be motivated largely by rapid urbanization surrounding the region's federal forests.

Adaptive Management Areas

In a move parallel to the Model Forests in Canada, ten Adaptive Management Areas (AMAs) were created in the Pacific Northwest (Washington, Oregon, and Northern California) as places to learn how to manage on an ecosystem basis in terms of both technical and social challenges.[38] The AMAs were an essential component of FEMAT's Northwest Forest Plan and were heralded as a new way of doing business in forest communities. These areas were to be models for community-based management approaches where agency and citizen stakeholders could experiment with local forest conditions and learn from the experience.

In chapters 11 and 12, Ryan offers a description of the AMAs and Shindler provides a quantitative assessment of the initial outcomes of this ecological/sociological experiment. A more comprehensive evaluation[39] completed in 2002 largely outlines the demise of the AMAs, attributing their failure to a host of problems internal to the Forest Service and the Bureau of Land Management, the agencies responsible for their implementation. Although several AMAs reached a modest level of success, most were plagued by lack of agency support including inadequate direction, financial resources, training, and skilled personnel, as well as a bureaucracy that greatly constrained innovation. The larger problem, however, was the inability to reach consensus about what it means to manage under an adaptive, ecosystem-based approach. At the time of this writing, the AMAs were largely defunct because all funding for these enterprises had been withdrawn and many personnel were reassigned; however, some areas continued to function loosely because remaining personnel recognized value in what they started and had been able to turn their efforts into small, local achievements.

Criteria and Indicators and National Reporting

Both SFM and EM rely on a system of criteria and indicators. Criteria are the goals, and indicators are measurable signs that the goals are being achieved.[40] As described in the previous section, the Canadian government has taken seriously this responsibility and been a key international player in promoting the need for such action (witness the Montreal Process). The U.S. government also complied with the Montreal Process; the Forest Service and the U.S. State Department assembled a group of forest stakeholders to provide an ongoing forum for sharing information and perspectives on SFM. These discussions resulted in a set of U.S. criteria and indicators in 1995.

The stakeholder forum has continued to meet periodically to review progress, and eventually this group evolved into the Roundtable on Sustainable Forests.[41] Participants have included other federal agencies, state and local governments, scientists, tribal organizations, environmental groups, private forest landowners, and forest product companies. Their focus thus far has been to advance the use of criteria and indicators and share information about SFM. Although several government and NGO research organizations are reportedly testing the criteria and indicators approach in the United States,[42] most of the effort appears to be confined to paper. There is little evidence that federal forest agencies are paying much attention; on-the-ground practices make little mention of these guidelines. In addition, the U.S. government pulled out of the recent Kyoto agreements, leaving our national role and commitment to the criteria and indicators process in question.

On balance, the United States has been much less successful than Canada at both embracing the SFM/EM model into its management institutions and transferring EM policies into forest practices. One of the major problems for forest agencies in the United States has been the failure to recognize fully that knowledge of organizational structure and behavior is fundamental to implementing EM.[43] This step has little to do with getting the science of EM right, something natural resource professionals are pretty good at. Instead, the complexities of incorporating the human element—including expectations and responses—into the management process have befuddled many forestry leaders and practitioners. Although the Forest Service and the BLM have acknowledged the importance of partnerships between stakeholders (managers, scientists, and citizens), they seem largely unaware of the radical implications of creating such partnerships.[44] As a result, there has been a broad-scale underestimation of the policy implications of changing power relationships. EM represents a significant change in how business is conducted and such changes must permeate agency actions (see chapter 15 by Stankey et al.). Managing this transition will require patience and leadership, and involve changes in deeply rooted beliefs and behaviors among forestry professionals.[45]

Key Traits of SFM/EM Today

The points mentioned thus far serve to outline the more formal arrangements our two nations have undertaken in pursuit of SFM and EM. If we look back over the past decade or so, we can chart notable progress (and some remaining questions) in on-the-ground efforts by forest practitioners. The following discussion reflects important traits that help characterize sustainable forestry in Canada and the United States today, and help propel us into thinking about the next decade of North American forestry.

Range of Forest Values Explicitly Considered in Management

Forest management in North America, regardless of the national origin, has always been dominated by the pursuit of timber. This is still the case, but the array of non-timber values now under consideration has grown dramatically. Wildlife habitat was probably the first non-timber value seriously brought onto the forest manager's agenda. In Canada this began with site-specific guidelines for the management of species-specific habitats,[46] progressing to species-specific habitat supply or suitability analysis,[47] and recently moving toward a broader spectrum of ecosystem, landscape and wildlife-habitat analyses in support of biodiversity conservation.[48] In the United States the regional assessments described earlier were often driven by concerns about wildlife and the need to protect essential habitat for fish and bird populations. For example, a key component of the Northwest Forest Plan (a result of FEMAT) requires researchers to survey populations and manage habitat for a lengthy list of species. However, recently this component has come under a good deal of criticism because it "interferes" with most all other forest management objectives.

The movement to protect wildlife has been taken up at the provincial and state level as well and, in certain cases, even been projected onto private forestlands. For example, in Nova Scotia mandatory wildlife-habitat restrictions had previously been implemented on Crown land, but in January 2002 regulations governing timber harvests on all land in the province were implemented.[49] Now, for the benefit of wildlife landowners must leave residual clumps in timber-harvest blocks, create special management zones in riparian areas, and leave coarse woody debris and standing snags intact as much as possible. Similar provisions have begun to appear in the United States in the various state forest practices acts.

By now we have gone much further than simply incorporating wildlife habitat in forest planning models. Terms such as *integration* and *managing for multiple values* are commonplace as management teams come together to consider the range of resources on forest landscapes including recreation uses, water and air quality, carbon storage, special forest products, scenic quality, and traditional land uses of Aboriginal peoples. The primary forest agencies in the United States (Forest Service and BLM) now conduct most planning activities using interdisciplinary (ID) teams to be sure all values are considered in the process. In this same regard, guidelines for SFM[50] are a part of the Canadian Standards

Association's national SFM standard. A key element of this trend in both countries is the slow but discernably progressive move to undertake quantitative predictive modelling for the newly incorporated forest values. This too is a requirement of the CSA SFM standard, as well as long being considered central to implementing the concept of adaptive management.[51] Some values lend themselves more easily to formal models, particularly certain features of forests such as tree growth, soil composition, distribution of plant species, and the like. Other considerations—such as social values and human uses—are much more difficult to incorporate into a modelling program, although scientists in both countries are now attempting to do this as part of decision support systems.[52] There is little doubt that greater understanding of ecosystems will require gathering and linking large volumes of biological data with social scientific information to establish critical relationships between human activities and ecological conditions.[53]

Extent and Quality of Public Involvement

Serious public participation in any forest-sector decision-making, whether local in forest management or national in forest policy, was hard to find even through the 1980s. There were indeed attempts made and events held, but too often the high level of formality kept participants away, and feedback as to the use made of public inputs was non-existent. Through the 1990s we have observed that much greater attention has been given to this aspect of planning and policy-making. In the case of the Canadian forest sector, recent assessments indicate the country has moved from among the worst to among the best at involving stakeholders and the general public in natural resource decision-making.[54] Perhaps this turnaround has come from watching its neighbours to the south. In a recent message to Canadian foresters, former U.S. Forest Service Chief Jack Ward Thomas suggested that "forestry stuff" seems to happen first in the United States and there were lessons to be learned lurking south of the border.[55] His cautionary words primarily targeted the changing social context of American forestry and the inability of forest agencies to predict and respond adequately to public concerns. As a result, citizens frequently have sought other solutions that better reflect their values, and these steps—invoking the courts, seeking congressional remedies, mounting public demonstrations, using the popular media—usually circumvent traditional agency authority. Indeed, these actions are not limited to residents of the United States; inhabitants of British Columbia, for example, have been especially effective with similar tactics. In any case, it is fair to say that governments on both sides of the border have (painfully at times) learned a good deal about managing in an era of social change.

All in all, the decade of the 1990s was a period of growth and learning in Canada and the United States regarding how to come to terms with public expectations about increased involvement in forestry decisions. Recent applied research in this volume (chapters 7 and 8 by McFarlane and colleagues and Stedman and Parkins) and elsewhere[56] has uncovered many cases where forestry personnel have successfully engaged their publics and reached decisions that enjoy consensus. Ontario provides a good example of trends in the development of public involvement in forestry in Canada. At the level of managing Crown forests, the forest-management manual of 1980 contained no references whatsoever to public participation in the planning process, while the 1986 manual,[57] curiously reverting to the concept of "timber management" rather than the broader "forest management," contained a four-page section on public consultation. The 1996 manual[58] has a thirty-six-page chapter on public consultation that describes the details of a substantial process that includes the operation of a standing local citizens committee and a special consultation program for Aboriginal People.

Consequently public participation has been prominently featured in most forest-policy and planning developments at the national, state, and provincial levels. In the United States the NEPA (National Environmental Policy Act) process is a major guiding force among federal jurisdictions, and most states attempt to satisfy some similar format in reaching decisions. Again, Canadian consultations typically occur at the provincial level and comprehensive approaches have been rapidly in development. For example, early in the 1990s the Ontario Forest Policy Panel developed a policy framework for sustainable forests using a process that included public newsletters, press releases, a tabloid discussion paper, stakeholder meetings, an invitation for submissions or phone calls, community and Ontario Ministry of Natural Resources staff workshops, sub-sector hearings, an interministry coordinating committee, and review opportunities on draft reports.[59] Successful examples in the United States are also emerging, most recently around plans that attempt to treat forest health problems or implement fuel reduction programs for protection of property at the urban-wildland interface where wildfire concerns are the greatest. Throughout the west, active partnerships have sprung up between forest agencies and citizen organizations such as watershed councils, homeowner associations, and "friends" groups to take on local problems.

Such assemblages of techniques are commonplace within the forest sector in both Canada and the United States today. While numerous positive examples have been revealed, the jury is still out on the success of these programs over the long-term. On balance, assessments of these processes tend to be kinder to Canadian efforts. In the United States practitioners have discovered that while having a plan in place—including information devices for public communication—is important, execution of participatory strategies is often another matter. A number of institutional barriers have contributed to procedural problems; Cortner et al.[60] provide an insightful discussion about how these relate to implementing EM. Specific examples at the interface level demonstrate the need for improved selection of agency staff in outreach positions, proper training of these individuals, and a genuine, well-defined role for citizen participants to achieve more successful interactions. Perhaps this is another case of certain events unfolding first in the United States; in any case, Canadian foresters would do well to pay attention to developments down south.

Time and Space Horizons for Consideration of Effects of Forest Management

Planning and forecasting horizons for timber-supply projections in our two countries have gone in opposite directions. Forecasts for Canadian forests in the 1980s were typically in the range of 80 to 100 years. By the middle to late 1990s, under the SFM regime, the horizons virtually doubled and tripled. For example, Ontario's forest-management planning manual[61] calls for forecasts to a 160-year horizon. Millar Western Forest Products Ltd. of Alberta recently submitted a forest-management plan[62] for its licensed public land that used a 200-year forecast horizon. The CSA SFM Standard[63] defines "long term" as double the normal life expectancy of commercial forest trees, up to a maximum of 300 years. The definition is provided to give guidance regarding wood-supply and forest-structure forecasts.

The U.S. situation is much different and one that merits description, if only to help explain differences in the ability of our two countries to establish planning frameworks. In the United States, current management horizons are tied to a forest planning process that was reshaped by the National Forest Management Act (NFMA) of 1976. The Forest Service gave planning responsibility to the 155 individual national forests and established 10-year planning cycles. They handled the task so poorly (see Stankey et al. in this volume) that the agency has struggled ever since to define its planning process. Indeed, political scientist Shepard wrote, the result has been "a nightmare. Billions of dollars were spent on planning; an agency that rightly prided itself on 'getting the cut out' missed its initial planning deadlines by a decade; and when plans finally hit the street, the street turned out to run straight to the courthouse."[64] The first decade alone yielded over six hundred judicial appeals to the first seventy-five plans. Most of these appeals, and the thousand or so that followed, were based on inadequate public involvement in the plans. In 1998 the Forest Service appointed a Committee of Scientists to assess the agency's planning processes. A primary conclusion placed heavy emphasis on the need for greater collaboration with stakeholders,[65] a finding that was later confirmed—along with a tendency toward excessive analysis and other management deficiencies—by the agency itself.[66] This internal review re-emphasized the agency's commitment to planning on an EM basis and called for re-tailoring the process to establish a modern planning and management framework that includes more long-range forecasting.

Why would anyone bother to make forecasts for such a long future period—as in the Canadian experience—when obviously the fiction-factor increases dramatically the further one moves from the present? Well, it is clear that people are concerned about the long-term future, and want to have evidence—any kind and quality of evidence—about whether a continuation of near-term planned management regimes would perversely lead to a resource situation deemed unsustainable, or at least one that is undesirable. Of course no one believes that such long-term estimates are the absolute truth (witness the U.S. situation), because so much can change between the present and a few decades into the future, let alone in a century or two. But long-term forecasts provide a sense of whether systems are pointed in the right direction, and both countries see merit in this view under SFM and EM.

From a spatial standpoint, forest-management plans of the past usually focused all their attention on areas where timber harvests and regeneration treatments were planned during the next twenty years or so. Now it is inconceivable that plans would not contain maps showing distributions of important timber and non-timber indicators across the whole forest.[67] In the United States, for example, spatial considerations are at the forefront of forest planning. Managing at the landscape-level, including entire watersheds or bioregions, is a central component of the EM concept. Forest managers and researchers are increasingly agreeing on the importance of looking beyond the immediate forest's boundaries regardless of ownership patterns. For instance, wildlife ecologists are particularly concerned with tracking the sustainability of key indicator species with huge home ranges such as caribou in northwestern Ontario, grizzly bear in Alberta, or lynx in Montana and Idaho. Given steady advances in natural resource databases, GIS capabilities, and remote-sensing technology, we expect it will become commonplace for forestry professionals to check routinely how their key indicators are behaving in neighbouring forests when plotting strategies to secure sustainability in their own forests.

One area of forest planning that gives us cause for concern is climate change. While uncertainties are rampant in any analyses of the effects of climatic changes on forests in Canada and the United States, we believe that, along with some desirable effects, there will be serious undesirable effects.[68] It is therefore curious that no forest managers (as far as we are aware) are accounting for climate change in the long-term forecasts of future forest structure and composition in their management plans. One would have presumed that the potential ravages of climate change on North American forests would at least attract some analytical attention, if not even some low-cost insurance by way of modest alterations to management operations to give forests greater chances to cope with the expected climatic shifts. Unfortunately, little has occurred so far, but we believe it will have to occur in the next few decades if our forest sector is to become well equipped for a range of possible climatic futures.

Responsibilities for Funding Forest Management Activities on Public Land

Arrangements for funding (timber) management activities on public lands also differ in Canada and the United States. In Canada's distant past, the most common form of timber tenure for the private sector operating on public land was the cutting license. Forest management—interpreted here as regeneration and all other non-harvest work—remained the responsibility of the provincial governments. In the 1970s there was a strong move across Canada toward evergreen licenses where the license holder would assume responsibility for regeneration and some other types of operations.[69] In some provinces, though, the responsibility for paying for such work remained with the government. In the 1990s, the private sector began to take full responsibility for paying for all its non-harvest operations in the woods. To guarantee the availability of funds for such operations, some provinces (most recently Ontario) instituted the concept of trust funds where the license holder is obliged to put a certain amount of money per unit of wood harvested into a trust fund for use the next year in funding the regeneration program.

In the United States, the federal forest agencies have long taken care of regeneration activities on their own lands. Requirements are typically spelled out in timber-sale contracts or fall to the national forest personnel where harvesting activities are conducted. These practices have come under fire lately as environmental groups have taken the agencies to task for under-value sales. Their complaint is that some forests are being unnecessarily harvested and that revenues do not adequately cover costs of regeneration and restoration. Most state governments have their own arrangements and in most cases these extend to private lands as well. For example, the forest practices laws in Oregon require landowners to replant within three years of harvesting, usually at the rate of at least two hundred trees per acre. One new twist on the federal funding scenario has occurred lately on national forests and involves recreation use. Under a fee-demo program national forests have been able to retain the recreation fees (largely from campground registrations and other permits) generated on their own lands to pay for local operations. This is an innovation, because previously all such fees went back to the federal coffers in Washington D.C. and funds were doled out through an independent budget process each year. Recreation fee programs on public forests are bound to expand as timber harvests are scaled back and managers look for other methods to cover operational costs.

Tenure Arrangements

Canadian tenure reforms in the name of SFM have certainly been an improvement, but mostly they have been minor variations on a common theme—public land ownership and private-sector timber exploitation through large corporate entities that own wood-processing mills.

Because of ownership arrangements in the United States, adoption of EM has meant something different: less timber removed from public forests and greater pressure on private lands to make up the shortfall. The next decade in both countries may well mean sorting out how our particular brands of SFM and EM can be implemented across landscapes and ownerships to achieve more socially acceptable forms of timber harvesting. Undoubtedly, these will include more alternative practices such as thinning to improve forest health, reducing the danger of large-scale wildfires, and providing logs for small-dimension lumber.

We doubt that there will be any appetite in Canada or the United States for the sale of publicly owned forestland to the private sector, although the subject does come up now and then.[70] But we do foresee a future that includes continued experimentation with corporate timber tenures (Canada), better cooperation for experimentation across forest boundaries (United States), and more implementation of the concept of community forests (both countries). There have been pilot projects in many provinces and on U.S. national forests, with British Columbia leading the charge in its community forest program.[71] Nevertheless, there are still calls for much more aggressive progress in implementing community control over public forestland.[72] What kind of progress is made may depend on whatever reforms are deemed necessary to bring resolution of the softwood lumber dispute between our countries.

One additional tenure point involves Aboriginal Peoples. In Canada, First Nations people have long been neglected in the use and disposition of public forests across the nation. In the United States, treaties of the past continue to hold up and any changes in government relations with the tribes occur in small increments. On the other hand, current politics of Canada are such that the Aboriginal agenda is high within the forest sector, and significant strides are being made to accommodate their interests and values. We believe that any growth in the concept of community forests will necessarily feature First Nations communities. A key challenge for the forest sector in both countries is to learn about Aboriginal values and how they can be incorporated into progressive forest-management regimes for SFM and EM. In our view, the best progress on these fronts will be made when we all move beyond blame for past injustices and chart cooperative pathways into the future.

Protected Areas and Harvest-Exclusion Areas

Environmental groups can well lament that efforts to establish a full-fledged protected-areas network across Canada have failed to meet self-declared targets. Most provinces have to date fallen short of their own expectations. This means that there is yet much work to do on the protectionist agenda. However, let us not lose sight of the accomplishments in this vein during the past twenty years. For example, during the 1980s logging was stopped in two large provincial parks in Northern Ontario (Quetico and Superior). In the mid-1990s, the largest protected area in the world's southern boreal forest was created just north of Thunder Bay. Wabakimi Park was expanded from some 160 thousand hectares to just under 900 thousand hectares. In the late 1990s, Ontario implemented its Lands for Life process that tried to settle on a significant expansion of protected areas across the public lands of the north.[73] The resulting agreements among government, industry and environmental groups (e.g., the Ontario Forest Accord[74]) represent significant breakthroughs in the further establishment of forested protected areas.

Another example that signalled a turning point in management approaches occurred in British Columbia's coastal rainforests during the mid-1990s amidst considerable citizen activism at Clayoquot Sound. As a result, the Scientific Panel for Sustainable Forest Practices was established and, through its influential reports,[75] heralded a new SFM strategy for the province in which the first priority was to maintain watershed integrity, biodiversity, cultural values including scenery, recreation and tourism. Second in priority was a sustainable flow of products. The order of these priorities indicates a (new) activist-driven, underlying theme throughout forest management in the province: wood products are no longer the primary force.

Similar stories can be told for other provinces as well as for the United States. The protection movement took hold in the United States with passage of the Wilderness Act in 1964 and over 100 million acres have been federally designated thus far, most with the original act and then sporadically through the 1970s and 1980s. A Wild and Scenic Rivers Act followed in 1968, and a number of designations have occurred since—with rivers in various conditions being classified as wild, scenic, or recreational in nature. Set-aside policies have diminished in recent years with little new land actually being protected under the Wilderness Act; additional classifications have been bitterly fought in the west. However, during his tenure President Clinton got around these congressional skirmishes by directly dedicating wilderness-like lands as national monuments. Several of these, such as the Siskiyou-Cascades National Monument in southern Oregon, came during his waning days in office and are still being debated. Final decisions about these lands by the Bush administration remain in question.

It may be noteworthy that citizen activism over natural resources appears to be on the increase in the United States, seemingly parallel to the Republican party's rise to dominance in national politics. In the Pacific Northwest, for example, things appeared to calm down during the 1990s, probably in response to the Northwest Forest Plan that focused on EM approaches to regional forests and promised to include citizen groups in the process. Now in the new decade, tree sitters have reappeared and timber-sale demonstrations are back in vogue. The difference in this round of activism is that plenty of other issues are at the forefront of American politics and environmentalists have to share the headlines with concerns about the economy, international terrorism, and wars in the Middle East.

Conclusion

We know that many foresters wish that the world would simply settle down, that things would return to normal so they could get on with the job they were trained for—managing forests. But the roles forestry professionals are being asked to play today are much different than in the past, when citizen participation was minimal and technical expertise was foremost. Yes, the world of forestry has changed as models for SFM and EM continue to be developed in Canada and the United States. It is likely that the guiding themes for our forestry professionals will be an ecosystem approach to planning, where the planning boundaries are ecologically based (not administrative), the plan is developed after interested parties are consulted (not before), and emphasis is placed on monitoring the management actions and outcomes, so that adaptive learning can take place. Recent experience has shown that policies cannot stem strictly from ecological interpretations or simply be mandated plans for large regions imposed by our centralized federal governments. Differences in biological, social, and economic characteristics also need to be respected and incorporated. Each localized management area must be able to develop management strategies and conduct research that reflects local concerns and conditions. This will require considerable flexibility, including a range of options for experimentation and assessment, a lot to ask in an era when forests (and forestry professionals) are increasingly subject to scrutiny, high expectations, and short timelines for demonstrating results. And yet, it is this same political climate that calls for improved understanding of SFM/EM in different geographic, temporal, and normative contexts. To say that forestry professionals live in interesting times is an understatement.

4

Beyond the Economic Model: Assessing Sustainability in Forest Communities

Solange Nadeau, Bruce A. Shindler, and Christina Kakoyannis

Introduction

There is a growing interest in both Canada and the United States not only in the sustainability of forests, but also in the ability of forest communities to sustain themselves through fluctuating political and environmental conditions. Interested parties of all stripes are attempting to understand how changes in forest management policies enhance or harm the future of these communities. Although many studies have historically used economic indicators as measures of community stability, more recently researchers have demonstrated that the relationship between a community and its surrounding natural resources goes far beyond economic dependency. Frameworks have emerged that attempt to include factors that more accurately reflect the broader range of sociopolitical influences on affected locales. For example, the existing set of skills and leadership present in a community, the propensity of residents to work together, and the available physical and natural attributes all contribute to how people might respond to changing conditions. In this chapter we discuss the notion of community from multiple perspectives and examine three conceptual frameworks for assessment: 1) community capacity, 2) community well-being, and 3) community resiliency.

Context

Over the last two decades forest-based communities have faced tremendous changes induced by economic, ecological, and political forces. Economic changes such as technological innovation, market differentiation, and more globalized economies have affected many communities. The results of these changes range from mill closures and employee layoffs to concentration of forest products processing firms to changes in traditional ways of life in rural areas.[1] Increasing concerns about environmental values have also driven forest policies in new directions. In Canada and the United States laws and resulting policies are now reflecting explicit mandates to consider more fully the sustainability of both biophysical systems and a range of human uses of forests. In Canada, a long-term commitment toward sustaining forest ecosystems represents a fundamental shift from traditional forest policies that emphasized sustaining commercial timber output. In the United States, where an ecosystem-based approach has been adopted for managing public forests, the goal is now on maintaining the health and integrity of ecosystems by balancing social, economic and ecological considerations.[2]

These changes in policy reflect a growing awareness in both countries of the social and ecological complexity of forest ecosystems, as well as a fundamental shift in environmental values nationwide.[3] These changes have also resulted in a growing interest among resource professionals, polititicians, and rural residents about the sustainability of forest communities, especially in identifying key factors that could explain why some communities have more facility to cope with change than others do.[4] Thus far, our ability to monitor and evaluate these community systems is limited due largely to a lack of sophisticated frameworks for assessing change across settings.[5]

In recent years, researchers have attempted to develop more comprehensive explanatory models describing interrelationships between forests and human communities.[6] This chapter presents a review of these efforts by focusing on the development of several emergent conceptual frameworks now in experimental use. First, however, we set the context for this discussion by briefly examining the historical perspective of forest policy regarding these communities. Second, we describe the notion of community stability, which traditionally has held an important role in forest policy and provides the initial basis for assessing the relationship between communities and forests. In the last half of the chapter, we examine the meanings of community and discuss the utility of the community capacity, community well-being, and community resiliency frameworks for evaluting forest communities.

Sustained-yield and Community Stability: A Classic Duo in Forest Policy

At the end of the nineteenth century, many questions arose regarding forest management in the United States and Canada. In both countries, a conservation movement was emerging, characterized by concerns that forests could not provide an infinite source of wood supply if past practices continued.[7] This movement had common roots in the United States and Canada that came together at the American Forestry Congress meetings held in Cincinnati and Montréal in 1882.[8] It was during this period of questioning that the concept of sustained-yield emerged as a policy goal to address fears of overharvesting.

During the 1920s in Canada, the concepts of sustained-yield and community stability became intertwined. Elwood Wilson, forester of the Laurentian Paper Company in Québec, and the first forest engineer employed in the Canadian forest industry, argued that forest industries needed to recognize that many of the communities developed in association with wood processing mills were highly dependent upon a sustained production of timber.[9] Accordingly, he believed forest companies had a moral obligation to the local workforce. Wilson and other foresters began promoting policies that would prevent the depletion of Canada's natural resources and the decline of its forest communities. After World War II, as the demand for wood products in North America skyrocketed, fears of a timber famine increased, and added to concerns about an eventual destabilization of forest communities. Under the assumption that a prosperous forest industry resulted in prosperous forest communities, natural resource policies in all Canadian provinces were dominated by the need to provide incentives for industry to practice sustained-yield forestry.

At about the same time, supporters for the need to stabilize forest communities in the United States inspired creation of the Sustained-yield Forest Management Act of 1944.[10] Under the Act, community stability became an official goal of the U.S. Forest Service, with sustained yield viewed as a means to ensure the prosperity and well-being of local residents.[11] These efforts were intended to guarantee economic and political stability through the maintenance of a non-declining flow of timber from public National Forests.[12] In short, the general conception of community stability held by Canadian and American foresters relied on the belief that a regulated forest would provide a steady flow of wood in a sustained and predictable fashion. At the time, foresters were largely unencumbered with alternative views about other values and believed a steady source of timber would ensure employment that in turn would lead to stable communities.[13] It followed that most research efforts in these locales were oriented toward evaluating sustained-yield and its effect on communities.

Community Stability: A Fuzzy Concept

While the term "community stability" has been commonly used in both forest policy formation and implementation, it raises considerable debate in the research literature. The discussion arises from the various meanings given to the concept of stability and the diversity of indicators used to measure that concept. Nevertheless, the notion and use of the term is important because it represents an attempt explicitly to recognize a relationship between forests and the people who both inhabit surrounding areas and derive their livelihoods from them, or otherwise depend on the values that forests provide. One of the earliest applications of the term "community stability" was in Kaufman and Kaufman's[14] seminal study of two Montana forestry towns. They used the term to describe a process of orderly change in rural areas. The authors recognized that stable community processes required economic diversification, community leadership, citizen participation, and a sustained flow of forest resources.

Over time, the concept of stability became associated with a notion of constancy and was usually measured by economic indicators.[15] A synthesis of research by Machlis and Force[16] indicated that measures of community stability have largely focused on the impact of forest industries and include the levels of harvest, production of forest goods, prices of wood products, and employment levels and salary within forest companies. As before, emphasis on economic measures for monitoring community stability arose out of the belief that sustaining timber production was an effective means for ensuring the stability of forest communities.

Over the years other researchers pointed out the limitations of the sustained-yield/community stability assumption. For example, Daniels et al.[17] demonstrated that a constant flow of timber was not a complete solution because more diverse factors such as the cyclic demand for forest products and constant changes in technology and fluctuating transportation costs also created uncertainty in these communities.[18] Other studies began to focus on the effect forest dependency has on the social context of communities, showing that many areas go through typical changes such as rapid shifts in population, employment, and prosperity as the industry upon which they depend follows a boom and bust cycle.[19]

The emergence of research regarding non-economic aspects of community stability is reflected in the positions taken by the U.S. Forest Service and the Society of American Foresters (SAF) in the 1980s. Both organizations proposed revised definitions for community stability that focus on the capacity of forest communities to cope with change.[20] A SAF task force noted:

> Community stability, as it relates to forestry is closely associated with jobs and economic benefits generated from the use of forest resources. However, the task force also recognizes that this topic cannot adequately be considered apart from several other related aspects, including: quality of life, environmental considerations, and the nontimber and noncommodity uses of forestland. Community stability concerns the prosperity, adaptability, and cohesiveness of people living in a common or functional geographic area and their ability to absorb and cope with change.[21]

Numerous studies conducted in the 1980s and 1990s adopted an approach inspired by the early work of the Kaufmans and acknowledge the complex dynamics of the community stability concept.[22] In particular, they aimed at developing a more complete understanding of the effect forest dependency has on communities. These studies usually refer to community stability as a process of orderly change observable both in the economic and social realms. This body of work often relies on a rural sociological perspective, and this broader conceptualization of the term led to the use of a more diverse set of indicators and provided new ways of thinking about the stability of forest communities.

In one example, Force et al.[23] attempt to establish a connection between community social change (community size and structure, cultural element, cohesion and anomie) and changes in local resource production, historical events, and poltical trends. Using a range of indicators to monitor variables, they found that timber dependency might only be a minor factor in influences on community change. Drielsma[24] used an even broader set of indicators to assess stability in forest, agricultural and tourism-dependent communities. He studied population flux, wholesomeness of family life, income, measures of prosperity and standard of living, community life, health, and a range of external influences and controls. His findings suggest that forest-dependent communities are among the least stable and prosperous because they tend to have a high population turnover and more social problems (e.g. divorce, suicide, low cohesion) than other communities. Furthermore, forest communities without major industrial facilities tend to have poor housing and public services, poor wages and earnings, and high seasonal unemployment. Similar studies[25] largely confirm these findings and indicate that stability of timber production does not necessarily result in prosperous communities. Drielsma[26] also noted that a sustained yield policy alone has little chance of leading to a stable community, in part because modern economic conditions encourage further processing of timber outside of forest communities. He recognized that in order to help forest communities support themselves on a long-term basis, policy options must be crafted that go beyond the economic aspects of a community.

The Emergence of New Concepts to Assess Forest Communities

Despite efforts made to clarify and broaden the meaning of "community stability," the term remains ambiguous and frequently leads to confusion. In response, researchers introduced several different frameworks for studying forest communities that allow for more thorough analysis of a range of factors. In addition, researchers also recognized the need to revisit definitions of "community;" indeed, the various ways in which "community" has been defined in studies of these areas has contributed to the confusion and emphasizes the need for clarification.

What Do We Mean by Community?

In past studies, particularly during the 1950s, "community" was defined as a human settlement in a given geographic area.[27] Under this definition, community assessments focused primarily on the economic dependency of the geographic community upon the surrounding resources and tended to view community narrowly as simply a source of labor for the local forest industry. This offered a rather reductionist vision of "community". Much earlier studies[28] used a more inclusive definition that integrated both social and economic components. They defined community as a human system with specific needs that should be included as a part of forest management. This emphasis on broader human needs in describing forest communities re-emerged in the 1980s.[29]

Certainly, this difficulty of defining the term "community" is not new. After reviewing the various meanings of "community" in sociological studies, Hillery[30] suggested three general approaches to clarify the concept. His typology, which is still widely used, refers to community by:

- Geographical location: a human settlement with a fixed and bounded territory
- Social system: the interrelationships between and among people living in the same area
- Sense of identity: emphasizing a group of people who share a particular set of values even if they do not live in physical proximity

Regardless of the type of community studied, we can begin to understand the linkages between humans and their natural environment at different levels of analysis. The first approach emphasizes a geographical analysis and suggests there is a relationship between social life and a specific, identifiable location. One strength of this territorial approach is that it allows the use of data that are collected on a geographical basis (e.g. county tax records, national census), to examine community demographics and the use of forest lands. However, from a social science perspective, this approach is limited in that it considers neither the nature nor the patterns of the relationships between people.

The social system approach provides a more in-depth examination of the interactions between individuals who are linked by geography. This form examines the network of relationships among people, including livelihoods, the local economy, community institutions and residents' uses of the forest; however, it may give little attention to the quality of those relationships. Finally, the third approach recognizes that people may hold shared values, but places no constraints on geographical proximity. It helps us understand the role that forests play in people's lives; how forests affect social and family life, and the values that people attach to forests. Thus, this approach is concerned with the quality of relationships between members of a non-territorially bound form of community.

Emergent Frameworks for Assessment

Beckley[31] recognized that the type of community studied is an important factor in how assessments should be designed and conducted. He proposed that three dimensions should be addressed in defining forest communities, suggesting that we specify the scale of the unit of analysis (individual, household, community, county, state, region), the type of dependence (timber, forest service, tourism/recreation, non-timber products, subsistence or ecological), and the degree of dependence (high, moderate or low). These three dimensions remind us that community can refer to a variety of diverse human settings. Therefore, special attention needs to be given to describing the major features of what we call a "forest community."

In parallel with these ideas, social scientists began organizing their study of forest communities around several conceptual frameworks that were first introduced by researchers in the sociology and ethnography disciplines. In particular three were adapted for assessing communities affected by recent turbulence over natural resource policies in both Canada and the United States In the remainder of this chapter we describe these frameworks and their intended use:

- *Community Capacity*: concerned with the characterization of a community's ability to face changes[32]
- *Community Well-being*: focuses on understanding the contribution of the economic, social, cultural and political components of a community in maintaining itself and fulfilling the various needs of local residents[33]
- *Community Resiliency*: concerned with capacity of humans and their institutions to adapt to changes over time while minimizing their effects on communities.[34]

These conceptual frameworks were first used in forest community assessments because they provided an opportunity to analyze the effects of forest dependency on various dimensions of communities. They were brought to the attention of many resource professionals with the advent of large-scale ecosystem studies in the United States. Although these assessments achieved mixed success, they provide an important means to supply policy makers and forest managers with information about the social, economical and ecological conditions in ecosystems for which management policy is being designed.[35] In the mid 1990s at least four such bioregional assessments were conducted and each included an analysis of forest-based communities:

- *The Forest Ecosystem Management Team* (FEMAT): appointed by President Clinton to identify alternatives to break the policy gridlock over the Pacific Northwest forests
- *The Interior Columbia Basin Ecosystem Management Project* (ICBEMP): launched to assess the natural resources and socioeconomic conditions of the Intermountain West
- *The Sierra Nevada Ecosystem Project* (SNEP[36]): modelling similar conditions and factors in California and Nevada
- *The Southern Appalachian Assessment:* targeting biological diversity, economic uses, and cultural values in the Southeast region of the country

Collectively, an essential contribution of these assessments is that they all illuminated the role and importance of forest communities. Not only did researchers provide details about the social, economic and political nature of affected areas, they also encouraged decision-makers to more fully consider the ability of these communities to respond to change. Because each bioregional assessment had a different mandate, these studies provided little agreement on a common model for assessing forest communities. Nevertheless, the three concepts—community capacity, community well-being, and community resilience—emerged as recognizable and useful ways of thinking about the human/ecosystem dynamics within these systems.

Community Capacity

The concept of community capacity emerged from a synthesis of research in human ecology, rural studies, and sociology.[37] In forestry the concept has been used to estimate the collective ability of residents to respond to external and internal stresses, to create and take advantage of opportunities, and to meet their diverse needs.[38] The major challenge of community capacity assessment is to identify the specific attributes of a community that facilitate or impede its ability to respond to problems or external threats. Various attributes have been assessed in studies of community capacity and can be generally grouped into four basic categories:

- *Physical and financial infrastructure:* physical attributes and resources in a community (e.g., water systems, open space, business parks, housing developments, schools, etc.) along with financial capital[39]

- *Social capital* (also called *civic responsiveness):* the ability and willingness of residents to work together for community goals[40]
- *Human capital:* skills, experience, education, and general abilities of residents in a community[41]
- *Environmental capital:* quality and quantity of the surrounding resources including water, air, soils, minerals, scenery, and general biodiversity[42]

The underlying assumption of community capacity is that the interactions between these elemental categories determine the ability of a community to face changes. Thus, positive and negative consequences of change are more likely to be balanced in a community with higher capacity, while communities with low capacity are more likely to be negatively affected. The level of capacity is influenced by the presence of each element and also by its quality. However, because complex interrelations exist among these elements, a change in one can affect the others in various (positive or negative) ways. For example, enhancing human capital by encouraging in-migration of highly educated people might lead to a reduction of civic responsiveness if those new residents tend to act independently and do not become involved in community affairs. It is not only the presence of an element that is important but also the effect it has on the others and, thus, on the overall community capacity.

Two researchers have been studying these ideas under the category of social capital, sometimes also referred to as civic responsiveness. Cornelia Flora[43] and Jan Flora,[44] who are working in rural communities in the midwestern United States, have been particularly interested in the source of norms, trust, and reciprocity that might develop within a community. They suggest that social capital at this level can be characterized largely by assessing three items: the respect for multiple points of view, resource mobilization, and the diversity of horizontal and vertical communication networks. They argue that while horizontal networks facilitate the inclusion of a diversity of groups, values and ideas, vertical networks reflect the interaction between the community and external organizations and institutions. Thus, these two types of communication perform different, but complementary roles in the development of social capital. This form of capital is also regarded as fundamental in sustaining or even enhancing democratic attitudes and practices in communities.[45] As such, social capital seems important to empowering communities to use local resources to meet their needs.

Community capacity was a major focus of the FEMAT social assessment. The assessment of three hundred communities in the western United States was conducted through workshops where panelists familiar with the local setting rated the communities on a capacity scale. Overall, the assessment team found that numerous factors such as community size, location, level of economic diversification, and leadership all contributed to a community's capacity to cope with changes in forest management. The team also concluded that many timber dependent communities were rated particularly low because of their sensitivity to changes in harvest levels and a low level of leadership.[46] Although the FEMAT study provides interesting insights about key influences on the ability of communities in the pacific northwest to adapt to changes in forest policy, it tells us little about how these changes might affect the welfare of residents or their quality of life. These concerns are embedded in another concept: community well-being.

Community Well-being

While community well-being has been used as an assessment framework in recent studies of forest communities,[47] the concept of "community well-being" is a difficult one to grasp. Although the term appears in many scientific papers, it is rarely defined. Wilkinson[48] offered an initial attempt by describing the concept as one that reflects and recognizes the social, cultural and psychological needs of people, their family, institutions and communities. This description reveals the complexity of the concept; and because of this complexity, studies on community well-being have adopted different approaches. Some studies of forest communities look at specific factors such as poverty or economic development[49] and rely mainly on social indicators. Others focus on more general well-being[50] by including a mix of social indicators, historical information, and something relatively new—primary data collected directly from community residents about how they evaluate different aspects of their life.

For example, in their studies of Alabama's forest communities, Bliss et al[51] compared the social, economical and environmental well-being of two forest dependent counties. Their assessment relies on a comprehensive analysis of social structures, ownership patterns, forest sector utilization and historical development patterns in the communities. They observed that a high concentration of resource ownership and product specialization posed problems for the overall well-being of residents. This finding was due in part to the fact that landowners had few, if any, incentives to participate in the improvement of the social well-being of the community in which their forests or mills were located. While these economic actors can make important contributions to the economic well-being of a community, their overall contribution to *community* well-being can be reduced by their negative impacts on the social environment. Other research[52] has also noted that concentrated ownership and control of natural resources can negatively affect the well-being of forest communities.

Aspects of community capacity have also been identified as important factors that can influence well-being.[53] For instance, in the Sierra Nevada Ecosystem project, Doak and Kusel[54] assessed well-being through an analysis of socioeconomic status and community capacity. Their strategy was to complement sociodemographic measures with self-reported measures collected in specific communities. They used indicators of housing tenure, poverty, education level, and employment to construct a scale to measure socioeconomic status. A series of workshops with key public officials were then used to assess community capacity. Results indicate that communities with a high socioeconomic status do not necessarily have a high community capacity. The authors attribute this weak correlation to the critical role of "social capital;" that is, a community's ability to work towards common goals. While socio-economic status provides information about the wealth of people in a community, community capacity informs us about the willingness of these people to share this wealth. Thus, these two concepts contribute in different ways to the general well-being of forest communities.

Community Resilience

It is not clear under what circumstances the concept of community resilience was first introduced in the community assessment literature. In 1990, Machlis and Force suggested resilience as an alternative to the concept of stability because it emphasized the ability of a community to cope with change. More recently, community resilience has been described as the capacity for humans to change their behavior, redefine economic relationships, and alter social institutions so that economic viability is maintained and social stresses are minimized.[55] In many regards, community resilience is similar to the concept of community capacity; however, the concept of resilience expresses a clear concern about the development and maintenance of a community's adaptability over time. In this sense it contributes a new element to the assessment of forest communities.

Unfortunately, one concern with the use of this term is that the social definition of resilience may be confused with the ecological meaning of resilience. Ecologically, resilience refers to the ability of a system to recover from a perturbation and the speed with which it returns to its original condition.[56] As a result, people may erroneously believe that community resilience is a concept for assessing how human communities return to pre-existing conditions after having responded to change. In contrast, the intended use of the concept emphasizes the prosperous evolution of a community and recognizes that a return to the status quo will not necessarily promote resilience.

Fortunately, the social science team of the Interior Columbia Basin Ecosystem Management Project helped operationalize the concept of community resilience. In their study researchers conducted workshops involving 198 communities and led participants through self-assessments of each town's ability to manage change and adapt in a constructive way. A resilience index was developed[57] to monitor communities by aggregating measures of residents' perceptions of certain community characteristics and conditions:

- aesthetic attractiveness
- proximity of outdoor amenities
- level of civic involvement
- effectiveness of community leaders
- economic diversity
- social cohesion among residents

Because researchers were also interested in how local people perceive their future, they assessed how communities thought about and prepared for future events. Results indicate that the most resilient communities are those whose residents have a clear vision of desired future conditions and have taken into account biophysical, social, and economic changes.[58] The researchers suggested that communities who have collectively considered change and approach it with a proactive attitude are better able to adapt and move forward. However, the ICBEMP assessment did not support FEMAT's conclusion that forest communities most dependent upon timber are the least adaptable communities. Unfortunately, this part of the ICBEMP assessment was abbreviated and provides few clues about the source of that difference. It therefore remains unclear whether the difference rests on the various aspects studied under the different assessment frameworks or if it is embedded in the community characteristics themselves. Further studies of community resiliency would help clarify the important elements that confer to a community the ability to maintain itself over time. Such research could also help clarify subtle differences between resilience and community capacity.

Conclusion

In this chapter we have retraced historical perspectives to describe essential terminology and provided a brief summary of several conceptual frameworks for assessing forest-based communities. We acknowledge there may never be a single definition for these concepts. Their value lies in providing ways to think about sustaining communities and remind us that the communities are continually evolving entities.

Because community stability initially evolved with an economic connotation, more inclusive terms such as community capacity, well-being, and community resilience are now more useful replacements. These concepts encompass the previous elements expressed by the notion of stability, but also provide a means to incorporate other current concerns and values held by community residents. These concepts also possess several common features—the most obvious being that community capacity is an important element of community well-being. Another important common feature is that each of these approaches requires researchers to go into forest communities because they cannot adequately evaluate these places on secondary data alone. Consequently, community members have come to play a larger role in assessments than under previous models. It is likely that these new forms of evaluation will be more meaningful to those affected and, as such, they have a better chance of being implemented over the long-term.

In addition, these frameworks recognize the contextual differences among communities and these tools help answer different questions about places and local residents. Community well-being is the most far-reaching concept as it assigns importance to the roles of historical background, quality of life, and concerns about people's capacity to adapt to change. Therefore, the notion of well-being is likely to lead to a more comprehensive description of a community than an analysis based solely on community capacity. Finally, while community resilience also shares certain elements embodied in the other two concepts, it is the one that prompts us to look forward, providing insights about the hopes and trust that residents place in their community's future.

Our intent in this discussion has been to help broaden our understanding of the interrelationships between humans and the forest settings in which they live and work. Ideally, further refinement of these conceptual frameworks will also lead to agreement on more specific criteria and indicators for evaluting the sustainability of forest communities.

Part Two

New Demands on the Forest—From Timber Values to Forest Values

There is mounting evidence that North Americans are undergoing a fundamental shift in the values they associate with forests. Now mangers are expected to produce a diversified range of goods and services, including pristine wilderness, habitat for endangered species, and picturesque vistas. Yet our appetite for pulp, paper, dimensional lumber, and other forest-based commodities is as strong as ever. This paradox makes the job of forest managers increasingly complex. We live in a finite world, and our forests may not be able to supply all that we demand of them.

Chapters in this section document these changing forest values. The groundbreaking research in this area occurred in the United States throughout the 1980s and into the early 1990s. Most of the five chapters in this section reflect recent Canadian efforts to measure shifts in their nation. Several of these build upon the original work from the United States. The cases involve national opinion survey research as well as provincial samples from British Columbia, Alberta, and Newfoundland. The evidence is consistent and largely irrefutable. Across Canada and the United States, citizens are placing greater emphasis on ecological services. In fact, in each of the studies presented here ecological services and environmental protection rank higher in priority than wealth creation, jobs, and recreation.

In many ways the new demands articulated through these studies of public opinion suggest that the institutional framework for forest management is antiquated. It was initially built as a scientific management program to service the sustained-yield model. Our current institutions for forest management were not created to incorporate public opinion, nor were personnel in these institutions adequately trained in how to receive input from the public and integrate it into management planning. This is part of the institutional reform that must occur if the public's new demand for a wider range of forest values is to be met.

5

Ecosystem Management and Public Opinion in the United States

Brent S. Steel and Edward Weber[1]

Introduction

The emergence of the United States as an advanced industrial society has led to an increasing array of social and political problems, which confound federal agencies' ability to implement effective policy decisions. One of these problems is what may be termed the "democracy and technocracy quandary," or the democracy and science quandary.[2] As an advanced industrial society, the United States faces many policy problems that are increasingly technical and complex in nature, such as the management of natural resources. At the same time, the United States is a democratic system of governance that has experienced a noticeable growth in distrust of government, traditionally conceived (i.e., elected representatives and expert bureaucracy), and increasing public demands for citizen involvement in governance.[3] As Frank Fischer in *Citizens, Experts, and the Environment* observes, "the increasing complexity of social problems, giving rise to increasing specialization and the expansion of elite 'public policy specialists,' puts Western polyarchies in the positions of being replaced by a 'quasi-guardianship' of autonomous experts, no longer accountable to the ordinary public."[4]

The concern is that the relationship between participation (democracy) and scientific expertise (technocracy) is mutually exclusive in character. On the one hand, placing too much emphasis on science and expertise as the ultimate determinants of policy outcomes risks the erosion of democracy.[5] Alternatively, too much democracy (i.e., direct involvement of citizens in policymaking and implementation) may relegate technical and scientific information to a peripheral role and increase the probability that complex problems will either be ignored or addressed in a suboptimal manner. There also is concern that the public's distrust of the scientific community[6] will translate into less public funding for the study of ecosystems and other complex phenomena, further hindering capacity to make effective policy.

Recent developments highlight and exacerbate the tension found within the technocracy-democracy quandary, particularly with regard to forestry and decision-making for sustainable policies. The growing embrace of Ecosystem Management (EM) across agencies and levels of government elevates the role of science as a tool for understanding environmental problems. Under this management approach, science and scientific experts help decipher the relationships within nature, and between humans and nature. In doing so, they also necessarily exert critical influence on policy choices and program design. There is a concurrent movement toward increasing opportunities for democratic process, as administrators and policymakers across the country attempt to reorganize and reinvent government. Central to this effort are innovative, decentralized institutional arrangements, which delegate or share the federal government's authority with private citizens or other agencies with similar jurisdictional and policy concerns. The fields of environmental and public lands policy have notably been affected by the larger "reinventing government" movement as the limits of top-down regulatory approaches have been promulgated by academics and practitioners alike.[7] Many now contend that effective environmental programs require complex, collaborative partnerships among diverse government, civic, and business actors at the state and local levels.[8]

Clearly, both EM and the movement to devolve (or share) authority with local citizens and other stakeholders "reinvent" the existing model for managing the environment. Yet are the twin efforts to reinvent natural resources management working at cross-purposes such that the technocracy-democracy quandary is becoming more difficult to resolve? Can the co-evolving trends of EM and decentralized control of environmental policy be reconciled one to the other? Moreover, successfully changing the way we manage nature is likely to require broad public support, especially from local stakeholders and citizens, each of whom will play far more substantial roles in policymaking and implementation. Does the American public support EM? What do citizens think about the devolution of decision-making?

This chapter examines the dual trends toward ecosystem management and decentralized, participatory environmental policy from the perspective of the general public. Using data from a national public opinion survey conducted during the summer of 1998, the paper examines factors associated with public knowledge and acceptability of ecosystem management and devolved environmental management in the United States. Before examining survey results, however, we present a brief historical perspective concerning the reinvention of public lands and natural resource management in the United States, followed by a discussion of the possible correlates of public acceptability of such new approaches. The next section examines public orientations and preferences concerning EM and the level of government most appropriate for implementing environmental policy. A concluding section further explores the questions posed above. Study findings indicate that significant numbers of citizens support the EM approach, the devolution of decisions (state and local government as the best levels to implement EM), and citizen participation in EM.

Devolving Authority and Reinventing Government

Administrators and policymakers across the United States are attempting to reorganize and reinvent government. Most of these efforts share three objectives: to improve program efficiencies, to harness resources outside government in the service of public policy goals, and to better facilitate the input of state-level interests, private sector groups, and the general public.[9] This move to decentralize public administration is occurring across a broad range of policy areas, from community policing to rural development and public health. It is clear that reinvention has not only taken hold and gained a substantial foothold at all three levels of American government, but it now enjoys considerable political support.[10]

The propensity to adopt alternative institutional arrangements premised on decentralization, collaboration, and citizen participation is especially pronounced in the environmental and natural resource policy world.[11] Regulatory negotiation, which actively involves a broad range of stakeholders in the specification and implementation of regulations, has become more widely used for federal pollution control programs.[12] The EPA has developed the Common Sense Initiative (CSI) in league with corporate America, state regulators, national environmentalists, and locally based environmental justice groups. The goal is to encourage innovation by providing flexibility in the use of a place-by-place approach to achieving pollution control standards.[13]

In the Western United States, more than one hundred coalitions of environmentalists, ranchers, county commissioners, federal and state government officials, loggers, skiers and off-road vehicle enthusiasts are cooperating in an attempt to improve ecosystem and public, as well as private, lands management arrangements.[14] Theoretically, these collaborative grass-roots ecosystem management arrangements work within the larger framework of national laws, not in lieu of them. The idea is to prevent degradation through long-term, holistic solutions to local problems, and to enhance the degree of local oversight and implementation expertise. These collaborations are viewed as a way to customize one-size-fits-all national laws to the particular conditions of individual ecosystems and communities.[15] Many state agencies recognized the need to coordinate their efforts with those of local governments, business, and private landowners and all have ample incentive to try something new. State environmental directors have become vocal about the lack of flexibility afforded by the EPA in the programs it delegated to the states. Problems the directors cited included excessive oversight, unrealistic standards, and lack of flexibility for implementing the programs, particularly rural and arid regions.[16] In New York, the Department of Environmental Conservation has conducted joint exploratory ventures creating integrated pollution management and prevention programs.[17]

Others like Daniel Kemmis (the former mayor of Missoula, Montana) endorse the burgeoning use of collaborative efforts by pointing to the limited capacity of centralized, federal management and the futility of the conflict-based "us versus them" approaches adopted by ideologues on both sides of environmental issues: "I do not believe that any solution coming from one end of the political spectrum or the other is going to have the capacity to do what this landscape requires. The danger is that one ideology or another will win a temporary victory because we did not work hard enough to find common ground."[18]

The use of decentralized collaborative methods is growing at other levels of government as well. Barry Rabe's[19] book, *Beyond NIMBY*, examines the phenomena in the United States and Canada within the context of local hazardous waste siting decisions.[20] DeWitt John,[21] in *Civic Environmentalism*, explores state-level collaborative games designed to resolve environmental issues associated with wetlands in the Florida Everglades, pesticides in Iowa, and energy conservation in Colorado.[22] A specific example involves the Ricelands Habitat Venture. Through arrangements with Ducks Unlimited, National Audubon Society, the California Reclamation Commission, and the California Rice Industry Association, rice farmers agreed to alter traditional practices so that their fields can be turned into wetlands during the winter months, thereby providing critical wintering habitat for migrating waterfowl along the Pacific Flyway.[23] Sirianni and Friedland[24] as well as several authors featured later in this volume document the growth of collaborative decision-making methods as part of a more inclusive and deliberative form of setting natural resource policy.

Managing Under an Ecosystem Approach

Over the last decade, EM has gained a growing legion of supporters in the political arena and the natural resource management fields.[25] Former President Clinton and Vice President Gore consistently supported EM through policy initiatives such as the Northwest Forest Plan (in response to the Northern Spotted Owl) and the Southern Appalachian Assessment, which involved fourteen federal, state, and local government agencies. Although with a different set of priorities than his predecessor (i.e., a much stronger emphasis on economic issues), President Bush has also called on management agencies to consider forest ecosystems, including the interests and sustainability of local communities. In the west, state legislators in Oregon, Idaho, and California have adopted policies encouraging the application of EM principles at the watershed scale. Steven Yaffee and his colleagues[26] have identified over six hundred cases of what they call cooperative EM, while Weber[27] describes the emergence of hundreds of grass-roots ecosystem management initiatives over the past ten years. In short, ecosystem-based management has rapidly become the basic management philosophy for many federal and state natural resource agencies.

While EM has many definitions, the former head of the U. S. Forest Service Dale Robertson described it as "a multiple use philosophy built around ecological principles, sustainability, and a strong land stewardship ethic, with a better recognition of the spiritual values and natural beauty of forests."[28] Robertson's description implies that management practices on public lands must meet certain requirements: they must be (1) ecologically sustainable, directing public lands toward a desired future condition which embodies the complexity of ecosystem interrelationships at a variety of spatial and temporal scales, (2) economically feasible, meeting societal demands for the myriad products of forests and public lands at a cost that does not exceed the priced and unpriced benefits gained, and (3) socially acceptable, reflecting a sensitivity toward recreational, aesthetic, spiritual, and other noncommodity values of public lands.[29]

The acceptance of an environmental *and* economic approach to managing nature is facilitated in part by the science underlying EM. This science borrows from the "impact" science approach and the chaos model of ecology. In the former case, the interactive effects of societal decisions are judged according to how they impact the whole of the ecological complex. As such, impact science evokes an overriding concern with the quality of nature—biodiversity matters—as opposed to a conservationist focus on the commodification and quantification of nature. EM explicitly recognizes the value of the ecological services provided by "healthy nature" (e.g., flood and water quality control by wetlands, waste assimilation, nutrient recycling) and the detrimental impacts of habitat fragmentation on biodiversity.[30]

But the tenets of EM reach beyond ecology. They also imply more collaboration among stakeholders, public and private, as established legal and administrative jurisdictions are blurred in favor of biophysical boundaries. The new focus encourages existing institutions to work together for the purpose of achieving ecological sustainability. From the perspective of a growing number of federal and state resource managers, EM is both a pragmatic attempt to solve increasingly intricate and complex problems of natural resource management and an opportunity to improve government performance by catalyzing as many resources as possible in support of public goals.

Determining Support for Devolution and Ecosystem Management

The embrace of EM by resource agencies along with accompanying trends toward the devolution of authority and implementation increase the importance of local communities (democracy) and science (technocracy) in the policy process. It is important to understand public perceptions and citizen acceptance of EM as it applies to public forestlands. It is also useful to test the (dis)comfort citizens have with different levels of government, particularly as these agencies attempt to share responsibility for managing ecosystems and working with communities. The following section describes public orientations toward these ideas and explains the methodology employed in our 1998 national public opinion survey.

Correlates of Support for EM

A number of authors have addressed various aspects of the relationship between social values and attitudes toward natural resources and the environment.[31] These discussions imply that the current debate about the disposition of ecosystems in the United States is, at heart, not only a professional and technological debate, but a debate about how forest ecosystems should be defined philosophically. The differences between the more traditional, anthropocentric view of forests and the emerging biocentric view thus cannot be settled by an appeal to facts alone.[32] Factual information does not speak for itself; it exists in a cultural context, within a set of assumptions about its relevance, and these assumptions include important value orientations.[33] It is a society's underlying values, to a large degree, that determine which facts will count as important. For these reasons, it is important to understand what those values are and to determine their connections with other relevant social, political and cultural factors.

Based on a number of recent social assessments, it is our judgement that public orientations concerning EM are influenced by a variety of factors.[34] Primary influences include sociodemographic characteristics, self or group interest, and value orientations.

Sociodemographic Factors. Group-based social attributes have been found to be important determinants of environmental values and behavior.[35] Among the most commonly employed measures are gender, age, and education. Age is a widely used variable in evaluating environmental orientation. For example, citizens in Western democracies born after World War II are considered to be more likely than older persons to focus on environmental concerns;[36] consequently, age (as an indicator of cohort) is an important background factor in any environmental study.

In addition, there may be a link between orientations toward EM and gender. There is some evidence to suggest that women are socialized to perceive moral dilemmas in terms of interpersonal relationships, and seek to resolve them by an ethic of care. Men, in contrast, may tend to perceive moral dilemmas in terms of more impersonal features of situations and attempt to resolve them by appealing to rules of justice and rights.[37] This differential socialization experience might lead women to take a more (personally) protective and biocentered view toward nature[38] and therefore be more supportive of EM, while men would tend to be less concerned about protecting ecosystems.

Level of formal educational attainment is included in this analysis because it is broadly associated with having a strong impact on environmental orientations.[39] Those individuals with higher levels of educational attainment are significantly more likely to have value orientations sympathetic to environmental concerns when compared to individuals with less formal education. Howell and Laska believe this relationship is not surprising because ". . . the evidence on both sides of an environmental issue frequently addresses a very complex etiology of causes comprehended more easily by the better educated."[40] We hypothesize this relationship to hold true for orientations toward EM, with higher levels of formal education associated with acceptance of EM.

The last sociodemographic variable included in this study is place of residence—rural versus urban. Some studies have suggested that urban populations are much more likely to have pro-environmental values as a result of better access to information and educational opportunities, and because they are more likely to experience environmental problems firsthand due to industrial activities and high concentrations of people.[41] If this relationship holds true for orientations toward EM, then we hypothesize more positive orientations in the less forested urban areas than in the countryside. This is consistent with the idea of wilderness as a desirable place, where interest in wilderness preservation has grown out of our urban culture.[42]

Interest Factors. An individual's orientation toward EM may very well be influenced by where he or she stands in relation to the productive arrangements of society.[43] Two factors that would affect value orientations are attachment to a natural resource extraction industry or membership in an environmental organization. Persons who rely on resource extraction or agriculture for their economic well-being, for example, are more likely to look at commodity interests as most beneficial, and may be cynical concerning EM because of its focus on sustainability and ecosystem protection. Environmentalists, on the other hand, may tend to view forests and watersheds in terms of broader public goals and to promote the preservation of natural resources.[44] However, because EM attempts to balance ecological sustainability with human use, it may be that environmentalists are cynical about the concept as well. A third interest factor we included in this study is the degree to which people use public lands and forests for recreation. We would expect frequent visitors to public lands and forests to be more understanding of the multiple benefits provided by forests and therefore have a more positive orientation toward EM.

Value Orientations. Orientations toward EM also are likely to be influenced by (or are a component of) general political and social values. For example, the liberal-left perspective has been identified with support for natural resource preservation[45] and higher levels of environmental risk perceptions.[46] Other research suggests that citizens on the left-liberal end of the political spectrum support policy proposals emanating from the environmental movement, while those on the right-conservative side of the spectrum have been found to be less supportive or even hostile to environmental concerns.[47] We hypothesize that those on the left would be more likely to have positive orientations toward EM while those on the right would be more negative. In part, this is due to conservative attachment to the status quo and use of the marketplace to allocate values. Liberals are more likely to critique the existing economic and political system and to support a wider range of noneconomic uses of public lands.

Another correlate of potential support for EM would be the growth of biocentric values toward the environment.[48] Many observers have argued that in contemporary America (and other advanced industrial societies), a New Environmental Paradigm (NEP) has emerged that emphasizes protection of ecosystems and the participation of citizens in the management of public lands.[49] This is a change in emphasis from more anthropocentric value orientations—often called the Dominant Social Paradigm (DSP)—toward public lands management which stressed the production of goods and services beneficial to humans. We expect that citizens who indicate support for the NEP will have more positive orientations toward ecosystem management than those who are more supportive of the DSP.

Methodology and Measurements

Samples. In order to investigate public views a national random digit dial survey was conducted during June and July, 1998.[50] The survey response rate was 54 percent, with 904 completed interviews out of 1,658 households contacted.

Dependent Variables. Questions used to assess public beliefs about EM asked respondents to identify their level of agreement with six statements (see Table 2). The statements were developed by Brunson and others[51] to cover various dimensions of EM. The five response categories ranged from "strongly disagree" to "strongly agree," with a neutral midpoint. After recoding items so that higher numbers reflected a positive orientation toward EM and lower numbers reflected a negative position, the responses were summed to form an indicator of support for EM ranging from 6 to 30. The reliability coefficient (Cronbach's Alpha) for the index was .74 suggesting that respondents were mostly consistent in their response patterns for the additive scale and that scale components were intercorrelated.

Independent Variables. The independent variables used to assess the impact of demographics, interest factors, and value orientations are presented in Appendix A. The demographic factors examined as predictors of value orientations concerning forests include age in years, gender, level of formal education,[52] and city size where each respondent resides (URBAN).[53] To assess an individual's perspective or interest concerning public lands and forests, three indicators were used. Respondents whose families depend on the natural resource extraction or agriculture for their economic livelihood were categorized as resource dependent (RESOURCE),[54] while those belonging to an environmental organization were classified as environmentalists (GREEN).[55] An additional interest indicator assessing the frequency of visits to public forests for recreation is also examined for its impact on orientations toward EM (RECREATE).[56] The factors used to assess value orientations include Van Liere and Dunlap's[57] indicator of the New Environmental Paradigm (NEP)[58] and a self-assessment measure of general political orientation (IDEOLOGY).[59] Summary measures for the various independent variables used in the forthcoming multivariate analyses are presented in Appendix A.

Ecosystem Management from the Perspective of Citizens

Univariate Findings: One of the first things asked of respondents was to indicate their level of informedness concerning EM. It was assumed that a lack of familiarity with the issue would lead to ambiguous responses to EM issues. Therefore, a brief statement describing EM[60] was provided and then respondents were asked how well informed they

Table 1. Self-Assessed Public Informedness About Ecosystem Management

In recent years many federal and state natural resource and environmental agencies, have increasingly emphasized Ecosystem Management on public lands. Ecosystem Management is a shift in the philosophy of managing America's public forests and undeveloped lands. It has been described as blending social, economic, and scientific principles to achieve healthy ecosystems and maintain biological diversity over long periods of time, while at the same time allowing production of the many valued resources our society seeks from its public lands and forests.

How well informed would you say you are concerning ecosystem management?

Not Informed	20%
Somewhat Informed	19%
Moderately Informed	32%
Informed	17%
Very Informed	12%
	N=868

were concerning EM as a management philosophy. The data displayed in Table 1 provide insight into general levels of informedness among our national random sample. Twenty percent of those responding to this question said they were "not informed" about EM; however, 61 percent were at least moderately informed about the concept. Overall, the level of subjective informedness appears to be very high for a relatively new and complex issue.

Table 2 reports the distribution of responses for six indicators of public beliefs concerning EM. The statements were designed to cover both positive and negative views attributed to EM by opponents and supporters.[61] Respondents who indicated they were "not informed" about EM are not included in these results because of a lack of familiarity from which to make judgements. Mean scores (3.68 to 3.95) for the first three statements indicates respondents were slightly more likely to agree than disagree with the positive aspects of EM. The mean scores (3.01 to 3.33) for the last three items suggest that the average respondent is more "neutral" concerning the statements that were meant to reflect skepticism about the EM concept. Standard deviations indicate a fairly normal distribution of responses. At the bottom of Table 2 an additive scale mean and reliability coefficient are reported (higher numbers reflect a positive perspective toward EM and lower numbers reflect a negative perspective). This indicator will be used in the forthcoming multivariate analyses.

Multivariate Analyses: Ordinary least squares estimates for the EM additive index are presented in Table 3. F-test results indicate that the

Table 2. Public Beliefs About Ecosystem Management *

Only respondents who answered they were "somewhat" to "very" informed about ecosystem management (Table 1 above) were asked to respond to the following statements (N=689):

Statements: ***[1=strongly disagree to 5=strongly agree]***	***Mean***	***s.d.***
Ecosystem management...		
helps us think about public lands as a whole instead of focusing on single resources such as timber	3.95	1.34
lets us protect endangered species while continuing to harvest resources from public lands	3.84	1.18
will enhance the long-term health of public lands	3.68	1.21
is being used as an excuse to extract natural resources from areas previously closed to resource extraction	3.33	1.07
is a misguided attempt to reduce public complaints without any scientific basis	3.21	1.18
is an attempt by environmentalists to stop natural resource extraction on public lands	3.01	1.01
Additive index mean/s.d Cronbach's Alpha=.74	18.97	4.22

* For the additive index assessing public orientations toward Ecosystem Management, the last three items were recoded so that higher numbers reflect a positive perspective toward Ecosystem Management and lower numbers reflect a negative perspective.

model is statistically significant, however the adjusted R^2 suggests that only 22 percent of the variation in public beliefs about EM is explained by our model. For the demographic variables, all four have a statistically significant impact on orientations toward EM. Younger respondents and women are significantly more likely to have positive orientations toward EM than their older and male counterparts. Education also has a significant impact with the more highly educated having more positive beliefs about EM than their less formally educated counterparts. For the variable URBAN, respondents living in urban areas are significantly more likely to have positive orientations toward EM than those from smaller cities or rural areas.

Table 3. Ordinary Least Squares Estimates for Support of Ecosystem Management

Variables:	*B*	*ß*
Age	-.07**	-.14
Gender	.32*	.09
Education	.37**	.13
Urban	.13*	.08
Resource	-1.34**	-.17
Green	-.04	-.03
Recreate	.44*	.09
Ideology	-.41**	-.11
NEP	.92**	.15

R-square =.23
Adjusted R-square =.22
F test =34.76**

* Significant at $p<.05$; ** Significant at $p<.01$

In regard to the interest variables, as expected those respondents who depend on natural resource extraction and agriculture for their economic livelihood are significantly less likely to have positive orientations toward EM than their non-resource dependent counterparts. However, environmental organization members are no more positive or negative concerning EM than nonmembers. Interestingly, when additional analyses were conducted on this group we found much skepticism concerning EM as a way to increase resource extraction on public lands. For the variable assessing the level of individual recreational activity on public lands, we find that those respondents who visit public forests frequently for recreation were more likely to have positive orientations toward EM than their counterparts who visit infrequently or never.

The final two variables concern dimensions of political orientation and are statistically significant in explaining orientations toward EM. When controlling for the independent effects of other variables, those respondents who identified themselves as liberal are more likely to have positive beliefs about EM than self-identified conservatives. The second value indicator shows that supporters of the NEP are more positive in their orientations toward EM than those respondents who identify more closely with the Dominant Social Paradigm. It is interesting to note, however, that when members of environmental organizations were removed from the analyses, the coefficient for NEP showed an even greater association with orientations toward EM. Strong supporters of the NEP who are not members of environmental organizations were very supportive of EM.

Preferred Level of Implementation and Public Participation

The data presented in Table 4 reflect public trust in government for implementing EM. Respondents were most trusting of local government and least trusting of the federal government. These findings are consistent with recent attempts to decentralize environmental policy decisions and implement practices through lower levels of government.[62]

A final question asked respondents their view about a realistic role for the public in ecosystem management. Results presented in Table 5 show that 3 percent of the respondents believe that resource professionals should make decisions without public input while another 14 percent feel citizens should be able to provide suggestions. The remaining 83 percent believe that the public should play a more prominent role in EM, from serving on advisory boards to actually making management decisions, with almost half (48 percent) selecting the last two categories, which would give citizens greater authority than current laws allow.

Table 4. Preferred Level of Government for Implementing Ecosystem Management

*Which level of government—federal, state, or local—do you most trust to implement ecosystem management?**

Federal	11%
State	29%
Local	38%
Trust all equally	22%

* Only respondents who answered they were "somewhat" to "very" informed with ecosystem management (Table 1) are included in this table (N=691).

Table 5. Public Participation in Ecosystem Management

In your opinion, which of the following would be the most realistic role for the public in ecosystem management?

None, let resource professionals and managers (U.S. Forest Service, Bureau of Land Management, etc.) decide	3%
Provide suggestions and let resource professionals and managers decide	14%
Serve on advisory boards that review and comment on decisions	35%
Act as a full and equal partner in making management decisions	36%
Public should decide management issues and resource professionals and managers should carry them out	12%

* Only respondents who answered they were "somewhat" to "very" informed with ecosystem management (Table 1) are included in this table (N=691).

Conclusion

These findings from our exploratory research shed light on the social landscape surrounding a move to a new model of forest management. We recognize their limits in explaining the full range of factors that may influence public acceptance of EM as well as more devolved forms of decision-making in the United States. However, the variables studied provide an initial snapshot from which useful observations can be made.

First, it appears that many citizens feel reasonably well informed about EM in general. Additionally, citizens in this study tended to respond favorably to the concept of ecosystem-based approaches, potentially viewing them as solutions to ensuring healthy forests over the long-term. However, the findings also document some skepticism among key constituencies for implementation of these strategies. Rural residents and those dependent on natural resource based economies were significantly less supportive of EM than their urban and non-resource dependent counterparts. Closer examination suggests that these key publics may become more supportive if devolved approaches that give state and local jurisdictions more influence are implemented. Overall, these findings are consistent with research in the western United States[63] and lend support to the notion that different constituencies will require different forms of engagement about resource allocation. One solution may come from an associated study where rural citizens were more willing to listen to and support the efforts of local resource managers as opposed to those from cities who preferred information from and interaction with the scientific community.[64]

Second, these findings suggest that mobilizing a broad array of citizens to embrace simultaneously the science of EM and enhanced citizen participation in the policy process may well be a matter of getting the institutional arrangements right. While these respondents may not have a great deal of confidence in national levels of government, it appears they do trust regional and local authorities. This would suggest that more decentralized approaches, as EM purports to be, offer an opportunity for building confidence and support between citizens and agencies. One key action seems to be for large federal organizations like the Forest Service to make good on their commitment to a more bottom-up management style. This may be problematic, where community constituencies may trust their local (federal) resource manager, but have serious reservations about the larger national agency allowing him or her actually to implement local decisions.[65]

A third point is that in spite of questions about their confidence in the bureaucracy, more of these respondents still believe that government agencies at all levels—but particularly the local and regional level—should exercise a good deal of influence in implementation of EM. Whatever their reasons for this support—either support for science-based systems or more citizen participation, or both—the EM model may provide a way to build a more collaborative constituency. However, there is also some evidence suggesting that the public will actively resist the use of EM if it is accompanied by top-down federal control.[66] One recent example is the Clinton administration's Interior Columbia Basin Ecosystem Management Plan, a massive effort to apply the EM model to 70 million acres covering parts of Montana, Idaho, Washington, Oregon, Nevada, and Wyoming. Public resistance, especially from local level officials and private property owners, was widespread and convinced Congress to cut funding for the plan. In any case, EM could provide methods to bridge the gaps between those who support one view or the other and, as a result, expand the number of citizens willing to try out these devolved management approaches.

It may be possible to construct situations that mesh technical expertise with public involvement in such a way that both sides come closer to being satisfied. With grassroots EM, science and democracy are each treated as essential to good policy decisions. Science is used to develop deeper understandings of complex problems and inform policy choices (alternatives). Democracy is served through broad participation in policy decisions and as citizens come to "own" outcomes they more willingly assist implementation.[67] In short, democratic processes are enhanced rather than eroded, science is given a privileged position rather than being ignored, citizens learn to trust both science and government, and, ideally at least, more of the complex, technical policy problems about the use of public forests can be resolved. Overall, it appears that our respondents are receptive to the concept of EM, but will be likely to wait to see how well it works before making final judgments. If EM is to succeed, it will most likely be accomplished in settings where citizens and stakeholders are included in the decision process—places where the consequences of choices, along with the attendant scientific uncertainty, are out on the table for all to evaluate.

Appendix A

Distributional Characteristics for Hypothesized Determinants of Ecosystem Management Support*

Variable Name	*Variable Description*	*Mean*	*(s.d.)*
Sociodemographic Indicators			
Age	Respondent age in years	42.3	15.20
Gender	Dummy variable for gender 1 = Female 0 = Male	.51	
Education	Level of formal education 1 = Some grade school to 8 = An advanced degree	5.27	1.21
Urban	Respondent residence 1 = Rural area to 7 = City of 250,001 plus	4.24	0.97
Interest Indicators			
Resource	Economic livelihood dependent on resource extraction/agriculture 1 = Dependent 0 = Else	.20	--
Green	Member of environmentalist organization 1 = Member 0 = Else	.11	--
Recreate	Frequency of participation in public lands recreation 1 = Never to 5 = Very frequently	3.01	0.95
Value Indicators			
Ideology	Subjective political orientation 1 = Very liberal to 5 = Very conservative	3.24	0.97
NEP	New Environmental Paradigm 6 = lowest level of support 30 = highest level of support	23.26	5.10

* Only respondents who answered they were "somewhat" to "very" informed with ecosystem management (Table 1) are included in this table.

6

The Cultural Context for Forest Policy Decisions: The Case of Western Canada and the United States

John C. Pierce and Nicholas P. Lovrich

Introduction

The neighboring countries of Canada and the United States share much. These two North American nations are joined by a common border, by an interdependent history of more than two hundred years, by a shared commitment to democratic institutions and values, and by a public policy agenda of similar challenges. Yet it is often said that there is much to separate this pair of North American democracies. Indeed, even a superficial acquaintance with the two nations would suggest that Canada and the United States are more like fraternal than identical twins. There are obvious contrasts in the institutional structures and practices through which democratic values are articulated in the two countries. The parliamentary vs. presidential systems of governance and the vastly different political party systems are the most obvious of those contrasts, but many other differences are present as well.

In recent decades considerable debate has taken place regarding the degree to which Canada and the United States have similar or different *political cultures*. That debate centers on the identification of the particular traits on which the two countries' cultures are distinctive, the sources of those differences, and the degree to which the larger forces of technology, media/information networking, and globalization of national economies have brought the two cultures closer together than they ever have been before.

Cultural differences in the politics of two countries are of particular interest to scholars because these differences may manifest themselves in how public policy takes shape in the two countries. The resolution of the policy disputes results in the authoritative allocation of values for a country; one would thus expect differences in the value bases central to political cultures to be revealed in concomitant contrasts in policy processes and policy outcomes. Consequently, our understanding the depth and the centrality of political culture differences surely will be enhanced by examining the degree to which political culture differences play out in salient shared policy areas. At the same time, though, for those observers whose primary focus is a particular policy arena, the importance of delving into the cultural roots of the distinctive and prototypical patterns, processes and outcomes cannot be overstated. A proper understanding of those policy roots will lead to the crafting of more effective, socially inclusive and legitimate policy.

This chapter weds the two questions of the cultural differences in Canada and the United States, and the deeper roots of variations in Canadian and US approaches to a common policy area. The geographic focus of our analytical attention is the northern Pacific Coast—namely Oregon, Washington, and British Columbia. The policy area focus is that of the management of forest resources. Forest resource policy strikes at the fundamental values of contemporary political cultures—how decisions about public policy should be made, the values upon which policy goals should be selected, and the policy mechanisms seen as legitimate means of achieving those values. This chapter begins with a brief overview of the historical, institutional, and cultural differences existing in the politics of the two countries. Given the extensive attention they receive throughout this book, forest resource policies and associated dimensions of public policy conflict are given but a brief overview.

The remainder of the chapter represents an empirical analysis of differences in significant dimensions of the political cultures of the two countries, and an explanation of how those political culture dimensions play out in the attitudes of the citizens of Oregon, Washington, and British Columbia. The forest policy focus is that of *ecosystem-based management practices*, a broadly framed approach seeking to balance the needs for the economic exploitation and the ecological health of forest resources; this is the common policy framework generally employed on both sides of the U.S./Canadian border. After comparisons of the citizens of the two countries on a range of cultural dimensions and on their attitudes toward ecosystem-based management practices, this chapter examines the degree to which attitudes about these forest resource management policies and practices are attributable to differences in Canadian and United States cultural orientations.

Political Culture in Canada and the United States

Political culture is taken to be "that mix of values, traditions, conventional actions, common symbols and prevailing expectations that guide the fundamental course of a society's characteristic approach to the collective resolution of conflict."[1] Political culture embeds assumptions and beliefs about who has what rights in society, how power is to be won and distributed, and where the location of sovereignty is held in regard to decisions about public and private life. Communities, regions, nations and groups of nations exhibit distinctive political cultures.[2] Historically, the principal focus of analytical attention by scholars has been on those distinctive cultural attributes of nations that divide one from another, and that make a particular country unique among the world's nations. More recently, though, heightened attention has been given to the effects of broad, sweeping global changes that are believed to be bringing historically different cultures closer together. Postindustrialism—in the form of globalization of the national economies, sustained economic wellbeing in large parts of the "developed" world, international communication and media systems, and new values in many locations—are seen to be breaking down pre-existing cultural differences among the highly developed countries.[3]

The dynamics of how the contrasting forces of Canadian and U.S. political culture either persist or disappear have received significant attention in recent years. While the two nations share a cultural origin in the British Empire and the American Revolution,[4] quite different sets of societal values, traditions and political institutions emerged out of that common origin.[5] The Canadian political culture has been described as deferential, organic, communalistic and particularistic.[6] Politics in the United States, in contrast, reflects the values of individualism, egalitarianism and entrepreneurship. Smith describes some of these fundamental differences in the following way:

> . . . The United States was to be understood as the liberal society par excellence, a community of free and responsible individuals bound together by the commitment of each of its members to accept their fellows. . . . Canadians, too, were able to devise a conception of the nation that could be used to assist its consolidation. . . . Their commitment to the idea that the nation was a thing of groups. . . a community of particularisms.[7]

Over the course of the past three decades a considerable literature has developed on both sides (and in the middle) of the questions of whether the political cultures of Canada and the United States truly differ, and whether whatever differences may have existed are persisting or are being dissipated by larger social and political forces.[8] This chapter investigates the degree to which citizens of Canada and the United States in the Pacific Northwest region of North America in fact differ on important indicators of political culture, and whether those differences are translated into differences in the area of forest resource policy preferences relating to ecosystem management and sustainable forestry practices.

Canadian and American Forest Policy

In 1990, in reference to forest resources in Canada and the United States, Thomas Waggener observed the following: "Given the interests or elements of public policy concerns and historical/cultural differences on each side of the forty-ninth parallel, it is perhaps not surprising that the two nations have implemented different policy instruments and institutional arrangements."[9] In reflection of these historical/cultural differences, in Canada over 90 percent of the commercial forest lands is held in public hands while in the United States approximately 70 percent are privately owned.[10] The much more diverse forest land ownership pattern present in the United States makes the creation of a comprehensive public policy on forest resources much more difficult. Nonetheless,

> British Columbia has constructed a system of expansive public ownership of forest land yet has delegated much managerial control to private entities. . . . In contrast, the United States has developed a system of mixed ownership and a tradition of public involvement, even in decisions about private forest land.[11]

Hoberg describes U.S. forest policy formation as "pluralist legalism,"[12] with these three principal characteristics: "(1) formal administrative procedures, with widespread access to information and rights to participation for all affected interests; (2) organized environmental groups with access to the courts; and (3) nondiscretionary government duties, enforceable in court."[13] In Canada, on the other hand, the principal locus of forest policy is situated in the provinces, and political authority lies with the parliamentary system in the executive branch. Indeed, Hoberg opines, "forest policy is made in profoundly different ways in British Columbia and the United States. . . ."[14]

Recent research also suggests that the opinion climate in regard to forest resources also differs across the U.S./ Canadian border. Washington residents, for example, are less likely to support clear-cutting as a policy than are residents of British Columbia, and they are more likely than their neighbors to the north to take the environmental side when considering trade-offs with economic values in forest policy.[15] Canadians from the Province of British Columbia remain markedly more trusting of governmental authorities than their American counterparts, and forest resources management issues are more center stage for them than they are for the citizens of either the State of Washington or the State of Oregon.

The Study

The findings reported here are derived from a combination of mail and telephone surveys conducted by trained university student interns working under the supervision of research faculty associated with the Program for Governmental Research and Education (PGRE) at Oregon State University. Funding for the various surveys was provided by PGRE, the Division of Governmental Studies and Services at Washington State University, and a Canadian Embassy Faculty Research Grant (partial funding for the British Columbia survey).

There were four principal sub-populations surveyed for this study: 1) a random sample of 559 households in British Columbia; 2) a random sample of 574 households in Washington; 3) a random sample of 600 households in Oregon; and 4) a random sample of 489 households in forest resource dependent Lane County and Linn County (as part of an NSF research project examining the role of science in natural resource management in Oregon's Cascade Mountains). The Lane County and Linn County results are not included in the analysis presented here. Both the telephone and mail surveys undertaken for this study were conducted during the period July 12 to September 20, 1999. The overall response rate for the combined surveys was 54 percent. The demographic attributes of the individuals surveyed in Washington, Oregon and British Columbia are reasonably similar. For example, the percentage female in the samples ranged from 47 percent in Oregon to 52 percent in both British Columbia and Washington. The average age among the British Columbia respondents is forty-six years compared to fifty in Oregon and fifty-one in Washington. Fifty-two percent of the British Columbia respondents had taken some college or more formal education, while the percentages are 55 percent for Washington respondents and 51 percent for Oregon respondents, respectively.

Findings

The initial question of concern in this chapter is the degree to which there are differences among British Columbia, Washington, and Oregon in the cultural context within which forest resource management policy is taking shape. Our analysis features two independent approaches to this question. The first analytical tack is to contrast the survey respondents in the three locations in terms of their overall positions on important cultural elements related to politics generally, and to the environmental policy arena in particular. We have identified six cultural orientations on which to compare the three samples—the New Environmental Paradigm, support for science and technology, support for public participation in policy formation, position on the post-materialism value index, self-identified liberal-conservative ideology and support for interest group involvement in the public policy process.

The second approach is to look at the relationship within each of the three sub-populations between positions on those six cultural dimensions and citizens' attitudes on *ecosystem management/sustainable forestry principles.* While the three survey sites may turn out to be quite similar in their cultural orientations, it is possible that the impact of those political culture orientations on forest resource policy preferences may be much more powerful in one location than in others. That is, forest resource management preferences may "engage" values in very different ways in different locations. We begin the analysis with a look at the degree to which the respondents in the three study locations differ with respect to their positions on the six cultural orientations.

Cultural Orientations

Self-Identified Political Ideology. The use of the liberal/conservative continuum has a long history in descriptions of western political thought. While the terms "liberal" or "conservative" or "left" or "right" have been employed to refer to a broad range of political differences (e.g., social norms, political tolerance, attitude toward change), most frequently they have been used to describe differences in orientations toward the role of government in the allocation of important values, economic or otherwise. Perhaps the primary political division in regard to forest resource policy has to do with perspectives as to the appropriate role of government in managing and controlling the use of that natural resource for the service of alternative values, whether economic or aesthetic in character. Thus, liberal/conservative identification may be one of those fundamental orientations that may define a political culture and that may provide the context for the development of and support for a particular forest resource policy.

The respondents to the study reported here were asked to place themselves on a seven-point scale, ranging from one (1) for "very left/ liberal" to seven (7) for "very conservative/right." The results, in the form of average placement, are shown in Table 1. The results reported in Table 1 indicate that there is a statistically significant difference among the three samples' ideological identification (F=5.2, p=.01). While the Oregon and British Columbia respondents are quite close to each other (means of 3.81 and 3.78), the Washington respondents are more conservative (mean of 4.03).

If intentional, planned and long-term strategies for the multiple use of forest resources (for timber production, recreational use, wildlife habitat, biodiversity) taps into the dimension of government involvement in the allocation of important values, then we would expect that the more liberal cultural context present in Oregon and British Columbia would lead to greater support for such strategies there than in the State of Washington.

Support for Public Participation in the Policy Process. Many public policy domains are seen as the arena for competition among dominant economic interests and largely outside of the window of oversight available for the general public. Issues dealing with natural resources and the environment, however, have been topics most central to the increasing demands for public involvement in policy formation and

Table 1. Self-Identified Liberal/Conservative Ideology*

	Washington	*Oregon*	*British Columbia*	*F*	*Sig.*
Mean Ideology	4.03	3.81	3.78	5.2	.01

*The self-identification scale is from one (1) to seven (7), with "very left/ liberal" at the 1 end and the "very conservative/right" position at the 7 end. The mid-point of the scale (4) is identified as "moderate."

oversight. Support for public involvement in policy formation, though, is one of those dividing lines alleged to separate the American from the Canadian political cultures. The American political culture is seen as egalitarian, focusing on the equal access of broad ranges of the citizenry to the formation of public policy; in contrast, the Canadian political culture is seen as more particularistic, more focused on the interaction among historic, traditional groups in the society. Again, two expectations emerge. The first is that the Oregon and Washington survey respondents would surface as more supportive of public participation in natural resource management policy setting than their Canadian counterparts. In particular, Washington State respondents—coming from a wide-open primary system and frequent use of the initiative and the referendum—would be expected to be the most supportive of public participation. Secondly, if the cultural argument is to hold sway, one would expect that those locations where the greatest support for public participation is found would also be the locations where there would be least support for either a forest professional determination of policy direction, or an open field for the interplay of interested parties.

Table 2 shows the results of an analysis of variance among the average sub-sample scores on an index of support for public participation. The index is created from responses to two survey questions: "Recently there has been considerable debate over efforts to increase citizen participation in government policy making. Where would you locate yourself on the following scale regarding these efforts?" The scale ranges from one (1) representing "no value" in public participation to seven (7) representing "great value;" and, "In your opinion, a realistic role for the public in natural resource management issues should be," ranging from one (1) signaling "no role" to five (5) reflecting the belief that "the public should decide management issues and resource professionals should carry them out." Responses to the two items are summed into a single score that

Table 2. Support for Public Participation in Natural Resource Management Issues*

	Washington	*Oregon*	*British Columbia*	*F*	*Sig.*
Mean Support for Public Participation	9.01	8.74	8.56	7.3	.001

*The index of support for public participation is constructed from two questions (see the text), with the low end being two (2) and the high end being twelve (12).

ranges from two (2) to twelve (12). The mean scores on the index are calculated for each of the three study locations.

The survey results reported in Table 2 show support for public participation to be greatest among the Washington State sample, and the lowest among the British Columbia respondents; the Oregon respondents fall in between. If cultural context is important, and if a professional or interest group-dominated planning or policy effort is on the table, then one might expect greater support from the British Columbia respondents for ecosystem management/sustainable forestry principles emerging from an elite-mediated source.

Post-material values. The last quarter of the twentieth century witnessed an unparalleled set of socioeconomic changes among Western democracies, loosely termed post-industrialism.[16] Those wide-ranging changes included sustained and large-scale economic growth, an increasingly central role for science and technology, globalization of media and information systems, and a decline in the threat of conflict among the world's major powers.[17] In turn, these changes have been shown to have restructured the sets of value priorities of the citizens of these post-industrial nations.

The development of much greater affirmation of "post-material" values, those focusing on the quality of life and the fuller participation of individuals in collective decisions affecting their own lives, has had several consequences. One effect is that policy preferences and political behavior concerning important issues relating to peace, the economy, the environment, and equality are increasingly driven by these post-material values. A second alleged consequence is the hypothesized convergence of political cultures among postindustrial nations, with shared post-material values supplanting the more nation-unique values that previously had characterized differences among countries. Thus, one important question in assessing the potential cultural foundations of differences in forest resource management policy is whether Canada and the United States (in this case, British Columbia and the states of Washington and Oregon) differ in the relative presence of these post-material values to a degree that would predict differences in policy preferences.

The respondents in the study reported here were asked a set of widely used questions (e.g., in the annual "Eurobarometer" survey, the periodic World Values Survey, and other cross-national studies) that allows the classification of individuals into three value types: post-material, mixed and material. The preface to the question reads as follows: "There is a lot of talk these days about what our country's goals should be for the next ten years. Listed below are some of the goals that different people say should be given top priority. Please mark the goal you yourself consider to be the most important in the long run. What would be your second choice?" The four options listed for survey respondents were these: 1) maintaining order in the country; 2) giving people more say in important governmental decisions; 3) fighting rising prices; 4) protecting freedom of speech. The second and fourth options represent post-materialist values, and the first and third options represent materialist values. The average scores on the post-materialism index are displayed in Table 3, with a 3 indicating post-material values (options 2 and 4 selected), a 2 reflecting mixed values (one post-materialist and one materialist option selected), and a 1 representing material values (options 1 and 3 selected).

The survey results reported in Table 3 show that no significant differences exist in average post-material values appear among the three samples. On the average, at least, the three policy locations are quite similar in this fundamental, postindustrial cultural orientation that may provide the context for policy formation. To some degree this is a mild surprise, given that the British Columbians are significantly lower than the Oregon and Washington residents (especially the former) on two other dimensions usually found to be associated with post-material values—namely, liberal political ideology and support for public participation in policy formation.

Support for Technology. The belief that policy issues can be resolved through the application of science and technology is no stranger to the natural resource arena. This belief in the efficacy of technology emerges strongly in the training of professionals in both the relevant natural

Table 3. Mean Post-material Value Index Scores*

	Washington	***Oregon***	***British Columbia***	***F***	***Sig.***
Mean Post-Material Values Score	2.08	2.01	2.02	2.2	.107

*The entry in each column is the average for the sub-sample on a three-point index of post-material values, with 1 being materialist, 2 being mixed, and 3 being post-material.

sciences and in the policy sciences.[18] In the former, of course, are the research investments made in identifying the appropriate techniques for sustaining healthy forests, maximizing their productivity over time, and the identification of other elements of forest management related to silviculture, plant pathology, social sciences, hydrology, natural resource economics, etc. In the latter the focus is on the management techniques and harvest and maintenance strategies that provide the administrative and political frameworks and structures within which policy priorities can be most effectively achieved. Implicit in this rationalistic approach is an underlying assumption that by focusing on the science and technology of forest resource management—be it a matter of administrative or biological science—the decisions and strategies developed by professional experts will be removed from the political arena, conflict will be reduced, and more effective goal-directed outcomes will be achieved systematically.

As the prototypical postindustrial country, one would expect the U.S. study sites to produce relatively strong support for technology, and in the United States one would expect Washington State, with the prominence of its high tech software and aerospace industry, to be more supportive than Oregon. Later in the analysis we would expect the level of support for technology to be associated with support for ecosystem management/sustainable forestry across all three research locales.

The level of respondent support for technology is measured through an index composed of two questions. The first questionnaire item asks for level of agreement or disagreement with this statement: "Technology will find a way to solve the problem of shortages of natural resources." A five-point response scale was employed, ranging from 1 (one) reflecting "strongly disagree" to 5 (five) representing "strongly agree." The second question asked respondents to agree or disagree with this statement: "People would be better off if they lived without so much technology." Again the scale ranges from 1 (one) for "strongly disagree" to 5 (five) for "strongly agree." Since the first statement is pro-technology and the second statement is anti-technology, we summed them into a single index after recoding responses to the second item in a positive direction. Consequently, the index of support for technology ranges from 2 (two) to 10 (ten), with 2 indicating the least support and 10 reflecting the greatest support for technology. The results contrasting the mean technology support scores in the three research locations are set forth in Table 4.

Table 4. Support for Technology*

	Washington	*Oregon*	*British Columbia*	*F*	*Sig.*
Mean Support for Technology	·5.86	5.66	5.78	2.1	.126

*The index of support for technology is constructed from two questions (see the text), with the low end being two (2) and the high end being ten (10).

The results displayed in Table 4 document the absence of any statistically significant difference among the three sub-samples with regard to their level of support for technology. While Washington State respondents do reveal a marginally higher average support score, a post-hoc Scheffe set of contrasts between Washington State respondents and the others showed that even these apparent differences do not reach statistical significance.

Interest Group Particularism. Contrasts drawn between the political cultures of Canada and the United States often focus on the more particularistic view of politics and society said to be predominant in Canada.[19] That is, Canadian society and politics are viewed as legitimately being organized around, and interests articulated on behalf of, segments of the population sharing important attributes such as class, national origin, or economic interest.[20] While in the United States there is clear recognition of such shared interests, in principle at least the fundamental building block of society and politics is the individual rather than the group.

Differences in the support for interest groups growing out of a more removed particularism or individualism may spill over into differences in support for forest resource policy based in ecosystem management/ sustainable forestry. One might expect that in those locations—or among those individuals—who support interest group activity more strongly one would find lesser support for the ecosystem management principles of contemporary forestry practice. Such a formal planning and policy practice might be seen as a structure designed to reduce the impact of interest groups present in more normal pluralistic politics and the exercise of interest group influence and power. In order to find out if the sub-samples differ in their support of interest groups, the respondents were asked the following question: "In complex societies such as the United States and Canada, interest groups emerge to represent different interests in society. What is your view of the role

Table 5. Support for Interest Groups*

	Washington	*Oregon*	*British Columbia*	*F*	*Sig.*
Interest Group Support	2.89	2.84	2.84	.31	.334

*The entry in each cell is the average score for the respondents in that sub-sample on responses to a question about interest groups being necessary for government to be aware of people's needs. The higher the score on the five-point scale, the less the support for interest groups.

these groups play in politics?" A five-point response scale was provided, anchored at one end by the descriptor "Interest groups are necessary to make government aware of people's needs" and at the other end by "Interest groups prevent government from being aware of people's needs." The results are shown in Table 5.

As the survey results displayed in Table 5 show, no significant difference emerged among the respondents in the three study locations with respect to their beliefs about interest groups. On the average, all three sets of citizens are slightly below the midpoint of the scale, in the direction of being somewhat favorable toward interest groups as necessary to make government aware of people's needs. Of course, while the summary statistic may show little difference among the three political cultures, it is possible that the centrality of interest groups may be reflected in the effects of variations in attitudes about them on policy preferences. That is, attitudes about interest groups may be strongly associated with preferences about ecosystem management in British Columbia, but very weakly connected in Washington State. That possibility will be assessed in a subsequence portion of the analysis.

New Environmental Paradigm. With some significant exceptions, until the middle of the twentieth century the prevailing view of the relationship of humans to natural resources was anthropocentric—that is, human-centered. Natural resources and the environment, while often to be valued, were important because of what they could contribute to the well-being and enjoyment of humans. Those human-centered benefits could be either economic or aesthetic, but still the dominant criterion for the evaluation of natural resource policy was its instrumental contribution to human well-being. However, the latter part of the twentieth century witnessed the rise of the environmental/preservationist movement as a major political force across the whole planet. The rationale for the considerate husbandry of natural resources and the natural environment became *biocentric*—that is, considered within a framework wherein humans, plants and animals all were seen as co-equal in an integrated, interdependent ecological system.

Linked largely to the work of Riley Dunlap[21] this all-inclusive way of looking at the relationship between humans and the natural environment has been labeled "The New Environmental Paradigm." Variations in support for the New Environmental Paradigm (NEP) and its biocentric tenets have been shown to be strongly related to variations in positions on important natural resource policy questions in prior cross-national research.[22] Moreover, considerable research has shown significant variations among political cultures in the relative predominance of the anthropocentric and biocentric views of the natural world among both citizens in general and natural resource policy activists.[23]

Support for the NEP world view is incorporated in this study for two distinct reasons. The first stems from the nature of the historic political cultures of Canada and the United States. The Canadian political culture has been said to be more communalistic and organic, with a sense of the interdependence and interconnectedness of all elements of Canadian society and the country's natural environment. In contrast, it is widely argued that in the United States a more segmented, instrumental legacy contributes to its culture, one wherein the well-being of different sectors of society (and their respective environments) are a consequence of the outcomes of self-interested behaviors structured by the marketplace. This difference is important because of the second reason—namely, that variations in levels of agreement with the biocentric NEP may structure support for the forest resource management policy issues under consideration here. We would hypothesize that individuals with greater agreement with the NEP would exhibit greater support for the ecosystem management/sustainable forestry principles featured in this study because of the assumptions in

Table 6. Support for the New Environmental Paradigm*

	Washington	*Oregon*	*British Columbia*	*F*	*Sig.*
Mean NEP Support	1.92	1.94	2.11	11.1	.000

*The entry in each cell is the average score on the adjusted NEP index (see the text), with a higher score indicating greater support for the NEP. The adjusted index ranges from 1 to 3.

the latter of *system* considerations, namely the interdependent effects of different decisions on other elements in the same (however complex) natural resource and human system.

The average scores of the three samples on the adjusted NEP index are reported in Table 6. The NEP index used in this study is based on a six-item scale, and each respondent was classified as being high, medium or low in agreement vis-à-vis the top, middle, and bottom thirds of all scores taken in the aggregate. That summary scale reported here ranges from 1 (one) for the low agreement end to 3 (three) for the high agreement end. As we had expected, there is greater support for the New Environmental Paradigm in British Columbia than in either Washington State or Oregon. This result seems consistent with dominant conceptions of the political cultures of the two countries. Later on in this chapter we will discover whether this NEP pattern translates into differential support for ecosystem management, either across countries or within them.

The Integration of Cultural Paradigms. To this point, the analysis has contrasted the three study sites in terms of their average position on each of the alternative ways of looking at the political, social and natural world. But cultures are characterized by more than simply the archetypical position of individuals on significant dimensions; they are also distinguished by the degree to which the views on the different positions themselves are interrelated. This interrelatedness represents an indicator of cultural integration, of a sense of the degree to which individuals see links between (or common elements among) different ways of framing their social and natural worlds. It also is an important indicator of the degree to which political and social conflict over important policies can be effectively organized and resolved. If there is a cognitive connection across different ways of looking at the world, then there is the heightened probability of effective communication among different segments of the society and individuals with different interests in outcomes from the policy itself. As a way of assessing this overall cultural integration within each of the three sub-samples we have conducted a factor analysis. That type of analysis reveals the degree to which there is cultural value integration within the political cultural frameworks citizens employ when they approach the natural resource policy arena. The results of the factor analysis are set forth in Table 7; separate factor analyses for the British Columbia, Washington, and

Table 7. Principal Components Factor Analysis Results for Cultural Variables*

Variable	*Factor I*	*Factor II*	*Communality*
Washington			
Technology	-.430	.592	.535
Participation	.521	.591	.540
NEP	.716	-.177	.545
Postmaterial	.515	.587	.610
Ideology	-.661	.217	.484
Groups	-.244	.194	9.709E-02
	46.78% of variance explained		
Oregon			
Technology	-.586	.472	.566
Participation	.622	.349	.509
NEP	.705	-.108	.509
Postmaterial	.477	.513	.491
Ideology	-.599	.411	.528
Groups	-.237	-.576	.388
	49.87% of variance explained		
British Columbia			
Technology	-.521	-.597	.618
Participation	.678	-3.09E-02	.460
NEP	.655	.386	.579
Postmaterial	.486	-.254	.301
Ideology	-.450	.451	.406
Groups	-.399	.542	.454
	47.11% of variance explained		

*The entry in each cell is the factor loading for that variable on that dimension produced by a principal components factor analysis.

Oregon respondents reveal both similarities and differences in how the political culture frameworks of these citizens are organized.

In all three sub-samples, two principal factors emerge from the factor analysis, suggesting a similarity in the relative complexity of the cultural belief structures at play. Moreover, in all three locations the amount of shared variation explained by the analysis is quite similar, with a slightly greater percentage explained in Oregon than in either Washington or British Columbia. This set of findings suggests that there is slightly greater coherence in the organization of the cultural beliefs in the southernmost of the study sites. In Oregon and British Columbia, the support for technology index is the most central to the cultural beliefs (see the communality scores) while in Washington State the post-material values measure holds that position. The centrality measure indicates the degree to which the other beliefs share a common relationship to the particular cultural orientation. While there is some variation in the factor scores associated with each of the indexes in the results, the overall outcomes are pretty much the same. The direction of the loadings for the variables is the same, and generally the highest loading is found among the same small set of variables. For example, on the first factor, in both Washington and Oregon the NEP index loads the strongest, and it is the second strongest in British Columbia.

In an overall sense the general structure of the political culture beliefs is about what we would expect—that is, significant similarities in structure and content, but enough distinctiveness in particular locations to account for some possible disparities in how citizens might respond to a particular policy question. At this point we turn to the question of whether the citizens of British Columbia, Washington, and Oregon differ in their policy orientations in regard to ecosystem management, and whether those differences—either across locations or within locations—are systematically related to the cultural dimensions we have examined to this point.

Support for Ecosystem Management Principles

Levels of Support. The surveys conducted for this study asked respondents four questions about their views about ecosystem management. The results for these four questions are reported in Table 8. Perhaps attributable to its historic political culture that sees society as more of an integrated, organic system, the British Columbia respondents are more likely to say that ecosystem management is a "responsible" approach, while Oregon is less, and the British Columbia respondents are more likely to say that this policy direction will contribute to the long-term health of public lands. No difference emerged among the respondents in terms of the perception that the policy is an excuse to extract natural resources, and very little difference in the perception that it is a strategy by environmentalists to stop extraction.

In order to create a summary measure of support for ecosystem management principles, the two general items ("responsible approach" and "enhance long term health") were combined into a single index score, ranging from 2 (low support) to 10 (high support). The results for this summary measure across the three study sites are displayed in Table 9. The findings reported in Table 9 reinforce the conclusion that British Columbia respondents express greater support for ecosystem management than do those in Oregon and Washington. While the

Table 8. Attitudes about Ecosystem Management*

% Agreeing	*Washington*	*Oregon*	*British Columbia*	*Sig.***
Ecosystem management is a responsible approach	62%	57%	72%	.000
Ecosystem management will enhance long-term health of public lands	57%	52%	66%	.000
Ecosystem management is an excuse to extract natural resources from closed areas	17%	23%	22%	.899
Ecosystem management is an attempt by environmentalists to stop resource extraction on public lands	24%	31%	23%	.030

*The entry in each cell is the percentage agreeing or strongly agreeing with the statement listed along the left of the table. Other response options included neutral, disagree or strongly disagree.

**The significance level is the result of an analysis of variance of the mean responses by each of the groups.

Table 9. Distribution of Respondents on Index of Support for Ecosystem Management Principles*

Support for Ecosystem Management	*Washington*	*Oregon*	*British Columbia*
Low			
2	2.1%	3.6%	1.8%
3	1.9	3.6	1.8
4	2.8	5.1	4.7
5	2.1	4.2	2.2
6	19.3	21.7	12.6
7	17.7	10.2	8.8
8	24.2	18.1	27.1
9	22.6	23.1	22.4
10	7.2	10.5	18.7
High			
Total	100%	100%	100%
N	(556)	(590)	(557)
Average	7.4	7.2	7.8

F=15.8, Sig.=.000

*The index of support is constructed from the first two items in Table 8, simply adding the respondents' scores on each item.

differences are not great, they are indeed statistically significant (p=.000) and in the direction that is consistent with broadly shared perspectives about how the political cultures of the two countries differ.

Cultural Foundations of Support for Ecosystem Management. There is significant evidence to this point as to the presence of political cultural foundations for attitudes about ecosystem management. That is, of the cultural variables we have examined the New Environmental Paradigm—the dimension that taps into the individual's belief in a more biocentric world—appears to resonate with the political cultures of the people in the Canadian and U.S. sites under study. The NEP perspective seems more consistent with the organic, integrated, communalistic view of society broadly attributed to Canadian political culture. Now we turn to the question of whether within each of the sub-samples those cultural orientations are related to support for ecosystem management—a result that would lend credence to the cross-national cultural interpretation. In this regard, Table 10 presents the results of regression analyses that predict positions on the ecosystem management support summary index using the six political culture variables (NEP, post-materialism, self-assessed political ideology, support for public involvement, support for technology, and support for interest groups).

The results set forth in Table 10 indicate that in each of the three research locations the group of political culture variables is able to provide a statistically significant prediction of the individual's position on the ecosystem management index, albeit the overall amount of explained variation is limited. Moreover, there is substantial similarity

Table 10. Regression Analysis Predicting Support for Ecosystem Management from Cultural Variables

Cultural Variable	*Washington*	*Oregon*	*British Columbia*
NEP	.204***	.150**	.241***
Postmaterialism	-.074	.064	-.059
Ideology	.057	.025	.006
Participation	-.030	.080	.033
Technology	.003	.136**	.117**
Group	-.086*	.008	-.040
R	.225	.206	.256
R2	.050	.043	.065
F	4.73	3.93	6.21
Sig.	.000	.001	.000

* p≤ .05; ** p≤ .01; *** p≤ .001

across the samples in the relative impact of the particular political culture variables. In all three samples, post-materialism, self-assessed ideology, support for public participation, and regard for interest groups have non-significant standardized regression coefficients representing their independent impact on ecosystem management support. In Oregon and British Columbia, but not Washington, support for technology is significantly related to ecosystem management support. Most important, though, is that in all three locations support for the New Environmental Paradigm has a significant relationship with support for the ecosystem management position. Thus, at both the aggregate level and at the individual level position on the ecosystem management policy index is related most strongly to support for the New Environmental Paradigm.

Knowledge and Political Culture. In policy areas that have significant levels of scientific and technical content, such as the environment and natural resources, some observers believe that the answer to resolving policy disputes can best be found in public education—in the enhancement of information available and the building of public knowledge. This belief raises the important question of whether knowledge level about the policy area makes any difference in the individual's policy position. Furthermore, a derivative question arises as to whether greater knowledge levels suppress the impact of the political culture variables on policy preferences.

This study did not feature a set of indicators measuring the level of knowledge of the respondents concerning forest resource ecosystem management policy. However, one item (admitted to be imperfect for this purpose) did ask survey respondents to answer this related question: "How well informed would you say you are concerning Ecosystem Management/Sustainable Forestry?" Response alternatives for this survey item ranged from 1 for "not informed" to 5 for "very informed." No difference among the three sub-samples obtained in the average level of self-assessed knowledge. However, Table 11 shows the effects of self-assessed knowledge on support for ecosystem management policy among and within the three study locations. In both Oregon and British Columbia, the low knowledge group is least likely to support ecosystem management; in sharp contrast, in Washington it is the high knowledge group that is least supportive. While no difference registered among the middle knowledge groups, at the high knowledge level the British Columbia respondents are much more likely than those from either

Table 11. Average Level of Support for Ecosystem Management: Controlling for Self-Assessed Knowledge Level*

Knowledge Level	*Washington*	*Oregon*	*British Columbia*	*F*	*Sig.*
Low	7.4	6.8	7.6	12.2	.000
Medium	7.6	7.5	7.8	1.4	.243
High	7.0	7.3	8.2	9.4	.000

*The entry in each cell is the average level of support on the Ecosystem Management index within each sub-sample, controlling for level of self-assessed knowledge. The level of self-assessed knowledge reflects a 5-point scale, ranging from 1 (low knowledge) to 5 (high knowledge). For this analysis, 1 and 2 are combined as low, 3 is medium, and 4 and 5 are combined as high.

Oregon or Washington to support ecosystem management. Thus, self-assessed knowledge does make a difference in the policy position, but that impact is not uniform across the three study locations, nor does it eliminate the greater support for ecosystem management among the British Columbia public, especially those who consider themselves well-informed.

The final question to be addressed is the impact of knowledge level on the effect of NEP, the most powerful of the political culture variables, on support for ecosystem management. The question surfaces because British Columbia exhibits the greatest support for ecosystem management, the highest level of support for the NEP, the strongest relationship between NEP position and support for ecosystem management, and the highest level of support for ecosystem management among the most well informed. In order to answer this question, we assessed the relationship between the NEP political culture variable and support for ecosystem management within each knowledge level within each of the three study sub-populations. The results of this analysis are reported in Table 12.

Table 12 reveals a startling difference between the Washington and Oregon samples on the one hand, and the British Columbia sample on the other. In Washington and Oregon, the relationship between support for the New Environmental Paradigm and support for the ecosystem management policy is the strongest among those with the greatest self-assessed knowledge. But in British Columbia, only among those with the greatest knowledge does NEP fail to have a significant impact on the policy preferences. Thus in the U.S. context knowledge provides the

Table 12. Correlation (r) of NEP and Ecosystem Management Support: Controlling for Level of Self-Assessed Knowledge[a]

Knowledge Level	*Washington*	*Oregon*	*British Columbia*
Low	.05 (202)	.05 (208)	.18** (193)
Medium	-.03 (234)	.09 (232)	.17* (252)
High	.2* (124)	.41** (130)	.01 (108)

[a]The entry in each cell is the correlation between respondents' position on the NEP scale and position on the index of support for Ecosystem Management, within each sub-sample and while controlling for level of self-assessed knowledge. The index of self-assessed knowledge runs from 1 (low knowledge) to five (high knowledge). For this analysis, low contains those at positions 1 and 2, medium contains those at position 3, and high contains those at position 4 and 5.

* $p \leq .05$; ** $p \leq .01$

vehicle for engaging NEP implications with the policy option, while in British Columbia that knowledge provides the impetus to override the predominant political culture orientations. As suggested in the conclusion, these very different findings on the interaction of political culture, level of self-assessed knowledge, and policy preferences may imply some very different approaches to reaching policy resolution in these political jurisdictions.

Conclusion

Public policy in all domains, and that of natural resources and the environment in particular, has the potential to engage the fundamental values that distinguish a political culture. Those values are embedded both in the outcomes of the policy process and in the character of that process itself. This chapter has examined the degree to which the Canadian and U.S. political cultures—as found in British Columbia, Oregon, and Washington—differ in ways that reflect broadly accepted historic differences in their relative content, and whether those differences have the potential to influence public policy in the area of forest resource management.

The conclusions reached here are rather straightforward. The citizens of British Columbia, when compared to their counterparts in Washington and Oregon, are less likely to identify themselves as liberal on domestic issues, less likely to support the importance of public involvement in policy formation, and more likely to support the concept of a New Environmental Paradigm. The citizens of British Columbia are more likely to support contemporary ecosystem management policy principles, and that support increases with greater knowledge there. In contrast, among the citizens of Washington greater self-assessed knowledge is associated with less support for ecosystem management policy principles. Perhaps most striking of all is the differential impact of knowledge on the relationship between agreement with the New Environmental Paradigm and support for ecosystem management. In British Columbia, high knowledge eliminates the relationship between NEP support and ecosystem support, seemingly overriding the impact of the distinctive cultural values. In contrast, in Washington and Oregon, only at the highest knowledge level is there a significant relationship between NEP support and support for ecosystem management.

These several results strongly underscore the importance of the political culture context in understanding public policy dynamics. They also suggest that different strategies should be taken in the two national environments studied here. Responding to the democratic ideal of broadly dispersed policy-relevant information and broadly inclusive policy processes will have different effects in British Columbia than it will in Washington and Oregon. Individuals managing the policy processes must recognize that increasing the information available in Washington and Oregon, as surely must be done both legally and politically, will enable individuals to focus much more directly on the value stakes at issue in the policy alternatives, and thus may serve to intensify the conflict. Policy makers will need to be sensitive to this outcome, and focus their attention on finding the routes that will maximize the shared values in the policy alternatives considered. In contrast, in the British Columbia setting the enhancement of public education and broader dissemination of policy-relevant information will likely serve to de-emphasize the value conflicts, at least as reflected in the New Environmental Paradigm. The consequence may be that the grounds for debate and dispute will be taken outside of the cultural realm, at least as indicated here, and focused more on such concerns as the economic implications of those policy alternatives.

These conclusions underscore the importance of the entire enterprise of probing deeply into the political culture foundations of public policy formation in general, and in the area of natural resource policy in particular. They suggest that the New Environmental Paradigm index is a particularly pertinent dimension of values for the study of natural resource policy dynamics, and that the anthropocentric versus biocentric perspectives on the mankind and nature nexus will likely prove a central dimension of value engagement in the years to come when issues such as global warming and biodiversity protection come to the center of the public policy debate. Finally, it is clear that Canadian political culture distinctiveness remains very much in evidence, though the signs of cultural convergence with the United States are also in strong evidence at the same time.

7

Public Values for Sustainable Forest Management in Alberta

Bonita L. McFarlane, Janaki R. R. Alavalapati, and David O. Watson

Introduction

The Model Forest Program was established in 1991 by the Government of Canada to develop and test nontraditional approaches to forest management policies and programs that are integral to sustainable forest management.[1] The Model Forest Network consists of eleven model forests, established across Canada, representing five major ecoregions. Each model forest is comprised of partners representing a diversity of forest values, working together to develop innovative region-specific approaches to sustainable forest management.[2] The model forests act as hands-on laboratories in which leading-edge sustainable forest management techniques are researched, developed, applied, and monitored. Ryan discusses the particulars of this program in greater detail in chapter 11. This chapter describes a case study of how one model forest, the Foothills Model Forest, has extended the analysis of forest values and demonstrated an innovative approach to incorporating a range of values into a valuation framework.

The Foothills Model Forest (FMF) is a nonprofit corporation that represents industrial, academic, government, and non-government partners.[3] The principal partners representing the agencies with vested management authority for the lands that comprise the FMF include a major forest company (Weldwood of Canada Ltd., Hinton Division), a provincial government department (Alberta Environmental Protection) and a federal government agency (Parks Canada Agency). Prior to the formation of the FMF these partners managed their respective lands in isolation from one another, and other stakeholders' input was largely through a referral process and after the fact. The FMF provides a venue to bring a wider range of stakeholders together to understand a broader range of issues relating to ecological, social, and economic aspects of forest management and to better accommodate a range of values associated with the forest.

The Foothills Region has the longest history of large-scale commercial forest use in the province, though it is short relative to other part of Canada and the United States. The pulp mill in Hinton was first established in the late 1950s. At the time, the community numbered less than eight hundred. Much of the wood for the mill was procured using horses in the 1960s, but these were quickly phased out in favor of more modern technologies. The community grew dramatically in a short period of time to its current size of around ten thousand residents. The region has at least two generations of residents who have grown up with the mill, though population turnover has always been fairly high in the community.[4]

The FMF land base comprises over 2.75 million ha of primarily publicly owned land in the Rocky Mountains and eastern slopes region of west-central Alberta (Figure 1). Resource extractive activities on the land base include industrial forestry operations, coal mining, and oil and gas developments. Jasper National Park, the Willmore Wilderness Park, William A. Switzer Provincial Park and several provincial recreation areas offer a variety of recreational opportunities and make the area a primary destination for many tourists and recreation users. Several resource-dependent communities are situated in or near the model forest the two largest of which are Hinton and Jasper with populations of 9,961 and 4,301, respectively.[5] Achieving sustainability of a range of values on a land base with such varied and sometimes conflicting uses requires an understanding of the values associated with the forest, the relevant stakeholder groups and their values and attitudes, and an assessment of how stakeholders will be affected by management decisions.

Measuring Values

Generally, economic values associated with commodity production have dominated forest values analysis and management and policy decisions in North America.[6] Reliance on economic values of commodities assumes that the only relevant values for management and policy are those to which we assign value, such as market price and employment. This has resulted in failure to adjust to changing societal values and provides only a partial account of what is valued in forests.[7] There has been a shift in societal values away from a focus on the utilitarian and commodity production to the less tangible non-utilitarian, such as spiritual, aesthetic, existence, and amenity values.[8] It is these values that have gained importance in the sustainable management paradigm and are critical to understanding why people care about forest management and are motivated to legal and political action.[9] In the late 1980s and early 1990s, the Government of Alberta aggressively sought out-of-province multinational investors to utilize the vast, untapped boreal mixed-wood resource in the northern reaches of the province. They met with strong local resistance due to the fact that non-commodity values for the forest were locally recognized, and local residents felt these were threatened.[10] These non-commodity values do not lend themselves to economic measurement and thus require other methodologies.

One approach that provides a more inclusive examination of values is methodological pluralism.[11] This involves using indicators from multiple methods across multiple disciplines to provide a more complete picture of the diverse values of forests and forest ecosystems. Reliance on one perspective or analytical framework such as the economic value

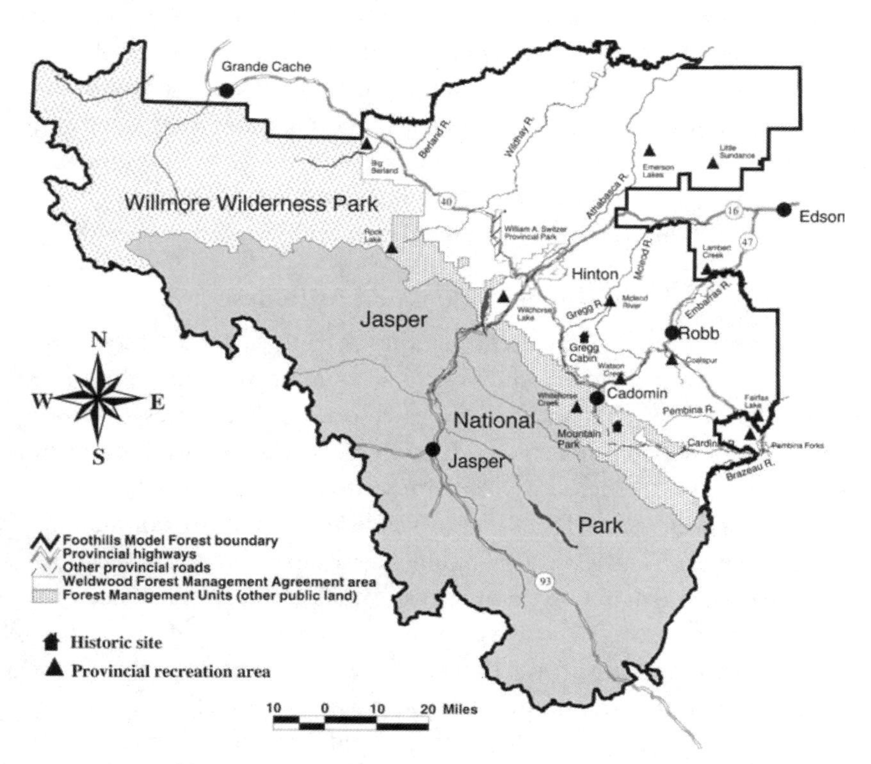

Figure 1. Map of the Foothills Model Forest, Albert, Canada

of marketed goods and services provides only a partial account of forest values and can result in increased conflict over resource management. Successful implementation of sustainable forest management requires an approach with a broad understanding of forest values and includes practices that are socially and politically acceptable.

With the implementation of the Model Forest Network, the FMF undertook an integrated multi-disciplinary social science program. One goal of this program is to include a more comprehensive set of values in management decisions and assess impacts of management and policy changes on a variety of stakeholders. Traditional economic analysis, social psychology, sociology, and valuation of non-market goods and services were used to identify the types of values associated with forests. An examination of held values among a range of stakeholders, the inclusion of non-market values in economic analysis, and improvements to traditional economic models were used to broaden the traditional economic valuation approach that has dominated forest management decisions in the FMF.

In the next section, we discuss research aimed at understanding the values and attitudes of stakeholders with varied interests in forest management. Next we discuss an attempt to assign value to some aspects of the forest that are usually ignored in economic frameworks by using non-market valuation techniques. Finally, we present the development of economic valuation models that include a more comprehensive approach to valuation than traditional economic models.

Stakeholder Values and Attitudes

The shift in forest management paradigms from sustained economic values and commodity production to sustaining multiple values is a reflection of changing societal values associated with forests. Values associated with commodity production focus on the production of goods and services that satisfy human needs and wants. These types of values have been referred to as instrumental, anthropocentric or human-centred values.[12] The less tangible values associated with the sustainable forest management paradigm such as spiritual, aesthetic, existence, and intrinsic values are referred to as non-instrumental or biocentric values.

Biocentric and anthropocentric values represent an individual's general beliefs about forests and have been defined as relatively enduring conceptions of the good related to forests and forest ecosystems.[13] They have been referred to as held values and provide an indication of what is valued in forests. A biocentric-anthropocentric continuum was used in the FMF as an indication of the management paradigm that might be acceptable,[14] for predicting management preferences and beliefs about forest management,[15] and categorizing stakeholders based on their value orientation.[16] By understanding the value orientations of various stakeholders, managers can predict how stakeholders might react to management practices and which groups will be affected by management changes.

In addition to broadening the range of values that are considered important to forest management, the FMF has recognized the need to broaden the range of stakeholders who have input into forest management and policy. In the commodity production paradigm professional foresters and other experts in governments and the forest industry were the dominant stakeholders. The definition of who is a stakeholder has been expanded under the sustainable forest management paradigm to include both users and nonusers of the forest.[17] In the case of the FMF, which is primarily public land, each Alberta citizen should have a legitimate voice in its management. The concerns of a broad range of citizens can be included at the philosophical approach and goal-setting stages of forest management.

In the FMF, stakeholder input has been represented through organized groups such as environmental organizations, labor unions, and chambers of commerce, and politically through local, provincial, or federal governments. Other more direct means of obtaining stakeholder involvement have included advisory committees, open houses, and workshops. These mechanisms have been criticized because they often elicit input from special interest groups who might not be representative of most stakeholders.[18] The FMF employed survey research as a means to complement existing methods of public involvement and to obtain an understanding of forest values and attitudes across a range of stakeholders. This approach helped in assessing the representativeness of the relevant populations of interest, and provided a common metric for comparing value orientations across stakeholder groups.

Six stakeholder groups were included in a study of held values associated with the forest. The groups included the general public of Alberta, members of environmental organizations, registered professional foresters (RPFs), members of public advisory groups (PAG), and campers and hunters. These groups were chosen to represent a variety of interests in forest management in Alberta. The general public was included because most of the forested lands in the FMF are public land. Environmental organizations are often viewed as having considerable influence in natural resource management. However, they are also viewed as an elite group whose interests are not representative of the general public. Registered professional foresters represent the forestry profession and thus have considerable influence in making and interpreting forest policy and recommending and implementing forest management practices. Public advisory groups advise the forest industry on forest management plans and activities and are the main forum for public involvement for the forest industry in the FMF. Recreation users were included in the economic development paradigm of forest management. In contrast, the public, environmentalists, and recreationists were similar to one another by being more biocentric and placing greater emphasis on environmental and social benefits because they represent non-timber users of the forest whose activities are often carried out in close proximity to industrial forestry activities.

A classification of stakeholders based on biocentric-anthropocentric values revealed disparate value orientations among the stakeholders (Figure 2). Of particular interest was the finding that RPFs and PAG members were more anthropocentric in their value orientations and placed greater importance on economic benefits than did the other groups.[19] The anthropocentric orientation reflects support for the values associated with commodity production. There was also substantial difference among the groups in terms of their attitudes toward the sustainability of current forest management. Generally, RPFs and PAG members had a more optimistic view of the sustainability of timber

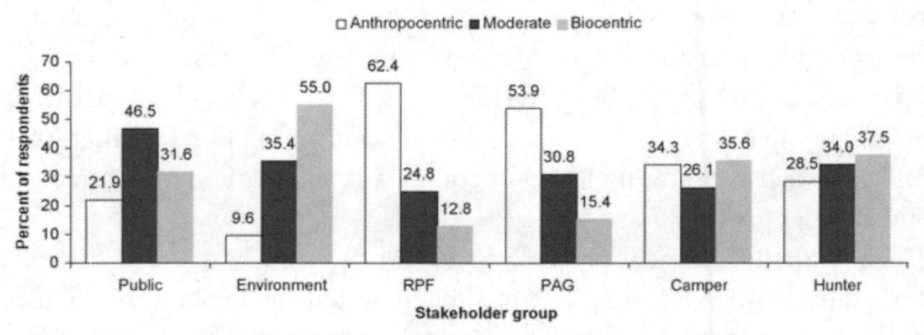

Figure 2. Distribution of Forest Values Segments among Stakeholder Groups. RPF = registered professional forester, PAG = public advisory group (Adapted from McFarlane and Boxall 2000b).

supply, the successful inclusion of multiple values in forest management, and the adequacy of public involvement. The other stakeholders believed that timber supply and the inclusion of multiple forest benefits was inadequate, and that the public does not have enough input in forest management decisions.

One important consideration for managers in the FMF is addressing the values of local residents and assessing the impact of management and policy decisions on these residents. Because of their dependence on the forest resource for their economic livelihood it is generally assumed that local residents will be the most affected by management practices and policy. Thus, they will differ in their values and attitudes from residents in non-resource dependent communities. For example, residents in forest-dependent regions in the United States have been shown to differ in value orientations by being more anthropocentric than residents in other regions.[20]

To compare the values and attitudes of FMF residents and residents in non-resource dependent communities the general population sample was divided into three groups: FMF residents, residents living in Edmonton and Calgary (urban residents) and those living in other communities. The urban group tended to be more biocentric while the FMF residents and other resident group tended to be more anthropocentric in value orientations.[21] The FMF residents, however, shared similar attitudes with other members of the public. They viewed current forest management as providing an inadequate timber supply, and as not managing for a range of values, and public involvement as not providing adequate consideration to forest-dependent communities and citizens of Alberta.

Applying a held forest values assessment has demonstrated what aspects of the forest are most important to various stakeholders. This analysis has expanded the traditional commodity-based approach by including some of the less tangible aspects of forests and demonstrated that those who have the most influence in forest management and policy (RPF's and PAG members) have a more anthropocentric orientation and view forest management quite differently from other stakeholders. This suggests many of the values important to most stakeholders are not receiving adequate consideration in forest management. The challenge for natural resource managers in the FMF is to incorporate these values into management decisions while maintaining many of the industrial activities occurring on the land base. One means of approaching this challenge is to extend traditional economic models to include a broader valuation framework, and to develop models that reflect the local situation.[22] While it may not be possible to reduce all aspects of the forest that have value into a valuation framework there are valuation methods that can incorporate some of these values to provide a more accurate assessment of the economic value of the forest. The next step in our analysis was to extend the traditional economic analysis to incorporate some of the non-market values associated with the FMF.

Non-Market Valuation

The FMF consists of primarily public land and provides numerous goods and services. Many of these goods and services, however, are not traded in economic markets. The lack of a market price has resulted in an undervaluation of many non-market goods and services or their exclusion altogether from consideration in forest management decisions. The held values analysis of stakeholders presented in the previous section suggests that non-market goods and services are most important to stakeholders. Thus the inclusion of non-market values in regional and provincial economic models is necessary to provide a more accurate assessment of the benefits, trade-offs, and interactions between market and non-market goods and services.

In order to provide an understanding of the economic value of a full spectrum of unpriced services, we attempted to assign economic value to some of the non-market benefits associated with the model forest. Non-market benefits include use and non-use values. Use values involve an interaction with the resource. These include, for example, recreation activities and subsistence harvesting. Conversely, non-use does not involve interaction with the resource and includes bequest (the desire to maintain something for future generations), existence (the benefit from knowing that the resource exists), and option values (the desire to keep open the possible future use of a resource) and many environmental services. Environmental services of a forest include, for example, erosion control of streambanks, water quality, air quality, biodiversity preservation, and carbon sequestration.

An important non-market service provided by the FMF is forest-based recreation. Camping was chosen as an indicator of recreation because it occurs throughout the model forest, has a large number of participants, and campers generally engage in multiple recreational activities such as fishing and hiking during a camping trip.[23] A zonal travel cost model (TCM) was used to capture the non-market economic value of camping. The data for the model were obtained from camping registrations from provincial park campgrounds and recreation areas within the model forest. The estimated trip value was $58.14. Aggregating all trips resulted in an estimated value of $436,631 for the benefits associated with camping in the FMF in 1995.[24] This is in addition to the fees collected by the campgrounds, and does not include other expenditures the campers may have made in the region, such as food and gasoline.

To capture some of the non-use benefits in an economic framework a new environment sector was created for the FMF. Within this sector values were included for various components of natural capital. These included the non-market benefits associated with forest recreation, carbon equivalent emissions, and carbon sequestration and dissipation.

A Valuation Framework for the FMF

Resource sectors such as forestry, mining, and tourism in FMF are a significant source of wealth to the regional economy. Potential changes in those sectors whether institutional or through market forces may have significant implications on the stability of the regional community. Decision-makers would like to know about impacts on local communities and stakeholders in response to any changes in resource sectors before they make decisions relating to resource management. Quite often it may be that management decisions in one sector may not only affect that sector but also impact other sectors of the economy and other values either positively or negatively. Consider, for example, an increase in the exports of pulp and paper products. To produce more pulp, more logs and chemicals are required, and the logging and chemical industries have to increase their outputs to meet this increased demand. To increase their outputs, logging and chemical industries have to buy more inputs from other industries causing an expansion in those industries. Increased logging activity and chemicals can, in turn, affect the environment and values that are not traded in markets. This implies that impact analysis may lead to erroneous conclusions if it is done on only one sector of interest or if non-market values are not considered in the analysis.[25] Economy-wide models, such as input-output (I-O), Social Accounting Matrix (SAM), and Computable General Equilibrium (CGE) approaches provide frameworks to account for intersectoral linkages thereby capturing spinoff and tradeoff effects. The CGE approach offers the additional advantage of incorporating values that are not traded in markets and thus are overlooked in I-O and SAM valuation approaches.

The FMF has provided a unique laboratory to assess and refine various types of economy-wide impact analysis models and to apply those models to generate key socioeconomic information that aid decision-making. In this section a review of FMF economy-wide impact analysis research is presented in hope of providing a unique perspective to the regional economic literature. One of the problems that we encountered in conducting regional impact analysis in the FMF was the lack of appropriate data. In general, regional impact analysis for small regions has been conducted based on provincial or national data. Although this top-down approach is frequently used in the United States through IMPLAN, a similar program has not been initiated in Canada. Furthermore, when there is a significant variation between a regional economy and the provincial or national economy, bottom-up approaches, or the collection of regional level information, are considered superior. For example, the five major resource sectors in the FMF, the forestry, the wood sector, the mining sector, the crude petroleum and natural gas sector, and the visitor sector account for approximately 85 percent of the regional economy (Figure 3). In comparison, the same five sectors at the provincial level account for less than 19 percent of the Alberta economy.

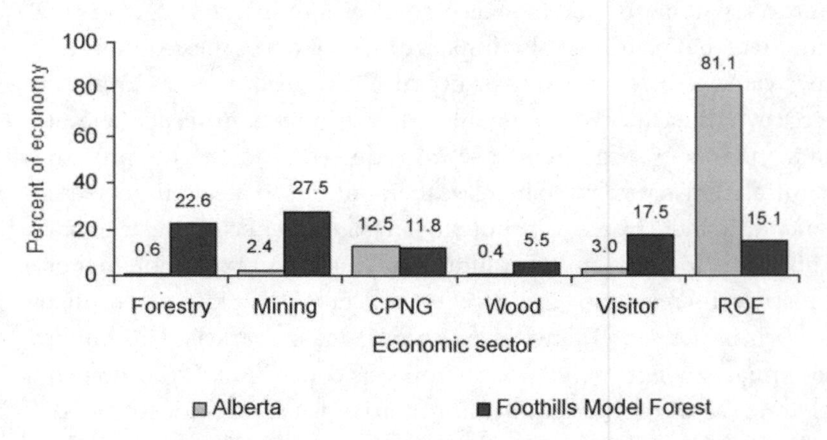

Figure 3. Comparison of Sectoral Contribution to the Economy of Alberta and the Foothills Model Forest. CPNG = crude, petroleum, and natural gas , ROE = rest of the economy (Source: Patriquin et al. 2001a).

Furthermore, there might be significant differences with respect to the source of inputs. For example, in the provincial economy, input requirements may be satisfied from within the province. However, a smaller regional economy such as FMF that uses the same mix of inputs may not be able to satisfy its needs locally. The estimation of imports and exports by industry is a major hurdle in generating a regional database from synthetic or top down techniques.[26] Synthetic regionalization techniques that do not adequately address changes in the source of inputs between two scales of economies result in biased impact estimates. In order to address this concern, a customized Social Accounting Matrix (SAM) was developed for the FMF. The first step in the process was the synthetic regionalization of the provincial SAM as a base with which to start. Second, regionally specific data gathered from a variety of sources were integrated and adjustments were made based on the nature of the regional economy. Third, minor adjustments were made to ensure data consistency in the SAM framework.[27]

In this process, three different types of information were collected to develop the regional database. First, dominant industries with small numbers of firms were contacted directly for financial information. In resource sectors with large numbers of firms, information such as total output and value of output specific to the region were obtained from various government agencies. Second, a visitor sector was developed using the results of two studies—a visitor expenditure study that obtained the total value of output for the sector[28] and a visitor sector employment study that identified the number of people employed in the sector and determined the visitor sector total wage bill.[29] The creation of a visitor sector in the SAM was accomplished by applying service sector coefficients to the value of total visitor sector output estimated from visitor expenditures in the region. Third, household expenditure data were obtained from a FMF household expenditure survey.[30] Total household expenditures were split into three income groupings: low income (less than $30,000), medium income ($30,000 to $59,999), and high income ($60,000 or greater).

The FMF SAM developed following the above procedure was used to simulate the impacts of various changes in resource sectors of the FMF. One of the simulations was the impact of a decrease in visitor-related activity of $15 million citing the Asian-Pacific economic crisis and the related decrease in tourism to Jasper National Park. The simulation results suggest that the visitor sector has an estimated multiplier of 1.7913. This implies that a $15 million decrease in the visitor sector results in an economy-wide decrease of $24.69 million. The results also suggest that the impact of the above shock on higher income households is greater than on lower income households.

A closer look at the underlying assumptions of the SAM model raises serious concerns about the validity of the information derived from this approach. For example, the assumption of fixed prices does not allow these models to capture the behavior of producers and consumers with respect to changes in prices of inputs and outputs. Second, the fixed-inputs assumption rules out the possibility of substitution between factors of production. Third, if sectors are not directly linked by interindustry flows of commodities, it is possible that they will still be interdependent because they compete for scarce primary factors (labor, capital, and land). All these assumptions may either overstate or understate the economy-wide impacts of any changes in the forest sector. Therefore, an alternative interindustry analytical tool, the computable general equilibrium (CGE) model, was used to conduct impact analysis in the FMF. This approach is thought to provide greater flexibility by allowing changes in prices and substitution among factors of production. Using this approach, Alavalapati, et al.[31] studied the impacts of a 6 percent decrease in the annual allowable cut (AAC) in the forest industry for the FMF. Results show that a 6 percent reduction in the AAC causes significant negative impacts on the forest sector. The reduction in the AAC and associated decrease in the supply of timber causes a 2.61 percent increase in stumpage cost. The reduction in the supply of timber also causes a 3.03 percent ($15.27 million) reduction in the forestry output. As a result there will be a 1.20 percent (14 jobs) and 2.47 percent ($6.94 million) decrease in the demand for labor and capital, respectively, in the forest sector.

A decrease in the AAC causes an expansion in other sectors of the economy. The output in the rest of the economy increases by 0.55 percent ($11.02 million). In the rest of the economy, the demand for labor and capital increase by 0.21 percent (9 jobs) and 0.92 ($6.06 million) respectively. Finally, a 6 percent decrease in the AAC cause a 0.53 percent ($2.24 million) decrease in household wage income. This implies that the increase in the demand for labor and wage in the rest of the economy cannot offset the decrease in the demand and wage in the forest sector. This type of tradeoff effect associated with any policy is difficult to capture through SAM analysis.

One of the problems with conventional general equilibrium impact analysis is the lack of explicit linkages between the economy and the environment or non-market values.[32] Few analytical models have been developed on a regional scale to integrate the economy and the environment. Therefore, an attempt was made to develop an environmentally integrated CGE model that reflected some of the non-market benefits for the FMF region.

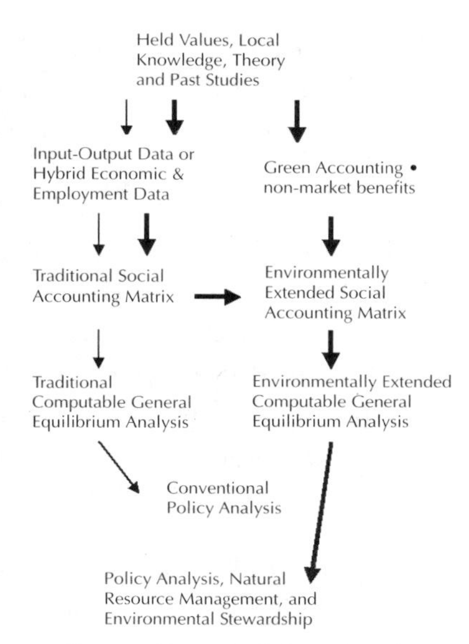

Figure 4. Framework for Integrating a Range of Values into Policy Decisions.

The conceptual framework that was followed in developing this model is shown in Figure 4.[33] The held values analysis discussed in the previous section provided an indication that environmental consid-erations are important to stakeholders and thus should be included in the valuation framework. The heavy black arrows in Figure 4 represent the path that is chosen to integrate an environmental sector into economy-wide impact analysis. Drawing on previous studies, various components of natural capital (non-market values associated with recreation, carbon equivalent emissions, and carbon sequestration and dissipation) were incorporated into a conventional CGE framework. The model thus developed was used to simulate two hypothetical scenarios.[34]

First, a reduction of 22 percent in mining exports is examined following the phasing out of an existing coal mine within the region without replacement. Second, a 7 percent increase in tourism that could result from a spillover of visitors into Jasper National Park due to increasing park use restrictions being placed on the adjacent Banff National Park. These economic changes will have impacts on output household income and environmental quality.

All sectors are shown to suffer in terms of output and employment in response to a hypothetical 22 percent reduction in mining sector exports. The net environmental productivity (NEP) is shown to increase by 133.85 percent. On the contrary the NEP decreased by 80.89 percent in response to a 7 percent increase in visitor activity. One explanation for this is increased vehicle traffic and transportation emissions associated with increased park visits. Furthermore, a 7 percent increase in visitor activity is shown to have a positive spinoff effect on all other sectors thereby improving the income of households in the FMF.

Discussion

Research on forest values in the FMF has extended the types of values included in values analysis, the methods used in the analysis, and the range of stakeholders being considered. The approach being taken by the FMF is developing tools to determine what aspects of the forest have value and, where possible, to include some of the non-market values into economic models. Although this approach is more inclusive than the traditional economic analyses of the past, it must be recognized that not all values are represented, and not all values can be reduced to a valuation framework. Many values associated with a biocentric orientation such as spiritual and intrinsic values are not captured in the economic models. Research using other methods, however, has developed tools (a values scale) that provide a common metric for measuring these types of values and comparing stakeholders' value orientations.

Socioeconomic impact information derived from applying the economic models is helping resource professionals make informed decisions for the FMF. Despite the progress made in terms of regional economic analysis, more research is sorely needed. The environmental adjustments made so far are quite modest. Only a few of the non-market values associated with the forest were included in the models. Better information on baseline environmental data and an accounting of changes in environmental quality would aid in the estimation of a damage function that links recreational use to environmental quality and sectoral output. In addition, there is an asymmetric impact of output on the environment. In other words, there are currently no environmental feedbacks in the general equilibrium system. Furthermore, the analysis could be extended to a dynamic setting by collecting relevant time series data on the economy and the environment. Despite these limitations, the approach adopted in the FMF represents state-of-the-art techniques for regional economic impact analysis and demonstrates how one model forest has embarked on a more inclusive values analysis in its quest for developing and implementing sustainable forest management practices.

8

Public Involvement in Forest Management: Toward a Research Program in Alberta

Richard C. Stedman and John R. Parkins

Introduction

This chapter describes a new research program in Alberta to assess the effectiveness of current public involvement (PI) practices using the *public sphere* as an analytical tool. Though there is much enthusiasm in Alberta for PI as a concept and as an essential component of sustainable forest management, critical analysis about what constitutes effective PI has been limited or absent altogether from discussion. Most research regarding the effectiveness of PI has focused on static criteria such as representativeness (e.g., do the characteristics of group members mirror the attributes of the general public?) or outcome-based measures (e.g., has a consensus-based decision been achieved?). Process-oriented criteria, such as the facilitation of information flow, flexibility of the process, and openness of the group to frank discussion and new ideas, are important but have been subjected to less empirical scrutiny.

We propose Habermas's[1] conception of the public sphere as a framework to guide further inquiry into the effectiveness of PI in Alberta. Our past quantitative survey research comparing public forest management advisory group members with the public at large has examined some elements related to the public sphere, including issues of trust, influence, and social status differences. But the research is not sufficient to engage the public sphere to its full extent. Therefore, we have undertaken in-depth case study research that examines PI groups specifically through the lens of the public sphere. In this chapter we describe the mandate for and current state of PI in Alberta and review existing research regarding PI effectiveness. We then discuss the concept of the public sphere, address the future direction of the research, and offer a number of preliminary insights based on our ongoing case study research.

The State of Public Involvement in Alberta Forest Management

Alberta experienced rapid forest industry expansion during the 1980s and 1990s,[2] stimulated by several concurrent developments. On the technological side, the development of pulping methods that could utilize previously non-merchantable tracts of poplar and aspen spawned large-scale thermo-mechanical and bleach kraft pulp mills funded by transnational corporations such as Daishowa-Marubeni International Ltd. and Mitsubishi Corporation. On the policy side, the Alberta government developed incentives to promote economic diversification in response to concerns about the province's heavy dependence on oil and gas revenue. This combination of modern pulping technologies and a business-friendly environment set the stage for industrial forestry at a scale never before realized in the boreal forest of Alberta. Along with the expansion in industrial capacity came guaranteed access to wood fibre on publicly owned land, although companies had to demonstrate sustainable management practices to renew their forest management licenses.

Public consultation was integral to the approval process for industrial expansion, but no formal process was in place for ongoing public involvement in the management of mills or in the associated forest management areas. Sensing apprehension from the Alberta public, the provincial government gave the forest companies an ultimatum for renewal of forest management licenses: develop a mechanism for public involvement. In response, Alberta's first public advisory group (PAG) was established in 1989 in Hinton with town council, Chamber of Commerce, ministerial, medical, Aboriginal, forest worker, university and college representatives. The group's mandate was to provide input into company forest management plans. In addition, it was intended to provide connections to the community, an arms-length perspective on company plans, and a reality check on potentially sensitive issues.[3] In the decade to follow, thirteen other PAGs formed in association with forestry companies across the province.

Each advisory group is unique, having its own set of organizational procedures and guidelines along with its own set of local issues and concerns. Although the groups are company-sponsored, autonomy from company control varies quite dramatically. Some groups operate at arm's length from the company, selecting their own members, defining their own issues of public concern, and challenging company and government policy pertaining to forest management. Other groups are more beholden to company interests, receiving information only from company-defined "experts" providing non-critical feedback to forest management plans, and addressing a limited range of public values and concerns. How companies respond to advisory group input is also highly varied. Officially, these are advisory groups with no direct decision-making control, but some companies state that forest management plans will not be forwarded to the government until approval is received from the public advisory group.

With so much variation in the groups' functioning and their relationships to sponsoring companies, some attempts are being made to formalize procedures. An umbrella organization was established to provide a forum for members of these groups to convene annually, discuss common issues of concern, and learn from the successes and failures of other groups. Thus far, attempts to normalize these advisory group procedures have met with limited success. Forest management companies have embraced the concept of input from groups consisting of a representative cross section of the interested public, and research suggests that the general public has endorsed this approach as well. What is missing is a critical understanding of what constitutes effective public involvement and how research from a theoretical perspective might begin to speak to this question.

Public Involvement Research

What are the criteria by which the effectiveness of public involvement might be evaluated? Researchers[4] generally agree that effective public involvement: (a) facilitates two-way information flow, (b) is flexible in scope, (c) is open to new input and participants, (d) provides guidance to managers, (e) allows for frank and open discussion, (f) is cost-effective, (g) gives something back to the participants, and (h) is representative of the desired target population.

In-depth investigations have examined process-oriented criteria for evaluating public involvement. Lawrence et al.[5] offer "procedural justice" as a criterion. This is the notion that "... the perceived fairness of public participation *procedures* can affect public satisfaction as much as the substantive nature of the resulting decision" (emphasis added). The authors contend that public involvement evaluations have over-emphasized outcomes relative to process. They suggest that many questions remain vis-à-vis the implementation of procedural justice, including the role of interest groups, the impacts on nonparticipants, and the role of historical mistrust of resource managers. Measurement remains elusive as well.

The work of Shindler and associates, especially that concerned with adaptive management, also proves informative regarding PI process.[6] By its very nature, adaptive management, or the implementation of policies as "experiments,"[7] must entail process-oriented criteria of effectiveness, as dialogue with involved citizenry unfolds over time and in response to iterative policy experimentation. More specifically, Shindler and Cheek[8] suggest a number of propositions consistent with the notion of process, including, but not limited to, inclusiveness, innovation, flexibility, and the use of participants' input in decision-making.

Research has also examined representativeness in some detail,[9] often finding that the participants' characteristics diverge widely from the public whom they purportedly represent. For example, Wellstead et al.[10] found that members of advisory groups are much more likely than the general public to be of high social status (income and education), to be linked to the natural resource industry for a living, and to have beliefs, attitudes, and values that are supportive of the resource industry status quo. Such findings are usually accompanied by concerns about these differences, such as whether the general public is being well served by groups that do not mirror its characteristics.

Theorizing Process-based Criteria: The Public Sphere

Although of great utility, most research that has addressed process-oriented criteria for assessing PI effectiveness lacks an organizing theoretical framework. In response we propose the public sphere, an idea that is deeply rooted in social history but is also catalyzing contemporary developments in deliberative democracy. According to Habermas, the public sphere is the realm of social life where citizens confer in an unrestricted fashion about matters of general interest, resulting in open debate and social learning. [11] Such a realm is said to find its most complete manifestation in the public coffee houses of seventeenth and eighteenth-century England, where bourgeois society met and discussed matters of trade, state government, and religion. As public spaces for information gathering and debate among private citizens over matters of public concern, these forums challenged the dominant institutions of the day. Like PAGs, the public sphere did not have decision-making authority; instead it emerged in opposition to institutional authorities because of mistrust between the public, the church, and the state. The public sphere stimulated new ideas and new perspectives on modern problems and exerted considerable influence on existing institutional authorities.

Calhoun[12] suggests that four elements or features of the historically defined public sphere are key to our theoretical understanding. First, Habermas refers to a kind of "social intercourse" in the public sphere that disregarded social status, or at least treated it as secondary to trust in the process of argumentation and to shared public concern. In this way, differences in social status, if not disregarded altogether, were *bracketed* and became secondary to deliberations over shared concerns. People of all educational backgrounds and socioeconomic status who shared common concerns were free to engage in debate.

Second, rational argument was the sole arbiter of any issue. Thus, any form of argument was to be guided by general procedures, or what is now commonly understood as "the force of the better argument." Essential to this debate was not so much the development of trust between individuals, but trust in a process whereby participants were afforded an equal chance to speak and to be heard.[13] Third, the public sphere allowed for the problematization of areas that hitherto had not been subject to critical debate. For example, issues related to church and state authority were brought into public discourse as matters of common concern for the first time. Fourth, inclusiveness, or at least the potential for inclusion, became an established principle. This principle was realized by a reading public with access to books and journals who were willing to engage in public discourse.

These features of the public sphere were produced by a unique set of social arrangements and intellectual energy in the early capitalist stages of Europe. While debates have turned on the extent to which Habermas correctly understood the cultural conditions precipitating the seventeenth-century public sphere, his work is now generally accepted as an attempt to constitute the historical category of public sphere and then draw it as a normative ideal.[14]

To the extent that modern public advisory groups are implicitly mandated as a realm of open discussion and are open to input from a wide variety of points of view, the normative ideals of the public sphere can be used to critique them.[15] These norms can be used as a yardstick to identify areas where power, politics, or ideology prevents the true formation of public consent under an ideal deliberative process. Simply because these ideals are likely never to be realized in full, it does not mean that the analytic concept has no practical value. The value comes in Habermas's description of these public deliberations as places where individuals may learn from each other in ways that are free from distorted communication and manipulation. This normative ideal can then be used as a lens for examining the effectiveness of modern processes of public discussion.

Building a Research Program

In this chapter, we posit that forest management advisory groups can be evaluated against the standards of a public sphere. This seems justified to the extent that PAGs are intended to achieve a practical discourse that is impartial, open to new input and new ideas, and based on the principles of rational discussion and debate. A research program designed to provide an empirical foundation for evaluating PAGs would explore the process citizens use to examine questions regarding public forest management. Such research would investigate how communication within the public sphere is systematically distorted and, as a result, decisions are manufactured in support of dominant political and economic interests as well as explore the preconditions for legitimate decision making within the public sphere. These preconditions are defined by ideal speech situations where sincerity, comprehension, legitimacy and truthfulness are consistently found to be the universal basis for effective communication.[16]

One approach to developing a research program around the concept of a public sphere is to conduct what Dryzek[17] called a "pure critique" or what Skollerhorn[18] called a "critical approach." This approach involves constructing a particular standard and then assessing real-world structures and processes to determine to what extent they fall short of the ideal. It is exemplified in both Kemp and Webler.[19]

Kemp investigated a 1976 public inquiry associated with British Nuclear Fuels Limited and plans to build a waste recycling facility at Windscale. According to Habermas, there are certain institutional preconditions for practical discourse. These criteria "are intended to ensure that the consensus that emerges from practical discourse serves generalizable, and not particular, interests."[20] Briefly, these include: comprehensibility (participants must prove understandable), truthfulness (participants are honest and sincere to themselves and to others), appropriateness (bureaucratic regulations must be applied without bias), and truth (all validity claims much be subject to scrutiny and truthfully grounded). Using the standard of Habermas's "ideal speech situation," Kemp asked a series of questions pertaining to the Windscale discourse. For instance, to what extent were the participants equally able to initiate and perpetuate discourse? Were participants equally able to raise issues and provide answers? Were the rules of the inquiry fair to all participants? Judging against these standards, Kemp delivered a stinging critique of Windscale by concluding "the communication process that occurred at the Windscale public inquiry was in fact systematically distorted and that the subsequent decision to allow the construction of a THORP [thermal oxide reprocessing plant] did not reflect a genuine consensus on the issue and was not reached solely due to the force of the better argument."[21]

Webler's work was similar in that he developed an "evaluative yardstick" or a "procedural normative model of public participation."[22] Like Kemp, Webler based his model on the ideal speech situation. He then extended it to include criteria for fairness and competence in civil deliberations. The final evaluative product is a complex set of criteria against which any deliberative process can be assessed.

A different approach to assessing the public sphere also involves similar ideas by Dryzek ("constructive critique")[23] and Skollerhorn ("empirical approach")[24]. This approach takes the burden off the researcher to construct a normative ideal and places it back onto the subjects involved in the deliberative process. Citizens are invited to discuss their own experiences within the public sphere and to contemplate institutional designs and procedures that might foster a discourse less subject to distortion. Tuler and Webler[25] investigated participant expectations of public processes using a grounded theory approach to identifying normative principles of effective public involvement with forty-nine participants of the New England Northern Forest Lands Council. Seven principles emerged from their analysis, including access to the process, power to influence process and outcomes, and access to information. These criteria support the empirical foundations for evaluating deliberative processes, irrespective of Habermas's theoretically derived ideal speech situation.

Building upon the constructive/empirical approach outlined above, we are embarking on a long-term research program on public advisory groups in Alberta. The research is governed by a subject-oriented evaluation of the deliberative process in question, but is informed by normative ideals of the (public sphere) ideal speech situation. It will include extensive case study research with three to four public advisory groups in the province. Through participant observation, document analysis, and personal interviews we plan to construct a picture of 1) how PAGs function; 2) who is involved in the process; 3) how members are selected; 4) the range of issues under consideration; 5) how the discussion agenda is established; 6) individual perceptions of manipulation or sincere talk; and 7) the variables that serve to facilitate rational debate, social learning, the spread of new conviction, and the general politicization of the deliberative process. To remain consistent with critical theory, we will also need to understand structural features, or the historical, political and economic circumstances under which the social interaction takes place. The views of those not directly included in the PAG process may become important as well. The dynamic tension between corporate, expert, and lay interests and the historical relationship between forest companies, the state, and local communities remain integral to our understanding. We anticipate that this research approach will yield practical understandings of how advisory groups might be improved.

Thus far we have examined the Alberta PI experience through survey research. We addressed conventional questions of representativeness as well as the adequacy of the PI activities, perceptions of influence, and the relevance of various information sources. Findings are described in the following section.

Public Involvement in Alberta

A survey was mailed to one thousand Alberta households in June 1999, followed by a similar survey sent to 158 members of eleven public advisory groups across the province in August 1999. Table 1 indicates the breakdown of respondent categories and the number of respondents in each category.[26] In the tradition of research on representativeness, we compared PAG members with the general Alberta public on a

Table 1. Categories of Respondents

Survey category	*Number of respondents*
General population	664
Foothills Model Forest residents	135
Urban residents from Edmonton and Calgary	242
All other respondents from smaller urban centers and rural areas	287
Members of public resource advisory groups	71

Table 2. Socioeconomic Characteristics of Survey Respondents

Characteristic	*General Public*					*Statistics*	
	FMF residents	*Urban residents*	*All other residents*	*Total*	*PAG members*	*F value*	*p*
Mean age (yrs)	44	45	46	46	48	1.3	.254
Women (%)	43	50	52	51	17	10.8	.000
Post-secondary education (%)	53	77	66	66	84	8.6	.000
Household income ≥$70, 000 (%)	37	35	27	31	49	6.3	.000

number of factors, including socioeconomic characteristics, attitudes towards public involvement, perceived level of influence over forest management by a number of actors, and information sources.

In terms of socioeconomic characteristics (Table 2) PAG members differ from the general public in a number of ways: PAGs have fewer women, are more educated, have higher household incomes, and are more likely to be employed in a resource-related industry than are the general public as a whole.

Respondents demonstrated near universal support for public involvement, broadly construed (Table 3). Nearly all respondents strongly agreed with statements such as "knowledge of public values

Table 3. Attitudes toward Public Involvement

Statement	*General Public*					*Statistics*	
	FMF residents	*Urban residents*	*All other residents*	*Total*	*PAG members*	*F value*	*p*
Knowledge of public values and attitudes helps to make better forest management decisions	4.10 84.4%	4.08 86.8%	4.06 88.2%	4.07	4.49 94.3%	1.29	2.76
Alberta residents have a right to determine how the forest is managed	4.33 92.6%	4.26 87.2%	4.21 88.9%	4.23	4.27 88.8%	.191	.903

1 = totally disagree; 5 = totally agree

and attitudes help to make better forest management decisions" and "Alberta residents have a right to determine how the forest is managed."

Respondents were asked to indicate the adequacy of public input into certain aspects of forest management and forest policy development (Table 4). Overall, respondents appear to be relatively satisfied with the level of public involvement specific to allocating forested land for logging, logging operations, and pulp and paper operations. However, they felt there was inadequate public involvement in nontraditional forest activities such as allocating protected land, recreation and tourism development, and the design and monitoring of public involvement activities. This was primarily a point of agreement between PAG members and the general public.

Despite these points of agreement, there were also fairly strong differences in attitudes and beliefs pertaining to some public involvement-related issues, including overall levels of support for PI, information used and trusted, and degree of influence among various forest sector actors. Several differences related to issues of process, including information used, the credibility of information sources, the perceived degree of influence among various forest sector actors, and

Table 4. Percentage Rating of the Adequacy of Public Involvement in Specific Forest Management Activities

Forest Management Activities	*General Public*									
	FMF residents		*Urban residents*		*All other residents*		*Total*		*PAG members*	
	A	*B*	*A*	*B*	*A*	*B*	*A*	*B*	*A*	*B*
Allocating forested land for logging	20.7	36.4	20.6	34.4	16.9	36.8	18.6	35.7	11.6	42.0
Allocating forested land for protection	5.0	55.7	1.4	59.3	3.8	56.4	2.7	57.7	12.9	41.4
Developing recreational and tourism activities in forested areas	6.7	51.3	12.4	31.2	9.8	38.1	11.0	35.0	5.7	38.6
Logging operations	25.2	28.6	21.8	30.1	24.0	29.2	23.0	29.6	11.8	35.3
Pulp and paper operations	23.3	23.3	23.4	29.8	21.7	27.0	22.5	28.3	11.8	32.4
Designing specific public involvement activities	1.7	50.4	0.9	54.4	0.8	50.2	0.8	52.1	7.2	55.1
Monitoring the outcome of public involvement processes	1.7	47.1	0.5	44.0	0.8	39.5	0.6	41.5	5.8	58.0

A = Too much; B = Not enough

Table 5. Sources of Information about Forest Management

Source	*% of respondents*				
	General Public				
	FMF residents	*Urban residents*	*All other residents*	*Total*	*PAG members*
Newspaper	81	82	80	81	48-
Television	68	82	79	80	30
Radio	47	55	53	54	32
Friends or relatives	48-	26-	37-	32-	25
Forest industry	69+	22	32	27	89+
Government agencies	24	17-	15-	15-	66
First-hand visits to forests in Alberta	67+	52	59+	55+	83
Environmental and conservation organizations	32-	43+	35	39	48-
Forest scientists such as biologist or ecologist	28	33+	25+	29+	68+

+ = considered most accurate; - = considered least accurate

normative assessments about which actors ought to have more or less influence. The general public was most likely to gain information about forest management from media sources and less likely than PAG members to gain information from the forest industry, first-hand visits to the forest, research scientists, and government agencies (Table 5).

Respondents were also asked to report their perceptions of the most accurate and least accurate sources of information about forest management, which is clearly an issue of credibility and trust. Overall, respondents perceived scientists and first-hand experience with the forest as the most accurate sources of information; personal contacts and government agencies were considered least accurate (Figure 1).

Again, some interesting differences emerged between the public and PAG members. The general public (predominantly urban in Alberta) was likely to consider environmental organizations as accurate sources of information, with PAG members largely considering these sources to be inaccurate. In contrast, PAG members tended to consider information from the forest industry as accurate, while the larger public tended to distrust this source.

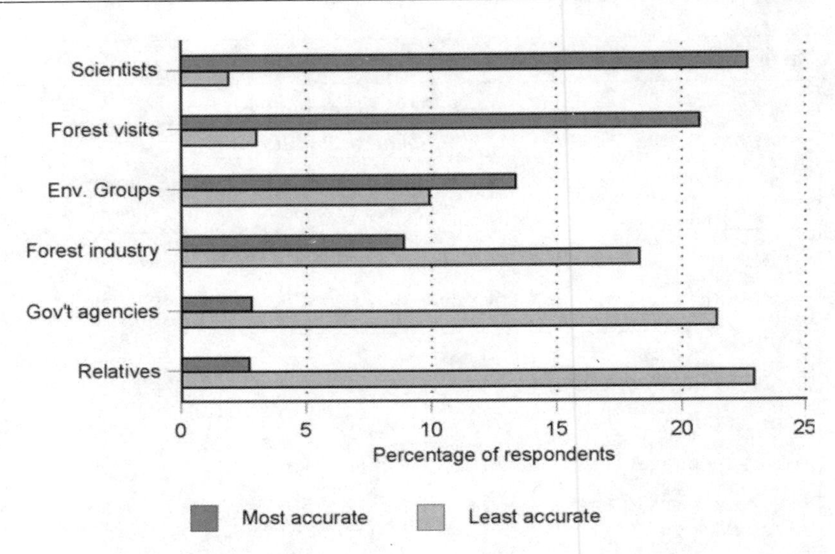

Figure 1. Perception of Accuracy of Information Sources (general public only)

We asked survey participants to indicate their beliefs about who influences forest management decision making (Table 6). Respondents agreed that forestry companies and the provincial government were highly influential. The general public believed the federal government to be much more influential than did PAG members. Survey participants were also asked who *should* have greater or lesser influence over forest management decision-making. The public indicated a desire for less influence by the oil and gas industry and forestry companies, while PAG members indicated a desire for less influence by government. Urban residents also indicated a stronger desire for conservation and environmental groups to have more influence over forest management decision-making, whereas PAG members were less likely to desire increased influence from these groups.

In summary, PAG members and the general public appear to differ quite dramatically in socioeconomic characteristics, and to a lesser degree on attitudes and values surrounding forest management. Although there are substantial points of commonality, such as strong support for public involvement, PAG members generally seem to depend on the forest industry for information (and to trust this information source), and are more suspicious than the general public of environmental groups and the public. Many of these comparisons, although reminiscent of work on representativeness, address perceptions germane to issues of process, including trust, influence, and social status.

Table 6. Perceptions of Levels of Influence in Forest Management Decision-making

Source	*% of respondents*				
	General Public				
	FMF residents	*Urban residents*	*All other residents*	*Total*	*PAG members*
Provincial government	75	70	71	71	94
Federal government	36	35	38	37	9
Forestry companies	80	80	75	78	88
Oil and gas industry	45	52	47	49	42
Forest scientists	25	19	22	21	21
Environmental and conservation organizations	27	14	22	18	17
First Nations and Aboriginal groups	26	17	25	21	25
Recreational users	6	7	8	8	10
Urban residents of Alberta	5	2	4	3	13
Rural residents of Alberta	8	5	5	5	7

Discussion: PAGS and the Ideals of the Public Sphere

From these preliminary findings we can draw some initial ideas about PAGs in relation to the public sphere. We examine the characteristics of Alberta PAGs using the four elements of the public sphere described by Calhoun: bracketing social status differences, inclusiveness, rational argumentation, and the problematization of new issues.

Bracketing Social Status Differences

The quantitative findings showed few status differences *within* PAGs: members are generally of high socioeconomic status and are well connected to the forest management industry for their economic livelihoods. Therefore, within the confines of the PAG groups, the issue of bracketing social status differences is relatively moot. Still, participants who do not share these attributes may find themselves quite alienated if steps are not taken to foster their participation. Perhaps more significantly, it appears that people of low socioeconomic status do not typically make it to the public involvement table, so it is unlikely that the "bracketing" of social status vis-à-vis the larger public is occurring.

We also need to think more widely of the bracketing of social *difference* as well as just status. For example, our survey research demonstrated that PAGs are comprised primarily of pro-forest industry representatives. Given that PAG groups are commonly organized and moderated by the forest industry, it is quite likely that agreement with forest industry ideals confers a type of status as well. Under the lens of the public sphere, a broader conception of social *difference* beyond socioeconomic status is important to bracket.

Inclusiveness

Previous research asserts the importance of two-way information flow in the public involvement process, where public involvement participants are not just passive recipients of forest industry information, but provide feedback and articulate their own attitudes and beliefs.[27] Though PAGs have no formal decision-making capacity, they may still have an important role to play in shaping discourse and in placing certain expectations on the company. If they are more closely to approximate the ideals of the public sphere, PAGs can and ought to serve as information conduits, where information they receive at the public involvement table is distributed and re-circulated throughout the community. In turn, interested citizenry ought to be able to provide feedback to their representatives such that their concerns eventually find their way back to the public involvement table for discussion. In this sense, the responsibilities of PAG members do not end when the meeting is over.

Including the broader public in this communication will help to bring the views of a diverse public into the discourse. This expansion has implications for other elements of the public sphere. For example, the conception of the PAG extending beyond the table to include wider discussions with other community members poses a far greater challenge to the bracketing of social difference. Given the pro-forest industry orientation that seems to characterize these groups, it is reasonable to ask whether the views of environmentalists in the wider community would be solicited and given due consideration by PAG members, even under an expanded model of involvement. Another crucial social difference quite difficult to bracket is the issue of local versus non-local interests. Being "local" often takes on a privileged status, while the views of "outsiders" are often considered irrelevant to local forest management. This is especially so when the local person has interests that align with traditional forest management. It is relatively unlikely that communication of PAG members with their "public" will include discussions with people who are outside the local community, but may still have a legitimate interest in forest management. This legitimacy is genuine throughout the province, as Forest Management Area licenses are allocated on Crown Land, which is held by the Province for public use and benefit.

Trust in the Process of Rational Argumentation

With procedural rationality, or trust in the process of rational argumentation, people are expected to set aside their initial positions or roles and listen to the logic and rhetorical strength of others' arguments. This is a hallmark of the public sphere. Ideally, participants will trust the process and acknowledge the rationality of someone's rhetoric (i.e., logical consistency, validity of the claim) rather than discount others' claims as a matter of course because they seem to represent different interests than their own.

Does trust occur in PAGs as currently constituted? Our survey research demonstrated that some sources are viewed as more trustworthy than others. Further, in our field study we have observed that PAGs are often constructed along the lines of quotas of "interest," for example, "to accurately represent the community, we need someone from the mill, an environmentalist, someone from the chamber of commerce. . ." This method of filling seats around the table is not conducive to fostering procedural rationality, but rather engenders feelings of "turf-protecting" among participants, where environmentalists need to be true to their cause (represent other "environmentalists"), people employed in the forestry industry feel they need to reflect the opinions of their co-workers, and so on. In a public forum, these positions may become exaggerated, intentionally or unintentionally, by their representatives to prevent the pendulum of opinion from swinging too far in an undesirable direction. There is the potential for representatives to become caricatures of their positions, which inhibits trust in those who hold contrasting views. Moving PAGs in the direction of procedural rationality would require strong efforts to reduce the "interest emphasis" of the de facto quota system so common in existing groups.

Several other considerations need be raised with respect to procedural rationality. First, this ideal becomes more difficult to attain when considering the wider community as potential participants, where PAG members act as conduits for a wider array of community sentiments. It would be difficult to provide a more diffuse set of participants with the same information sources as PAG members, and it would be hard for them to trust in the logic of arguments to which they are not directly exposed.

Another consideration is the importance of exposure to information. Duinker[28] emphasizes the need for good information as important to the public involvement process. Organizers of PAG groups tend to agree, often speaking of the need to "bring members up to speed" regarding the complexities of forest management. They achieve this goal by bringing in speakers pertaining to areas of interest such as pests and disease or regeneration rates. The need to continually bridge the knowledge gap between new and continuing members is often used as a rationale for seeking low rates of turnover on committee memberships. The strong emphasis on the provision of information poses problems vis-à-vis the public sphere, as it may be at odds with procedural rationality and inclusiveness. It is difficult to divorce the process of argumentation in a public forum from the information underpinning the arguments. Although the public sphere might suggest a process where the quality of one's argument was determined via criteria independent of technical information provided (e.g., logical consistency), it appears that this is rarely the case in practice. For example, arguments that are at odds with the information provided by "experts" are likely to be viewed with skepticism, especially by those who agree with the substance of the information.

A final aspect of procedural rationality involves power and extends beyond the characteristics of the PAG participants themselves. Recall that PAGs have no formal decision-making authority; in this sense, they lack power. However, given the earlier discussion of PAG capacity to influence forest management even without decision-making capacity, other manifestations of power also come in to play. Powerful outside interests (i.e., the forest management company sponsoring the public involvement forum) could be manifested in either direct or indirect (implied) forms. For example, direct control of discourse may occur when the managing forest company or government agency tries to keep the discussion within acceptable policy-relevant bounds (e.g., "we know that not meeting the Annual Allowable Cut is not a possibility because it would result in the local mill having to drop a shift"). Implied control, which might foster the perception among members that open disagreement or redirecting the agenda is inappropriate, is also important. For example, PAG meetings are often held in forest company meeting rooms, with meals, beverages, and even financial recompense provided by the company. A "home turf" effect can send the message that expressing disagreement with the agenda or goals of the forest management company is inappropriate.

Engaging New Areas in Discourse

The public sphere is intended to support the "problematization" of new areas and issues through the process of public deliberation. This might be manifested in moving beyond strict adherence to traditional values (production forest management and Annual Allowable Cut) to include alternative strategies for a sustainable community such as maintaining "special places" or managing for nontraditional forest products and uses. Another way to expand the range of topics on the table is by opening the process to the wider community. PAG members could share ideas and solicit input from their friends and neighbors. However, we have witnessed a tension in PAGs, where the perception exists that these ideas distract the group from "the real issues."

Summary and Conclusion

As the importance of the forest industry in Alberta has grown, so too has the awareness that the public has a stake in the way forests are managed. Public involvement is now recognized as important to sustainable forest management, and there is a formal mandate that PI must be included to receive or renew forest management licenses. Public advisory groups have proven to be a popular technique, but there is unease about whether they are truly effective in the sense of contributing to sustainable forest management, or whether, in their current incarnation, they are merely fulfilling the letter of the law.

Research into the effectiveness of these and other public involvement mechanisms in forest management is in its infancy. Most has focused on whether these groups mirror the characteristics of the population they are assigned to represent. Similar to previous research, our findings indicate that these groups differ from the general public in potentially important ways, including socioeconomic characteristics, connectedness to the forest industry, and attitudes towards components of public involvement.

Rather than asserting the need for these groups to become more representative, we argue that member representativeness is hardly sufficient to determine the effectiveness of these groups. Other process-based factors could render these criteria less important. We used Habermas's conception of the public sphere as a theoretical foundation to examine the participation process identified in previous public involvement literature. Under this conception, issues of trust in the process of rational argumentation are expected to supersede the substantive content of interest group-derived positions. New issues outside the narrow confines of "business as usual" are expected to be taken seriously, irrespective of social status differences.

Our preliminary examination of public advisory groups suggests that they fall short of the idealized notion of the public sphere, as they are characterized by relatively circumscribed communication along the lines of interest rather than procedural rationality. The role of information appears to be crucial to the process. The emphasis on informing members tends to limit inclusiveness and discourage procedural rationality as well. This is not intended to be a stinging rebuke to the conveners of such groups; not all ideals of the public sphere are readily attainable by providing information to ongoing forest management efforts. However, these ideals should not be readily dismissed as academic irrelevancies. These principles are worth striving for if practitioners are serious about improving the effectiveness of public involvement and convincing a sometimes skeptical public that effective civil deliberations can take place. Moving towards the inclusion of a wider public via communication with PAG members, de-emphasizing the initial positions of participants (e.g., "environmentalist" or "logger"), and broadening the legitimacy of different information sources will enhance the groups' political and practical effectiveness.

9

After the Fall: Perceptions of Forest Management in Western Newfoundland

Thomas M. Beckley and Brian Bonnell

Introduction

Canada is a vast and diverse country. Its forests range from temperate rainforests to Carolinian hardwoods to boreal mixed wood. Several chapters in this volume describe sustainable forest management in the western part of the country (McFarlane et al., Parkins and Stedman, Pierce and Lovrich, and Ryan). In contrast, this chapter addresses public perceptions of sustainable forest management in Newfoundland at Canada's eastern extreme. In many regions of the west, resource exploitation is in an early phase—the communities are young, the mills and processing facilities relatively new, and the history of resource exploitation is short. Europeans have been present in Newfoundland for over four hundred years. Its earliest habitation by Europeans was based almost entirely on resource extraction, though not of forest resources, but rather fish. Nevertheless, use of the forest has been important throughout Newfoundland's history, and industrial forest use has a century-long history there.

This chapter also reviews data about the traditional values Newfoundlanders have held for forests, and how that may be changing. Newfoundland is noted for its active reliance on its forest resources for basic needs.[1] The survey research that we review in this chapter covers a rural portion of Newfoundland. This area has a higher reliance on the forest industry than the rest of the province. All these things taken together might suggest that our survey respondents would express a utilitarian and anthropocentric value set. In fact, our results show that biocentric values for the forest are fairly strong, and likely growing. Part of the explanation for this change, and for the high demand for public involvement in forest management, comes from recent experience in another resource sector.

Within the last decade, the mainstay of the Newfoundland economy, the cod fishery, has experience a much-publicized collapse. The collapse of the cod fishery is one of a very few cases in North America of an economically important natural resource industry being "overshot" [2] through overzealous exploitation and a failure of policy, regulation, jurisdiction, and scientific capacity. Many people cite mismanagement by the federal government for the cod collapse, and many Newfoundlanders rue the day in 1948 that a slim majority (51 percent) voted to join Canada in Confederation. That act paved the way for federal intervention in the management of the fishery. The whole story, however, involves a complex web of international relations, technological development, competing sub-sectors within the industry, and a frustrating inability on the part of scientists to prove irrefutably (at least to policy makers and a substantial portion of the concerned public) that the cod stocks were declining past the danger point. The story of the cod collapse has been told in great detail elsewhere[3], and others have documented the human cost of that event.[4]

Not long after the cod fishery collapse, a new resource giant appeared in Newfoundland in the form of off-shore oil. The opening of the Hibernia platform in 1997 signaled a new era for Newfoundland. This development is largely credited for an economic renaissance in the province in recent years. The two industries, fish and oil, could hardly be more different, with the former expanding while the latter may have declined beyond a point where it will recover in our lifetime. The oil development is highly geographically concentrated, and much of the economic benefit from that industry is similarly concentrated in the St. John's region. The cod fishery, in stark contrast, was dispersed in hundreds of small harbors and settlements around the entire island. As well, the oil industry is capital intensive whereas the traditional cod fishery was labor intensive. The fisheries that have emerged in the aftermath of the cod collapse (namely shrimp and crab) also tend to have a more concentrated industry structure. The result in both instances is greater wealth but less widespread distribution of that wealth than was the norm in the cod fishery.

Newfoundland's Forest Sector

The distribution of land cover types of Newfoundland tell a story in themselves of a challenging and unproductive landscape relative to most of the rest of North America. Thirty eight percent of the land area (9.4

million acres) is classed as productive forestland. Just over six million acres, or 24 percent, is classed as non-productive forestland. This land, while forested, produces only softwood or hardwood scrub that will not reach merchantable size. Thirty seven percent of the land is bog or barren land, and just under 1 percent of the land area is agricultural.[5] With this sort of profile, it is not surprising that for four hundred years, the wealth of Newfoundland was gained from its fertile seas, not its rock-bound interior. Nevertheless, by the 1930s, the forest sector accounted for 50 percent of Newfoundland's exports by value.[6] Just two decades earlier, the forest was still utilized almost exclusively for domestic fuel, domestic construction and flakes (wooden platforms) for drying cod.

The history of Newfoundland's forest sector is quite interesting, though not entirely unique. Newfoundland was a colony of Britain until 1949, when its residents voted by the narrowest of margins to join Canada. The colony was developed in the way most colonies are—with the external capital of a few powerful families who were given extremely favorable concessions and who were required to return little to Newfoundland in the way of reinvestment in the local economy.[7] In the late 1800s, the Reid family was given nearly four thousand square miles of land in exchange for building a railroad across the island. The intent was to open up the interior to industrial development. After some fits and starts, this is exactly what occurred. Crown timber licenses were initially granted to sawmills, but with the arrival of two major pulp mills in the 1920s, the licenses were consolidated. Newfoundland gave some of the most favorable timber concessions in the twentieth century. Licenses for the big newsprint mills were for ninety-nine years. The ground rent was $2.00 per acre, with no royalty assessed for pulpwood and a $.50 per thousand board feet levy on wood processed into lumber.[8]

While the use of the forest resource is widespread throughout the province (for firewood, wild game, etc.), industrial forestry remains concentrated in a few processing facilities. Corner Brook, on the Bay of Islands on the west coast, is home to a pulp and newsprint mill. Grand Falls/Windsor, in the interior, is home to another such facility of similar vintage. In the 1980s, Stephenville, also on the west coast, became host to a third newsprint facility. The vast majority of the volume of wood harvested on the island is processed in these facilities. In actuality, the reach of the forest industry for its wood supply is quite wide. Much of the interior of the island is crisscrossed with roads and accessible by the forest industry. In recent years some of the large pulp and newsprint manufacturers have even gone offshore for their supply of timber. Currently, the provincial government and industry are looking to Labrador to feed the island mills, while residents of Labrador are lobbying to have that wood processed locally.

There are nearly 380 census subdivisions in Newfoundland. Of these, 14 are heavily timber-dependent census subdivisions on the Island and another 19 census subdivisions that are moderately timber-dependent. Heavily dependent communities have over 50 percent of their employment base in forestry activities while moderately dependent communities have between 30 and 49 percent.[9] Province wide, over four thousand individuals are employed in the forest industry and the provincial payroll in the sector was 114 million dollars in 1997.[10]

Newfoundland has the highest proportion of publicly owned forestland among all the Canadian provinces. The Government of Newfoundland and Labrador owns 99 percent of the province's forest and large portions of it are leased in long-term renewable leases or limits to major industrial processors. As mentioned, two leasees on the island have ninety-nine-year leases on their timber limits, both of which are coming up for review and expected to be renewed. A full 95 percent of the fiber harvested in Newfoundland is used to produce newsprint, the majority of which is shipped to the U.S. market. In 1997, there were 158 wood processing establishments in the Province; however, the vast majority of these are small sawmills.[11]

This chapter reviews survey research that was designed to elicit the views and opinions of the general public toward forest management in western Newfoundland. Despite the press about the cod collapse and the rise of the petroleum industry, forestry was and is a significant player in western Newfoundland. Twenty-one percent of our random sample survey reported that they or someone in their immediate family derive some income from timber related activities (work in the pulp mill, a sawmill, timber harvesting, selling firewood, etc.). Nine percent derive income from non-timber forest activities (trapping, guiding, selling berries, rabbits, etc.).

Sustainability in natural resource contexts is often described as having three dimensions or elements—ecological sustainability, economic sustainability, and social sustainability. These have been characterized in a number of ways, including "as three legs of a stool," implying that the stool cannot stand with any one leg compromised. Others have suggested that the ecological foundations are the most critical element and that it is the ecological pillar upon which the social and economic benefits rest. We do not wish to engage in this debate, or privilege one element of sustainability above the others. The survey was quite comprehensive and portions of it address all three facets of sustainability. For the purposes of this chapter, we are interested in the following questions: How do people in Newfoundland view the forest and forestry today? Do they feel that Crown (i.e., public) forests are being managed sustainably? Do they take a more skeptical view of government management, or are they more vigilant due to their recent experience with the collapse of the cod fishery? Is there greater recognition of social and economic dimensions of sustainability in Newfoundland compared with other jurisdictions, either after extensive use of forests or because of the cod collapse? Is there greater demand for public involvement in forest management in Newfoundland as compared with other jurisdictions? After a description of our methods and the study site, we will answer these questions in turn.

Methods

A mail survey was conducted in the spring of 1999 in the Western Newfoundland Model Forest (WNMF). The Model Forest program is described in more detail elsewhere in this volume (Ryan in chapter 11, and in chapter 3 Duinker et al.). In short, a model forest is an area where people with a direct interest in the forest, supported by the most up-to-date science and technology, participate in decisions about how the forest could be sustainably managed. The WNMF was the primary sponsor of the survey and provided the funding for the project. The WNMF is located roughly in the center of the west coast of Newfoundland and covers an area of 923,000 hectares of boreal forest. It contains a national park (Gros Morne) and dozens of small towns and villages in addition to one reasonably sized town (Corner Brook, population 22,000). The population of the entire region is only 46,000, and vast expanses of the model forest area are uninhabited.[12] The geographical boundary of the survey approximated that of the Western Newfoundland Model Forest. In addition, two towns located just outside the boundaries of the model forest were included (Stephenville, population 7,700, and Deer Lake, population 5,200).

The sampling frame was obtained from the NewTel Communication telephone directory. Survey respondents were first recruited by phone, using a two-pass systematic random sample. Telephone recruitment was necessary in order to obtain addresses of potential respondents. A total of 633 surveys were mailed, all but nine of which were delivered. The Dillman[13] "total design method" was used, and subsequent post-card reminders and additional surveys were sent to the sampling frame. The total response rate for the survey was 74.4 percent. We believe that the high response rate, in itself, suggests that people in the region are concerned with forestry issues. The survey covered a vast range of subjects with fifty-six questions. Included were items about forest use, perceptions of the model forest, perceptions of sustainable forest management, questions related to where people obtain information about the forest, questions about the pine marten (a locally endangered species), as well as the usual demographic questions.

For the purposes of this chapter, we will focus on a subset of the findings. These will cover the three widely cited aspects of sustainability: ecological, economic, and social. Ecological sustainability will be examined through the questions regarding regeneration rates and endangered species. Economic sustainability will be examined through respondents' rankings of forest use priorities and their perceptions of future wood supply. Finally social sustainability will be treated with responses to questions about public involvement in forest management. First, however, we will describe the region, with particular emphasis on forest use.

Forest Use in the Study Site

Western Newfoundland has been host to a pulp and newsprint industry since 1923, but the history of forest use predates the mill by a couple of centuries. That use of the forest, however, was primarily for subsistence purposes. Most of the history of Western Newfoundland is based on fishing and subsistence activities. The sea provided a tradable good, though often the fishermen were kept in debt by the merchant class of St. John's. The forest provided most everything else. People relied on the forest for building material, for food (berries, moose, caribou, birds, small game), rudimentary medicines, and of course fuelwood. Many of these traditional forest uses remain popular in Newfoundland. There is a strong tradition of doing for oneself, doing for one's neighbors, and when both those options fail, doing without.

Some unique customary laws, such as the right to harvest timber from Crown land for domestic consumption, have evolved from the traditions of forest use on the island. Historically this was done via water access, though today the prevalence of ATVs, four-wheel drive vehicles and an improved road network means that most people access the forest for fuelwood harvesting from the land.[14] Hunting also remains a popular activity in Newfoundland, particularly in rural communities. Results from a national survey conducted in 1996 show that 15.1 percent of Newfoundlanders reported participating in hunting; nearly three times the national average of 5.1 percent. As well, nearly twice as many Newfoundlanders reported gathering nuts, berries and firewood (20.2 percent) as compared with the national average (11.0 percent).[15] There is clear evidence that subsistence activities remain very important in contemporary Newfoundland culture.

Our survey results corroborated the national results. To document the persistence of these historical traditions, we asked our survey respondents about their use of the forest and use by members of their household over the last three years. As we suspected, a majority of our respondents use the forest for gathering wild berries (82.1 percent), harvesting large game such as moose, caribou or black bear (72.4 percent), fish (60.6 percent), rabbit (56.7 percent) and firewood (53.7 percent). While there is a recreational element to all these activities, there is an important subsistence component as well. In addition to these uses, 32.8 percent harvested their own Christmas trees, 22.2 percent harvested peat moss, black earth or soil, 16.8 percent harvested saw logs, and 15.3 percent harvested living trees, shrubs or flowers. Fewer than 10 percent identified craft materials, mushrooms, furbearers and fiddleheads as items procured from the woods.

Patterns of work, ease of access, and demographic characteristics of the island facilitate the active use of the forest. The dependency ratio in Newfoundland is high. The dependency ratio is simply the proportion of residents that are below working age or above retirement age, to the remainder of working aged people. Many able-bodied persons of working age have left the island to seek employment elsewhere. In general, unemployment in Newfoundland runs about 25 percent and in rural Newfoundland the number can be even higher. Only 37 percent of our respondents indicated that they were full-time employed. Just over 7 percent were employed part-time, and over 10 percent were employed seasonally. Ten percent of our respondents were homemakers, 8.2 percent were unemployed, and 16.5 percent were retired. The remaining respondents were on disability, were self-employed, or were students. Around 15 percent of the sample work in the forestry sector, and an additional 40.6 percent work in retail, services or manufacturing. Some workers at the pulp mill may well have placed themselves in the latter category. Virtually everyone in the study area has ample opportunity to forest areas for both recreation and subsistence purposes. Given that 99 percent of Newfoundland is public land, nearly everyone has some Crown land in his or her backyard or in close proximity.

General Views of Forest Sustainability

Most of the survey questions in this project were closed-ended questions. However, we asked one open-ended, general question about what sustainable forest management meant to the respondents. This was asked at the beginning of the survey so that subsequent questions would not bias or influence their responses. We were curious as to how broadly or narrowly respondents conceive the issue of forestry sustainability.

For the most part, respondents appeared stymied by the question. While it did not produce much usable fine-scale data, the fact that people had a rather difficult time articulating a response is an interesting result in and of itself. Responses were grouped by theme and the highest response category had to do with the theme of future action. The third highest response category referred to benefits for future generations. These two responses represented approximately 22 and 12 percent respectively. Second in importance was reference to replanting, reforestation or silviculture by 16.5 percent of respondents in their interpretation of sustainability. These activities, while still critical to sustainability, are more reflective of priorities related to sustained yield of timber crops rather than sustainable forest management. Other elements of sustainability, as currently articulated in national and provincial policy documents, and as understood by resource managers, were mentioned by less than 10 percent of the respondents. Wildlife was mentioned by 9 percent, forest protection by 8 percent, employment by 4 percent, and recreation by about 4 percent.

In a second question regarding the general state of affairs with forest management in Newfoundland, people appeared to be similarly stymied. Over one fifth (21.9 percent) were not willing to offer an opinion regarding the degree to which they felt Newfoundland's management of its forests was currently sustainable. Of the remaining 78 percent, only about 18 percent expressed either a strongly positive (8.8 percent) or strongly negative (9.0 percent) response. By far the largest response category was persons who expressed that they felt Newfoundland's management of its forests was somewhat sustainable (44 percent). This suggests a moderate degree of satisfaction with the status quo. The remaining 16.3 percent expressed the view that the Newfoundland forest management regime is somewhat unsustainable. As we shall see, however, this lukewarm endorsement of the status quo does not hold up in subsequent questions regarding the disposition of Newfoundland's forest estate.

We obtained another general assessment of respondents' views of the state of the forest by asking whether they felt the woods are in better shape today than five years ago. Here the overall assessment was negative. More than twice as many (48.3 percent) either disagreed or strongly disagreed than those who agreed or strongly agreed (22.5 percent). Furthermore, the intensity of the negative sentiment was stronger, with only 2.9 percent strongly agreeing with the statement, but a full 18.1 percent strongly disagreeing.

Views of Forest Practices

While the general views of the sustainability issue in forestry suggest a lukewarm endorsement of the status quo, questions about specific issues in forest management present a very different picture, particularly with respect to explicit forest management practices and priorities. The message was clear that residents of western Newfoundland would at least like to see a more prudent or cautious approach to forest management, if not a more ecologically friendly management system. Later, data on their views of public participation will show that they do not trust government or industry alone to implement this new

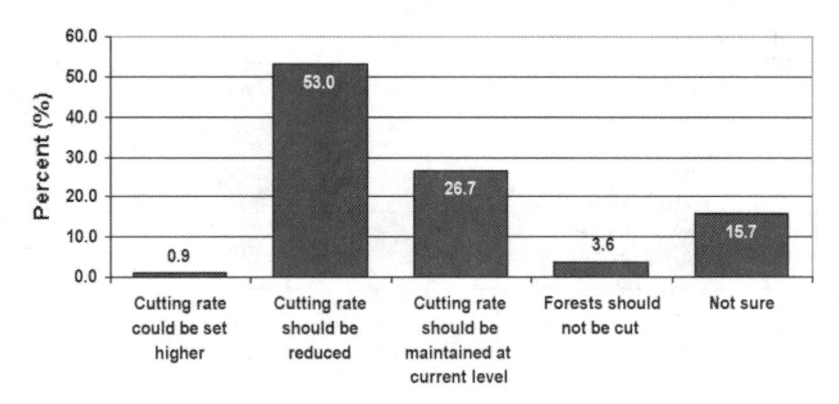

Figure 1. Opinions Regarding Harvest Rates for Newfoundland's Forest

paradigm. The changes desired are significant enough that they would like to see considerable citizen oversight in that process.

Most people were willing to express an opinion regarding the timber harvest rates for the province. Figure 1 shows less than one percent of the population believed there was room to increase the harvest and a true majority (53 percent) believed the cutting rate should be reduced. Newfoundland is noted for its utilitarian values and, not surprisingly, less than 5 percent of our respondents expressed the forest should not be cut at all. This result seems to contravene the view expressed by 53 percent who believed that forest management in the province was either very or somewhat sustainable. In response to the direct query on harvest rates, just over a quarter (26.7 percent) of the respondents suggested they were satisfied with the status quo.

It appears that part of the concern with forest sustainability has to do with effective enforcement of existing harvesting regulations (Table 1). Nearly 80 percent of all respondents strongly agreed or agreed with the statement "The provincial government should enforce commercial harvesting regulations more effectively." In contrast, just over 7 percent disagreed or strongly disagreed, with the remainder being unsure or neutral. This suggests that the level of confidence in the existing system to deal with infractions within the current management framework is low.

Table 1. Perceptions of Forest Conditions and Forest Practices (% of response)

	Strongly Agree	*Agree*	*Neutral*	*Disagree*	*Strongly Disagree*	*Not Sure*
The woods are in better shape than they were five years ago	2.9	18.6	10.7	30.2	18.1	19.5
The provincial government should enforce commercial harvesting regulations more effectively	27.5	52.0	6.8	2.6	1.3	9.8
Forestry practices have few long-term negative effects on the environment	6.5	18.2	12.3	33.4	15.2	14.3

Jobs Versus the Environment

We were curious about the relative priorities of western Newfoundland residents with respect to economic and ecosystem functions of the forest. We recognize that many consider this a false dichotomy, and that the jobs-versus-the-environment debate is one that many promote because it furthers their interests in an adversarial stakeholder process. Because this issue of trade-offs between the economy and the environment is the subject of debate in and of itself, our first question in this section provided our respondents the opportunity to reject the premise. They were asked to strongly agree, agree, disagree, or strongly disagree with the statement "If we were to protect the environment more than we currently do, jobs will be lost" (Table 2).

Not surprisingly, many did reject the notion that actions that improve the environment will result in a reduction of jobs. A significant portion of respondents were either unsure (11.1 percent) or neutral (9.1 percent) on this issue. However, 56.9 percent disagreed or strongly disagreed with the notion that greater environmental protection will result in the loss of jobs. Most people in both the affirmative and negative side of this issue responded in the moderate categories of agree and disagree, but the overall proportion rejecting this trade-off was significant.

In the absence of any forced trade-offs, an overwhelming majority agreed that society has an obligation to protect endangered species. These were some of the more dramatic results in our survey in terms of finding societal consensus. Fully 93 percent of respondents agreed or strongly agreed with the statement, "We have an obligation to protect endangered species" of which nearly 50 percent strongly agreed with the statement.

When forced to make trade-offs between jobs and the environment in subsequent questions, it is very clear where the priorities of a majority of western Newfoundlanders lie. Fifty-nine percent of respondents disagreed or strongly disagreed that jobs in the forest sector were more important than protecting endangered plants. Not surprisingly, even more disagreed or strongly disagreed (64.3 percent) when the same question was asked with regard to endangered animal species. In contrast only 15.5 percent considered that forest sector jobs were more important than endangered plants, and 12.8 percent felt forest sector jobs were more important than endangered animals. Interestingly, the percentage of persons disagreeing with these statements corresponds quite closely to the percent of our sample that is employed in the forest sector.

Table 2. Perceptions of Trade-offs between Jobs and the Environment (% of response)

	Strongly Agree	*Agree*	*Neutral*	*Disagree*	*Strongly Disagree*	*Not Sure*
If we were to protect the environment more than we currently do, jobs would be lost	2.2	21.6	9.1	47.2	8.7	11.1
We have an obligation to protect endangered species	49.5	43.5	3.5	1.5	1.3	0.7
Protecting jobs in the forest industry is more important than protecting endangered plants	3.3	12.2	21.3	43.9	15.1	4.2
Protecting jobs in the forest industry is more important than protecting endangered animals such as the pine marten	2.4	10.4	19.1	46.1	18.2	3.8
Forest companies must do more to protect the environment, even if it means a loss of some jobs	15.2	53.3	17.0	9.0	2.4	3.1

Finally, respondents were five times more likely to agree or strongly agree (68.5 percent) with the statement, "Forest companies must do more to protect the environment, even if it means a loss of some jobs" than they were to disagree or strongly disagree (11.4 percent). There were a fair number of neutral responses to this statement (17.0 percent), but overall it is quite clear that when respondents were forced to choose between the environment and jobs, they choose to maintain ecological integrity.

The Role of the Public in Sustainable Forest Management

A major difference between the paradigm of sustainable forest management and the sustained yield paradigm that preceded it, has to do with an increased awareness of social dimensions of forests and forest management. Furthermore, a major component of the social dimensions relate to opportunities for the public to participate in resource management decision-making. We asked a number of questions related to the importance of public participation in resource decision-making. In the wake of the cod collapse, we anticipated this issue would be important to our respondents. For years there was a rather public debate regarding the proper role of the public in participating in the allocation of fish stocks. The public effectively lost that debate and the management of cod stocks was a non-democratic, top down endeavor. The public is not so much concerned whether the scientists or bureaucrats were at fault. They believe the government was at fault, and they do not wish to see history repeat itself in the forest sector.

As expected there was a very high degree of value expressed for citizen involvement in decisions regarding forest use. A strong majority (56.8 percent) expressed that citizen participation is of great value. This was one of our strongest responses in the entire survey. An additional 31.7 percent expressed that citizen participation has some value. Clearly, the public does not want to leave resource management to industry, or to government, without some sort of direct citizen oversight. Less that 7 percent expressed citizen involvement has little or no value (see Figure 2).

We asked about the performance of current institutional players that have the mandate to ensure public values are incorporated into forest management. The island of Newfoundland is divided into eighteen forest management districts. In some, the industry that holds the Crown license for the area is responsible for engaging the public. For management units with a mix of Crown tenures, the government of Newfoundland is responsible for providing opportunities for citizen

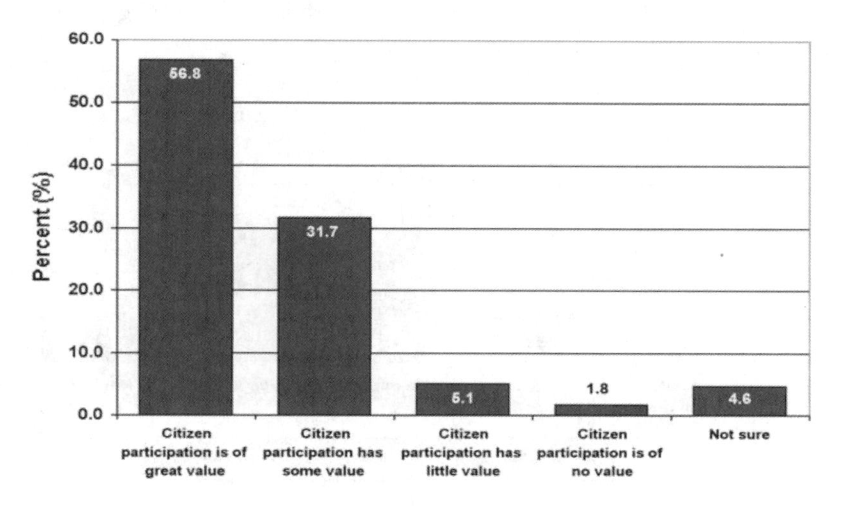

Figure 2. In your opinion, how valuable is it to have citizens participate in deciding how the forests of western Newfoundland are used?

input. The government faired slightly better than industry in our respondents' assessments of the degree to which these players provide opportunities for comment and consultation. Still, only 12.7 percent suggested the government provides some or a great deal of opportunity for public involvement, compared to 76.3 percent who felt there was little or no opportunity for public views to be heard. Less than 9 percent felt industry provides a great deal or some opportunity, compared to 83.3 percent who felt they provide little or no opportunity for public views to be heard (see Figure 3). In fact, these institutions do have systematic programs of public involvement. As well, the activities of the Western Newfoundland Model Forest, including the survey whose results we are reporting here, provide additional opportunities for the public to express their views regarding forest management. However, these efforts are largely unnoticed and unrecognized.

We also asked our survey respondents the nature of the role that the public should play in resource management. This is an area where the social contract between society and natural resource managers continues to be redefined. The response categories are included in Figure 4. Essentially, the options ranged from having no role (leaving all decision-making to resource professionals) to acting as a full and equal partner. While only 2.7 percent were comfortable with the public having no role, 29.6 percent preferred that the public be engaged as a full and equal partner. The same percent of respondents (29.6 percent) preferred what is essentially the status quo in which advisory boards review and comment on management goals.

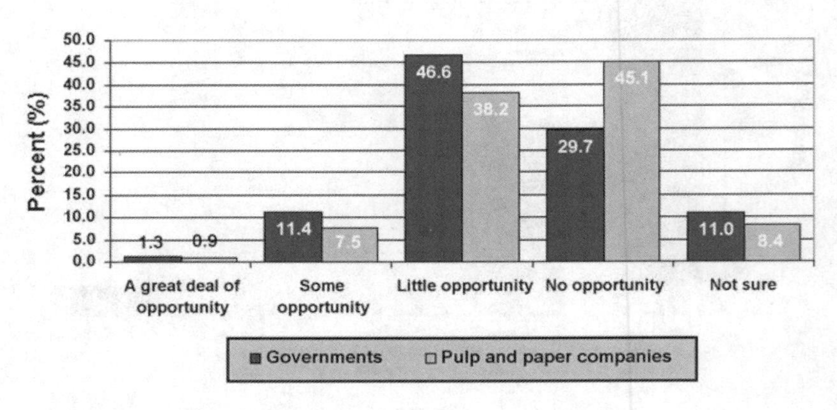

Figure 3. Respondent Views of Opportunities for Public Involvement Provided by Industry and the Provincial Government

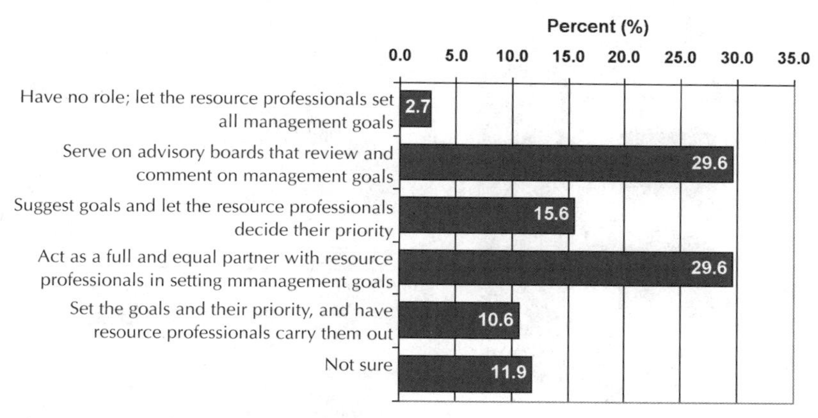

Figure 4. Respondent Preferences for the Most Appropriate Role for the Public in Forest Decision-making

Finally, we asked our survey respondents a more direct and personal question regarding public input in decision making in forest issues in Newfoundland. We asked how important they felt it was for them to have a say in how the forests of western Newfoundland are used. Over a third (34.3 percent) suggested it was extremely important that they have a say, and nearly half of the entire sample (47.6 percent) felt it was somewhat important that they have a direct say in forest management issues. Just over 10 percent were not concerned with having a say (see Figure 5). Clearly, the old paradigm of resource management being the purview of professional experts does not sit well with the general public. They wish to be involved directly and, despite the protests by industry and government that "No one comes to our open houses," there is strong evidence from this data that the public wants a seat at the table or at least some form of dialogue between decision makers and forest users. Low attendance at such events does not necessarily negate the evidence presented here. Rather, it suggests to us that our respondents crave meaningful involvement, not tokenism. They wish to be involved in setting goals for management, not in rubber-stamping plans that have been written without a significant degree of consultation.

Public involvement can be both expensive and slow. These traits are an anathema to industry stakeholders who prefer predictable outcomes, efficient operations, and manageable timelines. In fact, public involvement is an ongoing process that requires continual updating as preferences and values change. The exercise is not unlike a timber or forest inventory. One can take a snapshot of public sentiment but it

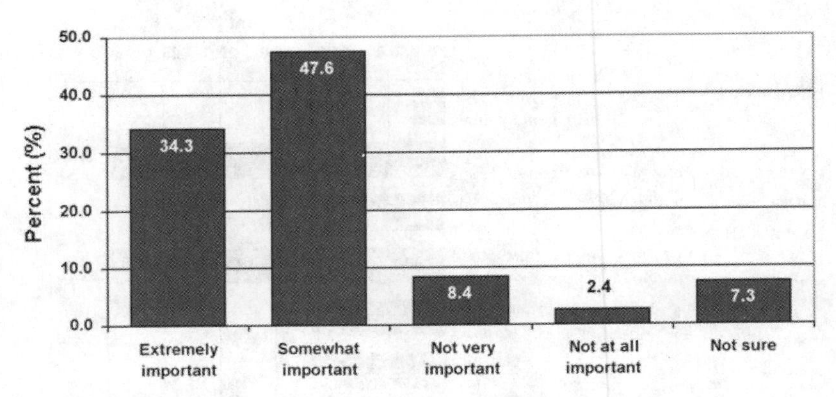

Figure 5. How important is it for you to have a say in how the forests of western Newfoundland are used?

requires periodic updating to reflect change and evolution in local conditions. At the same time, the scientific sampling principles that underlie a forest inventory also apply to a social inventory.

Ranking Dimensions of Sustainable Forest Management

In the final question to be reviewed for this chapter, we asked respondents to rank the importance of six forest uses. Our respondents' ranks for these uses, along with the mean responses, are listed in Table 3. Three of them relate to ecological dimensions of sustainability (a place for protection of water, air, and soil; a place for a variety of animals and plants; and a place for wilderness preservation), one clearly represented the economic dimension of sustainability (a source of economic wealth and jobs), one was clearly social (a place for recreation and relaxation), and one straddles the divide between social and economic dimensions (a source of non-timber products).

In this question, our sample of western Newfoundland respondents articulated a clear message as to where their priorities lie. The protection and maintenance of environmental services was the highest priority

Table 3. Ranking of Ecological, Economic and Social Importance of Forests

	Mean	*Rank*
As a place for wilderness preservation	3.27	3
As a source of economic wealth and jobs	3.87	4
As a place for recreation and relaxation	4.42	6
As a place for protection of water, air and soil	2.44	1
As a place for a variety of animals and plants	2.70	2
As a source of non-timber products	4.29	5

with a mean score of 2.44. Second and third on the list were the other habitat-related ecological dimensions of sustainability. The fourth most important use was the economic use of the forest, and the social use of the forest for recreation and relaxation was ranked last, a full two points on the six-point scale below the number one priority. The response category relating to non-timber use of forests, which we interpret as an intermediary response between social and economic factors, was ranked fifth. Taken as a whole, these responses clearly suggest ecological dimensions of sustainability are first and foremost the concern of western Newfoundlanders. Second most important are economic dimensions, with social uses of forests trailing rather far behind.

Comparison with Other Survey Results

Over the past decade there have been numerous other surveys relating to forest use, forest practices and forest sustainability in national, provincial and regional contexts. While all the studies produced statistically valid results, and many questions from our own survey were based wholly or partly on questions asked on these other surveys, we cannot make valid direct comparisons between our survey and surveys from previous years. However, some of the results are reviewed here to give the reviewer an idea of the consistency of results and to illustrate how our Newfoundland sample was similar or different to national samples.

In 1991, the Canadian Forest Service (then known as Forestry Canada) commissioned Environics Research Group Limited to carry out a national survey on forest values. The survey asked respondents to rank, by pairs, very similar categories to the ones we describe in Table 3. While the methods were different, the results were quite similar. A majority of the national public ranked a preference for protecting water, air and soil as the highest priority followed by global balance, habitat, wilderness, economic use, and recreation.[16] In most categories, Newfoundland respondents selected economic aspects over environmental. Either western Newfoundland has had a stronger environmental ethic all along, or there has been a shift toward more anthropocentric values in the last decade.

In the Environics survey, 49 percent of the national sample believed that overcutting and mismanagement were the greatest threats to Canada's forests. This mirrors our own results in which 53 percent suggested harvest rates should be reduced, and more than twice the number disagreed than agreed with the statement that the woods are in better shape than they were five years ago.

The McGregor Model Forest sponsored another national survey in 1996. It also elicited public values on a range of forest related issues including the appropriate role of the public in forest management. We replicated their categories and asked the question in a similar format. The results from the McGregor study's national sample and our own local sample were remarkably similar. The top response categories for the appropriate role of the public (serve on advisory boards that review and comment on management goals, and act as full and equal partner with resource professionals in setting management goals) were separated by less than one percentage point. Both these categories received responses of around 30 percent in both surveys.[17]

In 1991, TAWD Consultants Limited conducted a public survey on behalf of the Newfoundland Department of Forestry and Agriculture and the Protected Areas Association concerning the use, management and protection of forests and natural areas in the province. At that time, 48.3 percent of the public agreed forest management practices were ensuring adequate supplies of wood for future use. Furthermore, 29.2 percent felt that ample opportunities were being provided for public involvement in management decision-making.[18]

These results mirror our results quite closely. The priorities of Newfoundlanders appear in line with those the national public and consistent across time within the province. These survey results provide compelling evidence that there is a high degree of concern over environmental issues in Newfoundland.

Conclusion

The 1990s proved a decade of transition for the economy and culture of Newfoundland. Traditional fisheries were shut down to protect what little was left of cod stocks. New fisheries emerged and a major energy project came on line, which provided new wealth, though not a dramatic increase in the number of jobs for the province. Meanwhile, forestry remained an important sector in the economy, particularly in the west—the area from which we drew the sample for this study.

Newfoundlanders are often stereotyped as holding strongly utilitarian values toward the environment and natural resources. The 1991 national survey by Environics illustrated this to some degree. This stereotype is also widely attributed to rural residents in general, and our entire sample is derived from communities with less than twenty-five thousand people. However, our survey results did not bear out these generalizations. Residents of western Newfoundland appear as concerned as national samples that are predominantly urban (Canada as a whole is around 80 percent urban), with maintaining the ecological integrity of our forests. It is evident that there has been an evolution in public perception in western Newfoundland from an emphasis on economic wealth and jobs to more importance being placed on environmental values. This is probably the result of both increased education and awareness of environmental issues and a reduction in the economic dependency on the forest industry.

We were somewhat surprised at our respondents' inability to articulate clearly a definition or vision of what sustainability means for forestry in the province. We expected, with the cod collapse looming large in recent memory, that some basic, simple principles of sustainability would be in the forefront of our respondents' minds. Nevertheless, when confronted with specific questions about resource use and management, our survey respondents were clear in their preferences. First, they clearly value ecological integrity over economic or sociocultural uses of the forest. This was demonstrated consistently across a wide range of questions. Second, it is clear the public is skeptical of a top-down managerial style for forests. They desire substantial citizen oversight to government and industry management. This may well be a legacy of feeling left out of decision-making as cod stock declined past the point of being able to support an economical fishery. While reforms are taking place and both industry and government are substantially increasing opportunities for public involvement, the strength of the message contained in these results should make agency and industry officials realize that nothing less than a significant and sustained effort to engage the public will suffice in the emerging management paradigm. Perhaps we can learn from our mistakes. In the case of Newfoundland, this means placing higher value on ecological integrity and ecosystem function, and remaining vigilant with regard to how both public and private stewards interact with the public to set priorities and goals for resource use and management.

Part Three

Institutional Responses to Emerging Demands

The new language of sustainability has been adopted by federal resource management authorities in both Canada and the United States as well as in all provincial jurisdictions in Canada where resource management responsibility lies. What has not yet occurred to any significant degree is major institutional change in support of the new, more holistic paradigm. Agencies and external groups are in varying stages of responding to the emerging demands of ecosystem-based management. Contributors in this section report on these experiments in institutional reform as well as several new institutional arrangements. As these chapters attest there is no single path to sustainability. Where the chapters in section two were all about "demand," (the public's demand for emerging forest values), this collection of chapters is about current efforts to "supply" that demand through institutional reform.

Weber and Herzog champion the emergence of grassroots ecosystem management organizations (GREMs) in chapter 10. They review instances where local people, working together, have organized to lobby government agencies with land management authority. In some cases, citizens are actively involved in management themselves, and as such, they are working to supply their own demands. The authors view this as a positive trend in the pursuit of sustainability because key features of the movement include more direct citizen involvement (e.g. citizens taking responsibility for sustaining forests) and organizing on a scale that more closely matches ecological problems. Ryan (Chapter 11) and Shindler (Chapter 12) examine Model Forests in Canada and Adaptive Management Areas in the Pacific Northwest. These experiments are, in part, government attempts to supply the demand of greater citizen involvement, and a new mix of forest values. Ryan describes how these new approaches were created as experiments to transform traditional sustained-yield institutions into more responsive management agencies. Activities on these sites are to be informed by scientific information, but also by human values and stakeholder preferences from local communities. Shindler reports extensive survey research to evaluate the progress on the ten AMAs, finding that while intentions are good, agencies still have far to go in reaching these new goals.

10

Connecting the Dots: Grass-Roots Ecosystem Management and Sustainable Communities

Edward P. Weber and Christina Herzog

Introduction

As the limits of top-down approaches to environmental protection and natural resource management have manifested themselves in resource-dependent communities, the fields of environmental, natural resources, and public lands policy have all undergone significant shifts.[1] Many scholars now contend that creating sustainable environmental programs require complex, collaborative partnerships among diverse government, civic, and business actors at the state and local level—a dynamic similar to reform efforts in a multiplicity of other policy arenas, including education, policing, rural development, public health, and tax administration.

As part of this larger phenomenon, beginning in the late 1980s a new environmental movement began to take root primarily in *rural* areas in which local economies are inextricably tied to natural resources. These efforts began in the Western United States in places such as Willapa Bay (Washington), the Malpai Borderlands (New Mexico, Arizona), the Henry's Fork watershed (Idaho), the Blackfoot River Valley (Montana), and the Applegate Valley (Oregon). There is growing evidence that similar efforts are emerging in the eastern and southern United States.[2] The diversity of these efforts has caused scholars to apply a variety of labels to describe this new phenomenon, among them grass-roots ecosystem management,[3] community-based conservation, community conservation,[4] collaborative conservation,[5] and the watershed movement.[6]

Yet despite the variation, the new movements—identified as grass-roots ecosystem management (GREM) throughout this analysis—share several important characteristics. They organize on the basis of geographic "place," often according to biophysical boundaries such as watersheds or ecosystems. They are grounded in collaborative decision processes and active citizen participation. They reflect the belief that far greater common ground exists among traditionally opposed factions of rural communities than is commonly recognized, and that win-win-win outcomes for the environment, economy, and community are possible. Participants view the existing management regimes as ineffective, costly, lacking in accountability, and unable to sustain either the environment or rural communities. Finally, participants are taking a holistic approach to the notion of conservation and management with the expectation that doing so will promote sustainable ecosystems, communities, and economies. The Western United States is seeing the emergence of hundreds of broad-based coalitions of the "unalike"—citizens, government regulators, small businesses, environmentalists, commodity interests, and outdoor recreation enthusiasts—who are choosing alternative collaborative institutions for governing public lands and natural resources in an attempt to improve governance performance *and* enhance accountability to a broader array of interests.[7]

Of central concern to this analysis is the new movement's emphasis on a holistic approach to the notion of conservation and natural resources management. Participants are buying into these new arrangements because they expect efforts will be rewarded with sustainable communities. They seek communities "in which economic vitality, ecological integrity, civic democracy, and social well-being are linked in complementary fashion, thereby fostering a high quality of life and a strong sense of reciprocal obligation among its members."[8] The approach rejects the more traditional top-down, expert-led processes for deciphering and achieving sustainability because experts acting alone are inherently unable to deal with the complexity of the sustainability challenge and that long-term environmental policy success must involve the citizenry responsible for translating sustainability theory into on-the-ground results.[9] As the Willapa Alliance[10] notes:

> A sustainable community needs to be developed by the people who make up the community. It cannot be designed by a consultant. It cannot be implemented by experts hired specifically for the project. It needs to be implemented every day by the people who live and work in the community.

In order to explore more fully the relationship between these devolved, collaborative, and participative governance arrangements and sustainable communities, we rely on data from three exemplary cases—

the Henry's Fork Watershed Council in east central Idaho, the Applegate Partnership in southwestern Oregon, and the Willapa Alliance in southwestern Washington. Using primary documents, a review of the secondary literature, and field interview data from over seventy GREM participants,[11] we find that the search for sustainability in these communities involves a strategy to establish institutions, practices, and tools in support of sustainability, and to transform the way the community perceives itself, both in the relationships among community members and the community's relationship with nature. In fact, some of the participants in the three cases of GREM see their efforts as "a visionary project . . . [aimed at promoting a] paradigm shift in the way the whole community understands itself."[12].

Before exploring the six ways in which these community-based efforts connect some of the dots in the sustainability puzzle, the chapter briefly introduces the Henry's Fork, Applegate, and Willapa cases. The final section of the chapter explores some of the limits to the lessons for sustainable communities learned from our study of grassroots ecosystem management.

Three Cases of Grassroots Ecosystem Management

The Henry's Fork Watershed Council is a citizen-initiated, state-chartered watershed council in Idaho created in 1993. With broad support from Idaho's congressional delegation, Clinton administration officials, and Idaho's legislative and executive branches, the Council has been widely touted as a promising prototype for reinventing and improving natural resource management while re-engaging citizens at the watershed scale. The Council has been promoted at the national level as a new, proactive way to prevent species from becoming endangered.[13]

The Applegate Partnership, located in southern Oregon's Applegate Valley, was initiated in 1992 by a broad array of concerned citizens in tandem with the Bureau of Land Management and the U. S. Forest Service, who manage 70 percent of the land areas covered by the Partnership. Since its inception the Partnership has repeatedly been recognized as a potential model of future forest management. The Applegate Adaptive Management Area is one of the cornerstones of the adaptive management experiment in the Pacific Northwest (see chapters 11 and 12 by Ryan and Shindler respectively in this volume).

The Willapa Alliance is a grassroots effort in southwestern Washington, just north of the mouth of the Columbia River. Co-initiated by non-profits and private citizens in 1992, the Alliance's success in coordinating and catalyzing a comprehensive fisheries management plan for the area has prompted state legislative leaders, as well as Washington Governor Gary Locke, to propose the Alliance's administrative format as "the" template for rescuing threatened natural resources, like endangered salmon runs, throughout the rest of the state. Former Washington Governor Booth Gardner even has gone so far as to label the Alliance "an exceptional process of trying to create good jobs that can be maintained for generations to come . . . [and that] may serve as an example to the world."[14]

Mission, Norms, Process, and Management

The missions, participant norms, decision processes, and management approaches of the Henry's Fork Watershed Council, the Applegate Partnership, and the Willapa Alliance support the push for sustainability by infusing the communities with the value of sustainability and providing a different institutional framework for managing the interface between humans and nature.

The crosscutting "environment, economy, and community" mission statement, by valuing each element equally, emphasizes interdependence by recognizing that each of the policy sectors are linked in a way such that long-term success (sustainability) in one arena is dependent on success in the other two. The holistic ecosystem management approach recognizes that the watershed's "web of life" is interconnected and that individual, fragmented decisions by stakeholders often affect the health of other resources, habitats, and people. The mission and management approach both define environmental sustainability as inseparable from the sustainability of a healthy economy or a healthy community, directing decision-making to find of workable compromises that simultaneously promote the multiple policy goals.[15]

Participant norms reinforce the commitment to sustainability. Participants are asked to envision themselves as embracing a dual role, as community member and as representative of a particular interest (e.g., farmer, environmentalist, agency official). This obligates them to take a broader view of problems and to recognize the interconnectedness among people and policy sectors. The approach appears to lead to disapproval for decisions that only benefit one particular segment of the "place" or a single element of the three-part mission at the expense of others. The incorporation of sustainability as a critical public value adds legitimacy to the goal of sustainability. A broad cross-section of interests and community leaders are represented in GREM, making it for community residents to adopt sustainability practices on their own.[16]

The framework of the new governance arrangement attempts to promote sustainability by organizing according to ecological scale (e.g., watershed) as opposed to political-administrative scale, adopts a proactive orientation that seeks to identify and prevent problems before they occur, and more successfully mimics the flexible, adaptive, and cyclic character of ecological systems.[17] The key on the latter point is the adaptive management style practiced by participants. Adaptive management emphasizes learning through experimentation. Ideally, adaptive management increases responsiveness to ecosystem problems by promoting the values of continual innovation and adaptation in response to changing conditions, problems, and degree of success (or failure) enjoyed by solutions.[18]

Creating a Unified, Integrated Community-Based Network

The interactions in the three GREM examples are creating what some have called an essential component of a place-based community—a dense set of networks that can be utilized for communication, informal decision-making, and action.[19] In each case, the new dynamic connects together the series of individual, separate organizations and networks representing narrow, often self-contained segments of the population (e.g., irrigators or timber interests have one network, environmentalists another) to produce a more unified, integrated, community-based network whose primary purpose is a sustainable community. The new network strengthens the capacity of the community to act collectively in several ways (see discussion by Nadeau et al., in chapter 4).

First, the process of working together helps diverse elements of the community construct a common community history that becomes a way of defining the community by articulating the issues that matter. Such issues might be those that come up repeatedly, concern a wide number of diverse constituencies, or pose potentially catastrophic consequences for residents. By focusing on issues like changing patterns of land ownership, different uses of water resources, concerns over particular forest management treatments, or traffic congestion, for example, the common history provides a starting point for building new relationships and constructing a shared vision of what the community wants to become.[20]

Second, deliberating together and cooperating with others to provide community-wide benefits breaks down negative stereotypes and leads to new, positive working relationships grounded in trust. According to one HFWC participant, this is "absolutely critical to building trust within the community. . . . To the extent that we investigate and cooperatively pursue projects that help all watershed residents gain something, trust will follow."[21] A member of the Applegate Partnership agrees: "rather than being adversaries we've come to realize that we have a great deal in common. We still disagree on lots of things, but the Partnership helps us understand that we're also neighbors with a common stake in the health and well being of our community."[22] The newly minted trust encourages people to view others as partners and neighbors rather than adversaries, and to communicate in a direct, open and honest manner instead of hiding information for strategic advantage. Both developments unify the community, assist attempts to solve problems, and increase community capacity for effective self-governance.[23]

Third, the extensive networking means that institutions and decision-makers who used to become accessible only after "quite a bit of bitching and moaning or legal action . . . are now only a phone call away because of the trust that networking has created."[24] Citizens involved in the HFWC point to the example of Harriman State Park. When first approached about getting involved with the Council and cooperating to manage resources, park managers were reluctant. Now, however, they are very enthusiastic about the Council's collaborative format because they believe it has helped them more effectively to manage park resources, ranging from trumpeter swans to riparian restoration along the Henry's Fork.[25]

Fourth, the unified, integrated network strengthens the capacity for collective action by facilitating the creation of informal decision-making institutions to complement existing formalized arrangements. Two examples illustrate the point. An informal decision-making institution has developed in the Henry's Fork watershed to govern water releases from Island Park Reservoir. The Fremont-Madison Irrigation District (FMID) controls water releases and prioritizes them according to water rights claims by downstream irrigators. Starting in 1998, FMID has shown a willingness to be more flexible by releasing additional water to benefit the environment (e.g., to combat dangerously high water temperatures) at the request of Jan Brown, Executive Director of the Henry's Fork Foundation. There are limits to this arrangement—there must be "extra" water in the river. Thus FMID is unlikely to be very flexible during low water years. But no one interviewed for this project can imagine the institutional change *without* the years of working together on the Council and the creation of new relationships and trust among segments of the community who traditionally had no reason to communicate or cooperate.

In the Applegate Valley, responses by federal agencies to the 1995 Salvage Logging Rider[26] are also instructive. The rider redefined salvage timber so broadly that there was no limit to the volume of healthy "green" trees included in a salvage sale. Traditionally, no more than 15 to 25 percent of all trees in a particular salvage sale could be "green."[27] The rider expedited timber sale procedures, exempted salvage sales from standard administrative review and appeals processes, and severely restricted judicial review for such sales.[28] Further, the salvage logging rider assumed that all sales automatically complied with all federal statutes governing timber agencies (e.g., National Environmental Policy Act, Endangered Species Act, National Forest Management Act, etc.) and "all other applicable Federal environmental and natural resource laws."[29]

Many agency officials in USFS forests and BLM management districts took advantage of the enormous discretion offered by the rider to boost their timber harvest levels, often by resurrecting "old" sales that had previously been denied for violating one or more environmental laws.[30] Federal resource managers in the Applegate Valley, however, did not increase the harvest, despite a sharp decline in timber harvest volumes since the 1989 spotted owl court decision. Local agency officials "did not change the methods or timing of timber sale projects," continued with selective cut, thinning sales already planned or in process,[31] and did not permit salvage sales located in areas that Partnership participants had agreed were ecologically sensitive areas. In each case, the long-term cost of breaking the trust with community members was deemed too high to justify the short-term gain.[32]

Finally, the improved capacity of the community to act collectively does not occur in a substantive policy vacuum. Rather, it is effectiveness for a specific purpose. The unified, integrated, community-based network outcome is about cultivating a heightened sense of collective purpose centered on the sustainability message inherent in the tripartite "environment, economy, and community" mission. It is evident in the comments of "core" members that passion and commitment to sustainability drives the network forward.

> I was a skeptic at first. I didn't really believe in the Partnership's mission. But as I listened, I became more open to hearing what the dream could be like and seeing the opportunities for . . . experiment[ing] with the whole sustainability issue, treating communities and forests and lands together, and working across diverse people. It became for me an archetype of what we could do for the world. I mean it became a very, very powerful belief system for me personally, that if we can't pull it off here, we can't pull it off as a planet. And so I became . . . motivated by a genuine belief that this is absolutely the right thing to do for the Applegate . . . [and] that this will teach us lessons about sustainability for the future, about how we can solve [such] problems.[33]
>
> [T]he ecosystem is dynamic, it is constantly changing. . . . So what we're talking about is being flexible enough to cope with the changes . . . [a]nd so we talk about ecological sustainability. But we also need to talk about economic sustainability and community sustainability. . . . The fear is that we're now doing management on this land that over the years is not sustainable, which means that people—farmers, ranchers, loggers, our fathers—will be out of business based on the way that we have always done it. Yet people want to keep doing business the way they've always done it. . . . The reality is that we really need to reevaluate, to look at the ecological carrying capacity . . . the geologic, hydrologic carrying capacity of this watershed. Then just start to accommodate that. Ask what can the land support and work from there.[34]

A vibrant "sustainability" network can induce some behavior motivated by enlightened self-interest, on the understanding that what is beneficial for the collective community can benefit individuals. Self-interest is still treated as an important key for shaping behavior; since the mission has multiple emphasis on environment, community, *and economy.* The probability for success in a sustainable community is enhanced to the extent that economic activities (i.e., the profit motive) are aligned with the dynamic of ecological systems (see "Incentivization" section below). Yet participants in GREM accept that self-interest is malleable, shaped by social interaction, and that a unified network focused on sustainability provides a constant reminder that individual self-interest should be weighed against the collective, long-term interests of the community, including future generations.

Over the long term, then, the work of the new network will be eased as it reaches into all corners of the community and communicates the same kind of passion for and commitment to sustainability from more community residents. Participants know this will not be easy and that it will involve "a massive community paradigm shift. . . . We are talking about pulling rather than pushing, about moving whole communities systematically toward sustainable economic development. That is pretty heady stuff because we are talking about changing the way people live and act and react to their ecosystem, and develop their economy and think about community health. That is extremely complex. . . . But . . . getting to sustainability is absolutely essential. If we don't do it, we'll lose the pristine character and quality of life that make our place so special."[35]

Improving Citizen Capacity for Promoting Sustainability

There is a growing consensus among public managers, scholars, and development officials that success in achieving sustainable development, or any other public goal, requires more than simply the endorsement and pursuit of the concept by government officials and natural resource agencies. Instead, success requires the active consent and support of ordinary citizens.[36] Toward this end, the three cases of GREM seek to increase community capacity for sustainability through the creation of a more informed, skilled, and engaged citizenry. The belief is that such a citizenry will have both a greater interest in promoting sustainability for the long-term and a greater capability, thus increasing the probability that the interest in sustainability will be matched by results. The new governance arrangements do this by promoting a heightened awareness of what sustainable practices are, helping citizens recognize when something is amiss, and creating a cadre of citizens better able to monitor decisions affecting sustainability.

Education Through Community Outreach

The Henry's Fork Watershed Council, the Applegate Partnership, and the Willapa Alliance use a variety of methods to engage and inform the community about the status and potential of natural resources in the watershed. The three efforts also educate residents as to how such resources are connected to economic health and how they can be sustained. Active outreach involves purposive attempts to engage and inform community members, whether through the staging of events, meetings, and activities, or the distribution of information, written or oral, pertinent to the GREM mission. Outreach efforts remind citizens to adopt a different mindset when it comes to natural resources management, namely that management is the responsibility of everyone, rather than being the sole responsibility of public agencies.

Perhaps the most important public outreach effort by the Applegate Partnership is *The Applegator*, a community newspaper published by the Partnership on a bi-monthly basis. *The Applegator* is distributed free to all residents of the watershed and, if requested, to people outside the watershed. The paper tries to encourage residents to identify themselves with the Applegate community by adopting a non-partisan "us" instead of "them" approach for reporting community events and Partnership activities directed at environmental sustainability. The HFWC undertakes a similar effort, albeit using a quarterly newsletter format and a website Homepage. The newsletter, published by the Henry's Fork Foundation, a leading member of the HFWC, now reaches 1,800 people, up from 350 in 1992.

The Willapa Alliance reaches out to residents with their biennial "sustainable indicators" summits and several "conflict resolution" forums each year. The public forums typically draw several hundred people to explore policy issues of importance to the community. A good example is the October 1997 forum on the relationship between endangered species, healthy habitat, natural resource users, and regulations.[37] The Alliance also had both local newspapers, *The Chinook Observer* and *The Daily Astorian,* distribute over 9,000 copies of the executive summaries of the WISC Summits as inserts in order "to further engage the Willapa community at-large." Further, the Alliance maintains a membership list that numbered 9,300 as of March 1998.

In addition, all three governance efforts have published and distributed brochures highlighting the intricate linkages among the community, economy, land, and natural resources more generally.[38] Further, participants in these three efforts have also engaged in a number of attempts to get the message of environment, economy, and community out through presentations in local elementary schools, and increased interaction with the communities in question. School presentations in the Henry's Fork area are made throughout the watershed and occur, on average, once every two months, primarily in grades 5 and 6. The efforts have not only spurred "a lot more inquiries" on the part of teachers as to how the sustainability message can be integrated into their curriculums, they have led to "some really good feedback from the teachers as to how we could make the [materials] better."

The Willapa Alliance has focused its efforts on what it calls place-based community education. There are three main components: a school resource guide, The Nature of Home Program, and Student Institutes. In cooperation with local educators, the Alliance has crafted a 172-page school resource guide for grades K through 12. The guide, titled *Places, Faces and Systems of Home: Connections to Local Learning,* "is a toolbox to help [educators] learn and teach about Willapa by utilizing local resources. It is our hope that this Resource Guide . . . will encourage and support . . . learning opportunities [and] . . . a collective shift towards sustainable development."[39]

The Willapa Alliance's Nature of Home Program (NOH) is directed by a local community education specialist in cooperation with a local advisory task force. It involves the utilization of educational and scientific tools developed by the Alliance including a fisheries recovery strategy, the Willapa Geographic Information System, and indicators for a sustainable community. The group "strives to inform the present and future adult generations of their host ecosystem, to prepare them to be civic participants and leaders, and to aid the evolution towards stewardship-minded employers and employees. Most importantly, we utilize specific, local field sites . . . to make a connection that is not only intellectual, but personal as well, to foster one's internal stewardship values."[40]

Week-long Student Institutes in the Willapa area, utilized by more than one hundred area high school students to date, create opportunities for localized learning about sustainability through field trips, seminars with local experts from government agencies, schools, and the private sector, and projects wherein students themselves design and build hands-on, interactive, interpretive teaching tools.

The Applegate Partnership encourages similar efforts in local schools and is more active in reaching out to other community groups than the other two governance efforts. "[S]omething that we're particularly interested in . . . is building capacity for sustainability over time, first with the individuals involved in the Partnership, and then those individuals can network out to work with other community groups, whether they are focused on health care, or friends of the library, or the Apple Core looking at economic opportunities, or land use planning efforts. . . ."[41]

Skills, Expectations, and Sustainable Practices

Participants are convinced that the institutional dynamic improves citizens' skills and changes expectations in matters related both to governing and sustainable practices. By creating new opportunities for citizens to take control over their lives, citizens encounter new opportunities for empowerment, for building the citizenship skills critical to self-government, for accepting greater responsibility in governance, and for exercising local oversight and implementation.

The collaborative, deliberative, participative elements increase opportunities for citizens to engage in the primary political art of deliberating, or reasoning together. The active involvement reconnects citizens to government in a positive way by giving them a stake in governing and helps tame selfish passions through deliberation, information sharing, and a better understanding of the "big" policy picture affecting the community.[42] As part of this, the new governance arrangements remind citizens of the need for a new attitude toward the responsibilities of citizenship. The new attitude recognizes and accepts the connections between civic responsibility, active engagement in public life, and a community's quality of life. "Instead of sitting around and waiting for government to act, to solve the contentious issues tearing our community apart, we're here to tell people that sometimes local people can take care of their own problems, and their own facilities [e.g., local parks]. . . . We don't necessarily need government to do all of these things for us. It is our community. It is our responsibility."[43] According to another participant, it is also a matter of "convincing citizens to not let communities of interest from outside the community define the goals and aspirations of our community in absentia, by default."[44]

Field trips to inspect various facets of the watershed give citizens first-hand experience with what an (un)healthy forest, riparian area, or tallgrass meadow, for example, look like and the relationship between decisions and watershed health (sustainability). Regular presentations at meetings by scientists are designed to create a working familiarity with issues of importance to the watershed.

The Willapa Alliance has also enlisted community volunteers in the fight for sustainability. Through its Salmonwalk Training Workshops, 150 local citizens, including economically displaced fishermen, have been trained and certified as stream surveyors for habitat and salmon populations. The streamside survey information is utilized in the GIS system and the Willapa Fisheries Recovery Plan. Similarly, the Alliance has initiated local stream stewardship efforts modeled on the Adopt-A-Stream concept. Stream stewardship (e.g., Bear River Enhancement Association for Resources and Salmon) involves training volunteers in stream surveying methods, and bringing a diverse group of watershed stakeholders—government, land owners, fisher persons and timber companies—together through classroom and field education as well as field projects.[45]

In sum, the stake in governance creates ownership (i.e., added incentive to care about how a place is managed), while the field trips and additional knowledge make citizens more capable of understanding the impact of decisions on their place. Taken together, these elements create a cadre of citizens better able to monitor not only GREM decisions, but also the effects of federal and state agency decisions on sustainability. When combined with the extensive communication among community residents facilitated by public outreach efforts and the unified, integrated networks, the effects are even more powerful. "Instead of having just a handful of individuals interested in the design and implementation of a particular agency project, you've got hundreds of people now, that . . . are suddenly watching government actions" to see if promises are matched by performance.[46]

Incentivizing Sustainability

> To achieve [our] goals, ... [we] need to help others achieve their goals. The old way is a win-lose conflict, more for me, less for you. The new way is cooperation, a win-win deal, where there is more for everyone.[47]

Grassroots ecosystem management rejects the traditional coercive approach to natural resources management that "tend[s] to alienate the very people who can make good conservation happen—or who can block it through inaction, a never-ending search for loopholes, or just plain recalcitrance," and that "so few westerners—especially those living on the land—are apt to sit still for."[48] A classic case of resistance is the "shoot, shovel, and shut up" response by private landowners to potential Endangered Species Act listings. Others are afraid, hence unwilling, to improve habitat on their land because doing so may attract threatened species, thereby restricting their management options.[49] The responses confound the intent of the law by contributing to a perverse outcome—a hastening of the demise of endangered species, either directly or indirectly through fewer acres of prime habitat.[50]

Instead, GREM employs a cooperative, participative format not only because it can benefit nature or the community as a whole, but because they see it as better able to provide private benefits for individuals as part of the same bargain.[51] Dagget[52] aptly documents the logic behind many win-win deals in the "stories of ten ranchers who invited their neighbors, 'experts,' environmentalists, and others to work with them to find better ways to manage their rangeland. The [ranchers] speak with pride of revegetated lands, larger and more diverse wildlife populations—*and higher profits.*"[53] The potential for gaining individualized benefits through the HFWC, the Applegate Partnership, and the Willapa Alliance, incentivizes participation and consensus agreement on decisions. It is also crucial for convincing at least some, perhaps the majority, of *private* landowners voluntarily to adopt and support different, more environmentally beneficial land, water, timber, mining, and livestock management practices. Incentivization is thus another bridge to the private sector because to the extent that a series of initial bridges are built and management success ensues, the probability increases that more landowners will follow in their footsteps. In any case, whether one bridge is built or many, incentivization brings more private land into the management mix and increases the capacity to manage ecosystems as integrated, sustainable wholes.

Convincing private landowners to cooperate for the sake of sustainability does not always require higher profits, according to GREM participants, although the high profits/more sustainable practices combination is the optimal outcome. At minimum, they say, there must be enough certainty that cooperation and the adoption of more sustainable practices will not result in a loss of income, whether through higher operating costs accompanied by the same level of production, or the same operating costs accompanied by a lower level of production. In other cases, it is matter of autonomy, the right to control what happens on the land, or, at the very least, a guarantee of choice (versus diktat) from an expanded menu of more sustainable alternatives.[54]

The three cases of GREM are rife with examples of how the cooperative approach incentivizes participation and the embrace of more sustainable practices by private landowners. For example, efforts to identify and restore Yellowstone cutthroat trout populations in the Henry's Fork region, or to restore steelhead and salmon habitat in the Applegate Valley and Willapa Basin require access to private lands for surveying and habitat assessment purposes. Yet many landowners would rather not allow government officials onto their lands for "fear that the science and information might lead to more regulation, which usually means a loss of income. . . . We had an initial meeting with local landowners [in the Willapa], primarily tree farmers, egg farmers, and cattle owners, to gain their cooperation for our stream surveys for salmon. And a small tree farmer, who manages on a sustainable basis, spoke very passionately. He said, 'I have a couple of miles of excellent fish habitat on my property, and for years a county road culvert has restricted fish breeding in that stretch of stream. I want to open that habitat up for the sake of the fish, but I'm concerned that if I do, I'll pay too high of a price because I'll lose the ability to manage the area around the water.'"[55]

Through their actions over time, however, all three GREM efforts have established reputations as organizations that develop and share information with landowners for the purpose of *helping* them manage their property/resources in a more efficient *and* environmentally sensitive manner (e.g., GIS databases; stream surveys on fish populations and habitat).[56] The reputation as a helper, or partner, as opposed to a regulatory adversary who imposes added costs with less regard for the effects on economic viability, has "helped overcome lingering doubts about our mission and increasingly earned the respect and trust of [Willapa] Basin landowners."[57] The Willapa landowner concerned about letting Alliance members survey his streams ended up allowing surveyors access to his land and is now an integral part of the Willapa Alliance working to promote sustainability practices throughout the Basin.

Another example involves livestock management practices on the Diamond D Ranch in the Henry's Fork watershed. To the detriment of Targhee Creek's riparian ecosystem and water quality, the ranch had always allowed cattle to graze at will and to access the creek for watering purposes. However, in cooperation with the Council and in exchange for a partial subsidy totaling $10,000, the ranch made significant changes to cattle management practices. The rancher agreed to fence off key parts of the stream, install a watering trough with a float (to match water supply with actual demand), and adopt a different, more environmentally benign grazing practice ("hub" grazing focused on the water troughs rather than the stream). The net result was a more efficient usage of range grasses, a 70 percent savings in water usage, a healthier riparian ecosystem, and cleaner water.[58]

There also is the case of a stockyard business in the middle of the Applegate watershed. Instead of shutting down an established economic component of the community, the owner received a one-time subsidy to help restructure his cattle operations, volunteer help to protect and restore riparian habitat, and information about how different management practices impact environmental quality. The cooperative approach maintained the economic viability of this business, while concurrently ensuring progress toward greater environmental sustainability.

Others find that participation exposes them to new information regarding the interdependency between ecological processes and efficient business practices. This is what happened to some farmers in the Applegate Valley. They learned to recognize the interdependency between bats, which are a "night flying insect control" essential for maintaining crop health, and "old tree snags that we always felt were expendable. We found out that some [tree snags] house [the] bat communities [that help us]. Now I see snags in a different light."[59]

The Willapa Alliance, on the other hand, has nurtured a "strong relationship" with the Shorebank Enterprise Group (formerly Shoretrust Trading Group) based in Ilwaco, Washington at the southern edge of the Willapa Basin. Shorebank's goal is to create new markets and market opportunities for quality, "green" products from the Willapa seafood, agricultural, and specialty forest product sectors. The idea, according to Mike Dickerson, Shorebank's marketing director, is "to create a basic economic self-interest in protecting resources. [After all,] [w]hy would a business change its practices, and do things that may cost it more, if there's not an economic return."[60]

Further, the Applegate Partnership has played an instrumental role in linking certified organic farmers together to share best practices and to market produce directly to consumers in local markets. It also encourages the development of specialty wood products from surplus timber—small diameter trees, non-commercial grade wood, and brush remnants—much of which is already headed to the dump or the slash pile (i.e., burned because of limited commercial value). Both cases promote community sustainability by increasing the economic viability of local businesses, creating new jobs, and otherwise helping people make higher profits. Of equal importance, the increases in viability, jobs, and profits increase the opportunities and incentives for residents to make their living by engaging in business practices that are more environmentally sustainable.

Tapping a Broad Base of Knowledge

Grassroots ecosystem management expands the concept of expertise beyond scientific, bureaucratic and organized interest expertise, and seeks to engage and catalyze available community assets.[61] Participants, including some government agency representatives, expect that bringing "new," qualitatively different knowledge to the table will improve the effectiveness of the new governance regimes. Consequently, the likelihood improves that more of the primary mission, focused as it is on sustainability, will be achieved. In the first place, federal and state officials find GREM to be a "refreshing" forum for new ideas and potential solutions that typically do not see the light of day.[62] Second, the broad knowledge base tends to increase the probability of a more comprehensive understanding of problems, possible solutions, and consequences. Third, participants believe that the reliance on a broad knowledge base does a better job of incorporating the essential features of the real, functioning social, political, and ecological orders of the community in question. The new information increases the likelihood that the dynamic of human institutions—social, political, economic, and administrative—will be matched with that of ecological processes to the greatest extent possible.[63] Taken together with the ongoing, deliberative discussion format, the broad knowledge base tends to produce a more robust set of alternatives for solving problems, and a more reliable and realistic estimate of the parameters affecting program success.

The reliance on a broad knowledge base comes in part because many participants accept that real world problems typically do not fit neatly into the *singular* domains of traditional scientific disciplines, nor are they amenable to analysis excluding social impacts. From this perspective, while "natural and hard sciences are the key to unlocking natural resource problems, they don't come with the necessary instructions regarding how to apply them in human settings."[64] Thus social science is valued along with physical/ natural sciences (e.g., silviculture, biology, ecology, chemistry) and technical professional advice (e.g., engineering). In short, while technical expertise and "hard" science are important, they simply are not sufficient for understanding, much less solving, the sustainability conundrum, especially when it comes to a more sophisticated understanding of the self-organizing human systems that are essential to progress in the battle for environmental sustainability.[65]

At the same time, there is concern that professionals suffer from what the public administration literature calls the "trained incapacity" problem. Bureaucratic experts are taught to analyze certain types of situations in a specific way, using certain assumptions, procedures, and decision rules. They also tend to work from a restricted menu of possible solutions. Engineers, if given a water resource problem, will settle on technical, structural solutions—a dam to capture and control the flow of water, or a series of levees built high enough to prevent flooding. The range of possible solutions center on the technical mastery of the river rather than accommodating human settlement to natural flow patterns or employing non-technical solutions to the problem (e.g., adjusting insurance rates upwards to reflect the true risks and costs of flooding). High levels of trained incapacity increase the difficulty of learning a new set of premises and assumptions about how the world works. By extension, an expert suffering from trained incapacity will also have greater difficulty adapting solutions to fit changing circumstances, whether ecological, economic, or social in nature.[66]

It is as if GREM practitioners are putting into practice one of the chief lessons in Scott's[67] recent book, *Seeing Like a State.* Scott finds that the "formal schemes of order" favored by bureaucratic experts, grounded as they are in scientific management and imposed from above, *need* practical, local knowledge given "the resilience of both social and natural diversity and . . . the limits, in principle, of what we are likely to know about complex, functioning order."[68] Such formal schemes of order "are untenable without some elements of the practical knowledge that they tend to miss" and in many cases lead to tragic consequences, sometimes of epic proportion.[69]

Two examples of how local and/or non-technical, practical knowledge can help lead the way toward more sustainable practices involve the design and placement of concrete and steel head gates for managing water flows in the Applegate and the Henry's Fork area. In the Henry's Fork watershed, a fly fisherman with twenty years' experience reading the direction and speed of water flows on dozens of rivers pointed out how the placement of a head gate was "out of whack relative to the natural direction of the streamflow. In high water the stream will flow around the edge, causing heavy sediment flows [that are detrimental to fish and other aquatic species] and eventually rendering [the structure] useless. It needs to be shifted in this direction and made wider."[70] Within two years of being built, and as predicted by the fly fisherman, nature was starting to get the best of the engineer's design. Similarly, when a new head gate was installed on the lower part of the Little Thompson River in the Applegate, "the engineers decided to put a little wingwall on it that they thought would make it work better. [However,] [t]he ranchers looked at it and said, 'You can't put that in, it's going to cause a serious erosion problem.' The engineers said, 'No it won't.' Well, of course, come the first high water, what the ranchers said was going to happen happened. So it had to be cut out."[71]

Measuring Progress toward Sustainability

The common wisdom says that sustainability indicators are a necessary tool for solving the sustainable communities puzzle.[72] They provide the kind of feedback required for monitoring progress toward a predetermined, agreed-upon "sustainability" scale. The belief is that the richer, more accurate database improves the probability of sustainability by helping communities not only to make better decisions in the first place, but to fine tune and adapt programs over time as conditions change. Yet, precisely because GREM focuses on sustainable communities, indicators focus on more than just ecological sustainability. Instead, the emphasis is on integrating a set of standardized measures "of economic, social, and ecological health that are designed to gauge a community's systemic balance and resilience over long periods of time."[73]

Moreover, Heinen[74] points out, and many agree, that "[s]ustainability must be made operational in each specific context (e.g., forestry, agriculture), at scales relevant for its achievement, and appropriate methods must be designed for its long-term measurement."[75] The questions are: how and by whom? As noted at the beginning of the chapter, the conventional practice has been to let bureaucratic experts and scientists determine the specifics of sustainability indicators from the top down. However, as the case studies illustrate, GREM approaches these questions differently.

Participants in the Henry's Fork, Applegate, and Willapa cases recognize that operationalizing sustainability in their specific place and choosing appropriate measurement protocols are *political* questions as much as they are scientific because they presuppose a major shift in the behavior and practices of community residents. In addition, there are very few, if any, clear-cut answers to the technical dilemmas posed by the search for sustainability. There is often considerable controversy over appropriate monitoring protocols as well as the criteria (indicators) used for measuring sustainability. Finally, the act of melding the unique needs of each community together with national agency mandates is likely to benefit, for reasons discussed above, from a process grounded in broad participation. Given this context, GREM participants prefer a decision process that combines a bottom-up decision process involving as many parts of the community as possible with the top-down input from natural resource agencies' scientists and administrators.[76]

Professing allegiance to the value of sustainability indicators (SI) and changing the process by which they are determined, of course, are not the same things as actually constructing a viable SI framework. At present, only one of the three cases—the Willapa Alliance—has made much headway in defining sustainability, selecting indicators, and crafting applicable monitoring protocols. Yet Alliance participants are the first to admit that their efforts fall far short of being definitive.[77] Another case—the Applegate Partnership—clearly realizes the critical importance of SI to their overall efforts to achieve sustainability. To date, however, the AP has taken only a few tentative steps in the general direction of a workable indicators system despite having first discussed proposed criteria and indicators for measuring sustainable forest management in September 1993.[78] The third case—the Henry's Fork Watershed Council—has taken no steps in this direction.

Conclusion

New governance arrangements known as grassroots ecosystem management are an attempt to craft sets of institutional rules, processes and management practices aimed at fostering sustainable communities. Connecting the dots between institutions and sustainability requires an examination of both formal and informal institutions, and the roles played by education, incentives, different kinds of knowledge, and the capacity for measuring sustainability. And according to the practitioners and proponents of these new arrangements, connecting the dots also requires recognition that success ultimately demands a heightened level of commitment from those who live on the land and whose livelihoods have the greatest impact on the land. The efforts thus focus much of their energy on strengthening the bonds of community through the act of deliberating together over common problems, encouraging new relationships, and facilitating new cooperative networks.[79]

Clearly, creating a new community committed to sustainability is not a task for the faint of heart. It is complex. It is messy. It will engender at least some resistance from residents seeking to preserve the status quo. And it is likely to take a generation or more. There are concerns over the fragility of an approach grounded so heavily in informal institutions. How durable are these arrangements, especially if participant norms are a key part of the operational dynamic? Of the three cases, the Applegate Partnership has struggled to maintain collective commitment to participant norms, in large part because a particular segment of the Applegate—environmental activists from the town of Williams, Oregon—refuses to observe the norms either inside or outside of Partnership meetings. The activists' rigid ideology, behavior during group deliberations, and embrace of certain tactics, sometimes violent, all betray participant norms in some way. The result has been a decline in trust among participants, a decline in capacity to reach consensus, and an "[i]ncreased frequency of expression that involves blaming, fault-finding, [and] attacking."[80]

The behavior of the Williams' group helps to illustrate another potential flaw in collaborative arrangements. Participants acting strategically to protect a status quo that provides certain and identifiable benefits may well use the deliberative venue as a delaying device. They will engage in discussions, but will do what they can to ensure that decisions harmful to their interests are never implemented.

There is also the matter of leadership. Although it is not entirely clear, these collaborative efforts appear to require a special type of entrepreneurial, some might even say charismatic, leader who continually reminds participants of the benefits of collaborating for the sake of sustainability, while simultaneously providing them with assurance that their interests will be protected as the community transitions toward sustainability (i.e., makes clear the win-win-win character of the collaborative enterprise and its connection to sustainability).[81] In the specific context of GREM, leaders need to build trust among participants, forge lines of communication that eliminate (or minimize) information asymmetries, evince clear commitment to the holistic, sustainability-oriented environment, economy, and community mission, and exhibit a capacity for crafting workable compromises in support of sustainability.[82] Questions arise, however, as to whether there is an adequate supply of such leaders or whether training programs can produce more. Moreover, it appears that as the original entrepreneurial leaders affiliated with GREM have stepped to the side or otherwise moved on, the ability to deliberate and act in collective fashion diminishes, in some cases to the point of disbandment (e.g., the Willapa Alliance). Obviously, more research is needed to establish the importance of entrepreneurial leadership. Yet, to the extent that the prospects for success rest on such leadership, it will add one more degree of difficulty to GREM attempts to create additional community capacity for achieving sustainability.

Another concern arises in the area of bureaucratic discretion. The collaborative, intergovernmental character of GREM and the give-and-take of the deliberative dynamic suggest that success requires greater discretion for the agency line personnel who directly interface with the communities in question. Success also seems to require that affected line personnel have both the capacity and the opportunity to build the kinds of interpersonal relationships and individualized trust with the larger community that appear to be a key part of the glue holding the new governance arrangements together. Added discretion and the apparent need for long-term appointments raise the fear that agency employees will eventually compromise their loyalty to their parent agency, or even switch it to favor community over agency interests. The fear that agency personnel will "go native" and become captured by the community is, of course, an age-old one.[83] Will the efforts be commandeered by interests with hidden agendas that make a mockery of the sustainable community idea?

Beyond this, there are federal land tenure issues in the West that give federal officials control over huge swaths of rural land. In the Applegate watershed, for example, the Bureau of Land Management and U.S. Forest Service together manage 70 percent of all lands. In the Henry's Fork watershed, the comparable figure is 50 percent, including state agency lands. What this means is that unless there is government support for, active involvement in, and coordination with GREM efforts, even a high degree of community consensus and institution-building in support of sustainability may be for naught, or able to affect progress toward sustainability only at the margins.

Nor will it be easy to convince everyone that GREM communities are actually on track toward an environmentally sustainable future. This is because GREM fundamentally is about sustainable communities—humans are included in the equation and existing land use practices, extractive or otherwise, are treated as legitimate although in need of serious improvement in terms of their environmental impact. The sustainable communities approach is thus about balance among several goals and the competing values of a diverse society. It is pragmatic in recognizing that humans are an integral, legitimate part of ecosystems and deserve to coexist with nature. And ultimately, it is incremental rather than revolutionary. The question is: is a pragmatic, incremental approach to the environmental problematique compatible with environmental sustainability? Is there enough time, ecologically speaking, to experiment using an incremental approach? Specifically, some critics argue that the indeterminate, potentially long decision-making timeline is poorly suited to dealing with certain problems integrally related to environmental health (sustainability), especially in the area of endangered species.[84]

Others might argue that the institutional framework for ensuring sustainability is incomplete, especially in the case of the HFWC, because the attempt at sustainability is ad hoc—it occurs on a project-by-project basis. And although each decision is subjected to scrutiny in terms of its relationship to ecosystem health and sustainability, there is only a nascent infrastructure of sustainability indicators for measuring and monitoring outcomes against the desired "sustainability" outcome. Without such measures it is difficult to assess progress toward sustainability in a comprehensive fashion.

There are also legitimate concerns over the limits of an approach predicated on a small scale. The primary emphasis on ecosystems or smaller biophysical units (e.g., watersheds) as the appropriate unit for management, and the distinctiveness and complexity of individual ecosystems, favor arrangements focused at a relatively small geographic scale. In addition, the use of collaborative, consensus-based processes imposes logistical limits on scale. While directly involving dozens or even hundreds of local stakeholders is likely to be difficult enough, expanding the scale to the state or national level is likely to prove impossible in all but exceptional cases. There are also limits associated with "place," in particular the common embrace of physical landscape inspection as a management tool to help ascertain progress toward sustainability. The costs of catalyzing and coordinating such "walking tours" are likely to rise, perhaps exponentially, as scale expands beyond the local. As such, locally grounded GREM efforts are likely to be institutionally incapable of addressing environmental issues that occur on a regional, national or global scale (for example, global warming).

Nonetheless, if sustainable communities are about humans living together with nature, then it is plausible to accept that GREM is making progress toward sustainability in a number of ways. The ecosystem management approach is designed to mimic and adapt to the rhythms of nature. The unified, integrated community-based network is centered on the idea that ecosystem health is essential to long-term economic and community health. The collaborative, deliberative, participative institutional dynamic, along with outreach efforts, are making more people aware of the value of ecological services, biodiversity, and more environmentally sensitive practices. Others are being convinced to use fewer resources and engage in more efficient land use practices through projects that provide private benefits for individual landowners. Finally, the explicit reliance on a broad knowledge base facilitates innovation, problem solving, and effectiveness–all within the context of the sustainability-oriented environment, economy, and community mission.

What is needed now is more research devoted to connecting the institutional dynamic associated with these new governance arrangements to the actual outcomes. The logic of GREM suggests that individuals, and the communities they inhabit, will be more likely to adopt programs and decisions that promote a sustainable community. The question is: do they? Weber[85] offers a first cut at this by examining thirty outcomes produced by three exemplary cases of GREM and finds that "these new governance arrangements, rather than being only about process and relationship building, are also about developing new capacities for problem solving that promote environmental protection, economic concerns, and community sustainability all at the same time." The task now is to refine our understanding by conducting additional studies to see whether these results hold across a larger and more representative sample of cases.

11

The Ecosystem Experiment in British Columbia and Washington State

Clare M. Ryan

"Today, we do not know whether it is possible to achieve sustainability nor how to do it." Kai Lee

Introduction

The concept of sustainable forestry is not well defined. On-the-ground examples are elusive. Yet despite the difficulties, there are significant efforts underway in Washington State and British Columbia, aimed at achieving sustainable forestry. The case studies illustrate two different approaches to this shared challenge. This chapter describes and compares the evolution of four Adaptive Management Areas (AMAs) in Washington State with two Model Forests (MFs) in British Columbia. This is a companion piece to the empirical assessment of AMAs by Shindler in the following chapter. The MF and AMA experiments are barely ten years old, and it is premature to evaluate them in terms of success or failure. However, we can examine the evolution and early implementation of both initiatives. Both use an adaptive management approach toward achieving the goals of sustainable forestry.

Forest Management at the Crossroads

During the last decade, the challenges facing natural resource managers have substantially increased.[1] Industrialization, growing worldwide population, and increasing levels of consumption place greater demands on natural resources. In forestry, it is no longer acceptable in many parts of the world to harvest solely for economic gain. Economic considerations continue to play a primary role, but now they must compete with political and legal mandates, as well as social and scientific viewpoints.[2] As we struggle to reconcile all of these interests, we recognize that we know precious little about the ecological processes involved.[3] A good deal of uncertainty has surrounded most resource management decisions.

Forest resource management is in the midst of a shift away from what has been called custodial, traditional, or multiple-use management, towards a more holistic approach called ecosystem management. Ecosystem management embraces the philosophy that natural resources should be managed based on science, as well as social and political factors, with the understanding that resource management is largely a political process and that the ultimate purpose is ecological and socioeconomic sustainability.[4] Another fundamental principle of ecosystem management is the use of the best available scientific knowledge in planning, indicator selection, and monitoring.[5] Some critics feel there may be too many institutional barriers to implement ecosystem management effectively,[6] while others believe the concept is too vague and that the focus is largely on processes and procedures, rather than on clear identification of management objectives. Still others believe it is the best option available for achieving sustainability.[7]

The Challenge of Adaptive Management

The concept of adaptive management can be traced back many years to different disciplines, but recently the idea has attracted particular attention in natural resource management. With the publication of C. S. Holling's *Adaptive Environmental Assessment and Management* in 1978,[8] the concept began to attract scholarly attention and a growing appreciation of its potential utility in dealing with complex environmental management problems.[9] The subsequent publication of three separate volumes—*Adaptive Management of Renewable Resources*,[10] *Compass and Gyroscope: Integrating Science and Politics for the Environment*,[11] and *Barriers and Bridges to the Renewal of Ecosystems and Institutions*[12]—brought increasing sophistication, detail, and elaboration of the concept and its potential. Each contributed important discussions of the key elements upon which adaptive management is built; the importance of design and experimentation, the crucial role of learning from policy experiments, the integration and legitimacy of knowledge from various sources, and the need for responsive institutions.[13]

There are a number of definitions of adaptive management,[14] but most center on the concept of a structured process that facilitates "learning by doing." Adaptive management is a continuous process of planning, monitoring, researching, evaluating and adjusting management approaches which:

- views policies as hypotheses and actions as experiments
- accepts uncertainty and risk
- develops explicit models to include the assumptions and predictions of various management applications
- explicitly develops management alternatives and selects the one most likely to meet stated objectives
- carefully implements a plan of action
- monitors key response indicators
- compares information learned with expectations of results
- incorporates the new knowledge into the continuum of management experiments

The underlying premise of adaptive management is that knowledge of the ecological system is not only incomplete, but also elusive.[15] Thus, the experience of management itself as a source of learning. Adaptive management includes not only the use of scientific or expert knowledge in decision-making, but also the knowledge and values of stakeholders in an area.[16] As the challenges associated with forest resource management multiply, many look to adaptive management as an approach with the potential to reach the goal of sustainable forestry.

Adaptive Management Areas in the U.S. Pacific Northwest

It was the listing of the northern spotted owl and nine anadromous fish species under the Endangered Species Act that served as the Pacific Northwest's call to think about applying the concepts of adaptive management to forest resource management. After years of multiple lawsuits arguing about the effects of timber harvesting on the northern spotted owl, there was a general sense of gridlock and frustration about federal forest management in the Northwest. President Clinton convened a Forest Conference on April 2, 1993, in Portland, Oregon. A Forest Ecosystem Management Assessment Team (FEMAT) was created and given sixty days to formulate a sound, scientific strategy to help break the impasse. The FEMAT team included technical specialists and scientists from the USDA Forest Service, Bureau of Land Management, Environmental Protection Agency, U.S. Fish and Wildlife Service, National Park Service, National Marine Fisheries Service, and several universities. The mission was to work with an "ecosystem approach," giving particular attention to maintenance and restoration of biodiversity and long-term site productivity. The team was charged with maintaining sustainable levels of renewable natural resource outputs, including commodities and other values; it also emphasized maintenance of rural economies and communities.

Designation of AMAs

In June of the same year an encyclopedic report[17] was completed, proposing ten alternative management options for the region. Each option varied the acreage of land designated into categories of Late-Successional Reserves and Riparian Reserves (limited or no timber cutting allowed), Administratively Withdrawn Areas, and Matrix (most timber cutting occurring here). "Option 9" in the report recommended the designation of ten Adaptive Management Areas (AMAs) ranging from 84,000-400,000 acres each, spread across California, Oregon and Washington. The AMAs were designated to encourage "the development and testing of technical and social approaches to integration and achievement of desired ecological, economic and social objectives", with the overarching objective "to improve knowledge of how to do ecosystem management."[18] The AMAs were imagined as a set of "laboratories" where experimentation for technical and social learning could occur in order to better understand these systems and thus, fill the knowledge gaps needed to amend the Forest Service standards and guidelines and make more informed management decisions in the future.[19] Each AMA was linked to a particular "emphasis" in the Record of Decision. These emphases were derived from a variety of sources; past investigations and history, particular local circumstances, location within the region, connections with local communities, etc.[20]

Management and Decision-Making in AMAs

Management of the AMAs became the responsibility of the U.S. Forest Service or Bureau of Land Management offices already responsible for that land area. However, many of the AMA boundaries cross management jurisdictions, whether it is National Forest-National Forest, Forest Service-Bureau of Land Management, Forest Service-State Department of Natural Resources, or landowner boundaries such as public-private. There was a concern with establishing an organizational link between the areas and the management organizations. There was also recognition that establishing a clear and dedicated management presence for each AMA was important. The concept of an "AMA Coordinator" was already in practice in at least one of the areas.[21]

The establishment of the Provincial Advisory Committees (PACs)[22] was an important result of efforts to comply with the Federal Advisory Committee Act (FACA), implement the Northwest Forest Plan, and integrate various stakeholders. Provinces were designated across the region based on physiographic characteristics (geographic areas having a similar set of biophysical characteristics and processes, such as climate, soils, plants, etc.). Established by formal charters, the twelve PACs represent "communities of interest" in each province, and include representatives from federal, state, county and tribal governments, the timber industry, environmental groups, recreation and tourism organizations, and up to five other public-at-large members.[23] The PACs act as advisory groups only, and therefore wield no real decision-making authority. In addition, not all PACs oversee an AMA; only those containing an AMA within their province. This new, integrated forum for the sharing of ideas, values and knowledge was a novel approach to forest planning in the region. Participants hoped it would help build relationships and trust among stakeholders, and aid in the development of new plans based on consensus and, thus, less susceptible to litigation.

In theory, proposed actions within an AMA are presented by the AMA Coordinator or District Ranger (who might be one and the same) to the PAC, which discusses the options and makes consensual recommendations to the Forest Supervisor. The position of AMA Coordinator was intended to be a full-time responsibility devoted to the coordination of PAC activities, social integration, public education/relations and AMA plan development. However, these positions have since become more of an "add-on" to personnel's existing job responsibilities as budgets have decreased and positions are "underfunded" or downsized.[24]

Model Forests in Canada

The resource management issues facing Washington State do not stop at the Canadian border. Although subject to different laws and mandates from those of the United States, the Canadian government faces similar public criticisms and pressures, timber industry forest practices, and organizational issues within management agencies. A nationwide survey in 1990 by the Canadian government revealed that one in seventeen Canadians is directly economically dependent on the forest, and that 337 communities rely on the forest industry to support their economies.[25] British Columbia in particular has faced a number of contentious issues regarding forest management in recent years.

Although primarily the provinces retain land ownership, a few powerful timber companies have historically controlled forest management in Canada.[26] The long-standing system by which these companies retain power is known as forest tenure, which essentially means the Ministry of Forests delegates certain forest management responsibilities to private timber companies in exchange for long-term guarantees of timber supply. The specific balance of responsibilities varies according to the form of tenure.[27] This system was encouraged in the early years to help promote the young Canadian forest industry. Criticisms of the tenure system have grown louder in recent years.[28] They argue the system encourages inefficient timber use; the agreements are too short to allow trees to grow to a harvestable size, (therefore companies have no incentive to engage in reforestation or advanced silviculture); and the system lacks consideration for other forest values and attributes. The same public revolts in the Pacific Northwest over the northern spotted owl and old growth became prevalent in British Columbia, especially in Clayoquot Sound, where over eight hundred people were arrested for civil disobedience during a protest in 1993. This was the largest incident of its kind to date.

The climate of conflict and consultation that emerged in the 1980s put into place conditions ripe for the creation of the National Forest Strategy and the Model Forest Program by building a network of environmentalists, industrialists, and government officials committed to the task of exploring the question of sustainable economic activity.[29] In 1992, the Canadian Council of Forest Ministers released a national forest strategy report—*Sustainable Forests: A Canadian Commitment.* This report outlined the Canadian government's pledge to sustainable forest management. The objectives of the strategy were improved forest stewardship, public participation, economic opportunity, forest research, workforce diversity, and the inclusion of private lands in sustainable management strategies. The Model Forests were to emphasize management of forest ecosystems, management for a comprehensive range of forest values, public participation, and the demonstration and evaluation of sustainable forest practices.[30] The National Forest Strategy was signed at the 1992 National Forest Congress in Ottawa by members of the Canadian Council of Forest Ministers (save for Québec) and a cross-section of the forestry community.

Finally, in 1995, responding to criticism and public outrage both at home and internationally, British Columbia became the only province to adopt legally enforceable forest practices legislation. Known as the Forest Practices Code (FPC), it required companies logging on public lands to comply with prescribed rules and standards including road building, site preparation and silviculture.[31] Although criticized by environmentalists, government and industry, it is the most stringent of all provincial policies to date.[32]

Although the FPC dealt with the legislative problem of "soft" policy regarding the forest industry, it did not address the lack of inclusion in the decision-making process. The Canadian government traditionally has had a relatively closed-door policy with regards to forest management, but in recent years the public has begun to demand more accountability, involvement, openness and recognition of other forest values.[33]

Selection of Model Forests

The Model Forest Program in Canada shares the same fundamental goals as the Adaptive Management Areas in the Pacific Northwest—sustainable forest management and broader inclusion of citizens in decision-making processes. The Model Forest initiative is one part of a three-part strategy known as the Partners in Sustainable Development of Forests Program outlined in *Canada's Green Plan for a Healthy Environment* of 1990. The program was to create, by national competition, working-scale model management areas where a partnership of stakeholders would put ecological forestry into practice, where commercial forestry would coexist with wildlife, water, and fish, where research would be carried out, and the most advanced forest management practices applied.[34] Goals include shifting the forest industry from a sustained-yield to a sustainable development basis.

Model Forests were selected and designated through a system of proposals and competition, with specific criteria and guidelines for applicants, and weighted values. Timber or fiber was to be an "essential component" in a management philosophy that comprised other values and a variety of forest uses. The Model Forest program is skewed toward a "fine-filter" management approach; it is bottom-up, relying on individual species and stand management.[35]

The solicitation of proposals was announced in 1991, ninety letters of intent were submitted, and fifty proposals were received (with twelve submissions from British Columbia alone) in early 1992. A Technical Review Committee reviewed each proposal and the top five proposals were recommended for funding. Four others (the McGregor Model Forest in British Columbia among them) were recommended for funding on the basis of geography, and the Ministry of Forests added two additional forests, for a total of eleven. The Nova Forest Alliance was also added during the most recent (2002) round of funding. These Model Forests cover over 6 million hectares of land and represent five major ecoregions in Canada.[36]

Management and Decision-Making in Model Forests

The objectives of the Model Forest Program are: 1) to accelerate the implementation of sustainable development in forest practices, particularly integrated resource management; 2) to develop and apply new innovative concepts and techniques in forest management; and 3) to test and demonstrate the best forest practices available.[37] In addition, an implied goal is to bring traditionally adversarial groups to the decision-making table for capacity building and consensus processes. The program is also intended to provide the opportunity to experiment and test nontraditional approaches to implementing forest management policies in Canada. Model Forests in and of themselves do not have any decision-making authority regarding the use of forest resources and they are not considered legal entities with jurisdiction over an area.[38] Rather, they are stakeholder groups or "partnership" organizations that seek to identify common goals and objectives for the area and determine how policy makers can best implement them. Land use decisions are still subject to existing land use approval processes. However, in certain cases the Model Forest does have some jurisdiction, such as in the McGregor Model Forest, where the boundaries fall within a particular tree farm license held by the stakeholders and several Board members.

A Board of volunteer stakeholder representatives runs each Model Forest. Members represent industry, First Nations, government agencies at local, provincial and federal levels, environmentalists, recreation interests and local business. The goal is to develop a work plan of activities that comply with the overall goals of Natural Resources Canada.[39] The Board structure was determined by each individual Model Forest and allowed for appropriate representation of each area. Boards range in size from as little as four to over twenty members. The Model Forests themselves act as non-profit organizations and have small staffs that coordinate day-to-day research, education, technical and operation activities.

AMAs in Washington State

Four areas in Washington State were designated as AMAs as a result of the FEMAT process: Snoqualmie Pass, Finney, Olympic, and Cispus.

Snoqualmie Pass AMA

The Snoqualmie Pass AMA consists of approximately 212,700 acres straddling the Interstate 90 corridor in the Central Cascades of Washington State.[40] This east-west thoroughfare severely dissects the AMA, causing a number of problems and concerns for north-south animal migrations. The highway also brings hundreds of thousands of visitors and recreationists to the area every year from nearby Seattle. Several National Forests have been combined in recent years, and the Snoqualmie AMA currently falls within the jurisdiction of the Okanagan-Wenatchee National Forest in the north and the Mt. Baker-Snoqualmie National Forest in the south. This AMA also falls under the umbrella of two PACs rather than one. Of particular significance is the "checkerboard" pattern of land ownership in this AMA, whereby every other square mile is privately owned. This pattern has long been a problem for effective resource management and planning, but recently the Snoqualmie AMA has dealt with it by negotiating "land swaps" with private landowners to connect the acreage held by the agency. These land swaps are among the objectives for this AMA, which is managed for late-successional habitat and connectivity for north-south migrations. Without a critical minimum acreage held by the agency within the AMA, there would be no opportunity to manage for late-successional habitat required by the northern spotted owl and marbled murrelet.

Finney AMA

The Finney AMA is approximately 98,400 acres located 35 miles south of the Canadian border in the Northern Cascades. Nearby communities include Rockport, Concrete and Darrington, Washington. The AMA is primarily within the Mt. Baker-Snoqualmie National Forest yet extends into private and state lands. The emphasis of this AMA is the restoration of late successional and riparian habitat reserves. Unique to this AMA is the fact that nearly 90 percent of it is designated as Late Successional

Reserves (LSRs). With twenty recorded nesting pairs of spotted owls and the high percent of LSRs, the Finney faces extreme challenges in developing management alternatives and plans. Additional challenges are inherited. Since the 1920s, railroad logging occurred along Finney Creek valley bottoms and many roads were created to get from state or private lands onto federal lands for harvest. Increased timber production for homes following World War II meant more roads were needed in the drainage. In addition, roaded areas that did not meet wilderness criteria during this period suffered increased harvest and road density to compensate for designated wilderness areas taken out of timber production. However in the 1980s, following environmental outcries, concentrated harvests declined rapidly. The resulting landscape is scarred with roads, yet there is little to no income from timber sales to maintain them. The AMA Coordinator's position has changed hands several times since the AMA was designated.

Olympic AMA

The Olympic AMA is a unique configuration of disjointed parcels totaling 125,000 acres on the Olympic Peninsula.[41] The parcels surround the perimeter of Olympic National Park, lie completely within the jurisdiction of the Olympic PAC, and half of the total AMA acreage is classified as Riparian Reserves. Local communities have long depended on the forest products industry here and the result is a "preponderance" of young timber stands and a well-developed road system. Important to note is the fact the Northwest Forest Plan designated no Matrix lands within all of Olympic National Forest. This means that the primary source of timber is within the AMA. The emphasis for the Olympic has several components, including:

1. Create a partnership with the Olympic State Experimental Forest established by the Washington Department of Natural Resources.
2. Develop and test innovative approaches at the stand and landscape level for integration of ecological and economic objectives, including restoration of structural complexity to simplified forests and streams and development of more diverse managed forests through appropriate silvicultural approaches, such as long rotations and partial retention.
3. Survey and protect all marbled murrelet sites.

When the AMA originated there were four ranger districts, each with AMA parcels within their boundaries. Those four districts have recently been combined into two, the Pacific Ranger District and the Hood Canal Ranger District. The District Rangers of those stations make most of the daily decisions regarding AMA activities, but they also work fairly closely with the Olympic PAC during planning processes or when critical issues. The AMA Coordinator works out of the Forest Supervisor's office and acts primarily as an assistant for the District Rangers and interpreter of Northwest Forest Plan Standards and Guidelines.

Cispus AMA

The Cispus AMA is 143,900 acres located in the Gifford Pinchot National Forest in the southern Cascades. Nearby Mount Rainier and Mount St. Helens, both within moderate driving distance of Portland and Seattle, offer world-class recreation and tourism opportunities, making this an extremely valued and visible area. The history of relationships among tribes, industry-dependent local residents, federal/state agency personnel, and environmentalists has been tumultuous. The Gifford Pinchot National Forest has the historical reputation of being one of the largest timber-producing forests in the country up until recent years when numerous lawsuits from environmental groups succeeded in stopping or limiting the harvests. The impacts on local communities dependent on timber production, such as Randle, Morton and Packwood, were devastating.

The major emphasis for the Cispus AMA is the development and testing of innovative approaches at stand, landscape and watershed scales. Additional goals include the integration of timber production with the maintenance of late successional forests, healthy riparian zones and high quality recreational values. Several current studies on the Cispus involve efforts to learn more about these natural systems and develop effective management strategies. Many agency personnel affiliated with the Cispus AMA are quick to identify the Old Growth Remnants Study as the best example of adaptive management on the AMA thus far. Innovative funding methods, the motivation of the PNW researcher and the use of trained school children to conduct annual monitoring have highlighted this study and exemplify what type of experiments are possible on the AMAs.

Model Forests in British Columbia

As a result of the Model Forest proposal process in Canada, two sites in British Columbia were selected: the Long Beach and the McGregor Model Forests.

Long Beach Model Forest

The Long Beach Model Forest is comprised of 400,000 hectares of coastal temperate rainforest on the west coast of Vancouver Island. It extends from Hesquiat Peninsula in the northwest to Barkley Sound in the Southeast and includes Clayoquot Sound. This rainforest is both unique and rare, combining terrestrial, fresh water, estuarine, and marine ecosystems. There are only 30 million hectares of this type of forest in the world.[42] More than five thousand people live in the area, which includes communities within the territories of the Tla-o-qui-aht, Hesquiat, Ahousahat, Toquaht and Ucluelet First Nations, the District of Tofino, and the District of Ucluelet. About forty-five partners, including local communities, environmental groups, and the logging companies, are active in the area and take part in programs at the Pacific Rim National Park Reserve.[43]

The Long Beach Model Forest has been working to include industry and other stakeholders as partners in order to meet their primary social objectives. Activities often include the regional community to build capacity for sustainable forest management through education, youth internships, a Geographic Information System (GIS), and an evening speakers series.[44] The group is also in the process of implementing a number of applied forest practice experiments. Projects involve aquatic resources, the impacts of land use practices and natural disturbances, climate change, mapping of resources and cultural sites, and an examination of culturally important plants and how logging affects them. Considerable effort has been put into development of an extensive computerized database to serve First Nations and other communities in the area. Although numerous community-based connections resulted, the model forest did not meet the goals of the national program and was dropped from Phase III funding in 2002. The site has continued to operate less formally through a nonprofit organization, the Long Beach Model Forest Society.

McGregor Model Forest

Located 30 kilometers northeast of Prince George in central British Columbia, this boreal/montane/sub-alpine forest covers 181,000 hectares in central British Columbia on the western edge of the Rocky Mountains.[45] The tree species are quite diverse, and the forest produces an average of 386,000 cubic meters of timber annually, which is harvested year-round under a tenure agreement with the Ministry of Forests. In operation since 1993, the Model Forest's thirty-three partners have developed a three-pronged approach including scenario planning, strategic and operational planning support, and species indicators and adaptive management. A great deal of effort has gone into creating scenario-planning models that incorporates advanced modeling, forecasting, and visualization tools. The Model Forest also provides technical assistance in the form of workshops for forestland managers throughout Canada and Russia's Far East. In addition, The McGregor Group was created to consult on a for-profit basis using the computer technology developed by the Model Forest. The Model Forest is the largest shareholder of the Group, and hopes to become independent of government funding should further funding be cut back or evaporate.

Seeking Sustainable Forests through Adaptive Management

The Adaptive Management Areas and Model Forests share the ambitious goal of attaining sustainable forestry through adaptive management approaches. While we cannot yet evaluate these efforts in terms of whether sustainability has been achieved, there are several dimensions that can be examined and compared across the cases. For example, it is interesting to note that very similar programs with similar goals evolved through vastly different political processes, and are implemented under different conditions of authority and influence. In addition, the process of designating the areas was quite different, and all areas have struggled to improve participation by relevant stakeholders.

Most worrisome is the uncertainty of resources, both monetary and personnel, which leads to an atmosphere of tentativeness about the future of these management approaches, particularly on the AMAs. Funding has always been questionable and currently they receive no dedicated budget allocation. Plans establishing the AMAs acknowledged they would have substantial up-front costs.[46] These documents urged that adequate funding be provided to give the AMAs the opportunity to collect baseline inventory data, develop management plans and watershed analyses, determine criterion indicators and design monitoring plans, and provide for PAC meetings. However, actual annual funding was far less than required. For fiscal year 1995, the first year of allocation, the total budget for all ten AMAs was $493,000. In 1996, that figure remained relatively constant at $444,000. During these years, the funding was divided between the National Forest and Research divisions of the Forest Service. Spending peaked in 1997 and 1998 with figures of $754,000 and $715,000 respectively. In fiscal year 1999, the National Forest Regional Director decided not to fund the AMAs any longer, thus leaving full funding responsibility up to the Regional Research Director, who in turn withdrew funding as well. A decline in funding for regional ecosystem management activities triggered a decision by both the Region and the Research Station to terminate their support of the AMA program, a situation that prevails today. At present monies for AMAs are minimal, and projects are usually funded out of existing Forest or District level budgets

The Model Forests are substantially better off from a funding standpoint, with budgets approved in five-year increments. In 1991, the idea of a Model Forest Program as part of the Green Plan found approval, and garnered a $54 million commitment from the Government of Canada.[47] The intent was to implement the Model Forest Program in three phases. Phase I was under the authority of the Green Plan, with a budget of $5 million over a five-year period per Model Forest. Each Model Forest was given the autonomy to develop independently and uniquely in accordance with the particular needs and objectives of each region. Phase I was a period of self-identity and agenda setting for the Model Forests, and they were given the latitude to establish their own decision structures and techniques.[48]

With the end of Phase I in 1997 came the end of the first five-year budget. The Green Plan had no more funds to contribute, but the Canadian Forest Service (CFS) felt that the program had proven to be a worthwhile endeavor and chose to take on the responsibility of overseeing the Program.[49] This change in authority not only affected funding, which was promptly cut by 50 percent for the next five years (Phase II), but also changed the dynamic of the relationships between stakeholders/board members and CFS. Whereby CFS representatives had sat as fellow board members and peers, now there was an air of authority and overseeing of activities. Evaluations that had not been completed by the end of Phase I were requirements during Phase II, and a standard evaluation framework was developed by the Network and the CFS. Network representatives make site visits to the Model Forests to hold them accountable for evaluations and setting objectives in accordance with forest policies. Now the Model Forests are moving into Phase III, with the most recent five-year renewal occurring in May 2002.

Another oft-cited problem with implementation in both countries is the presence of institutional barriers. In the United States, frequently identified barriers stem from the operational culture of its management organizations. Policies guiding the Forest Service and the BLM often result in prescriptive approaches and standardized rules that blanket the entire forest system. This approach is a primary constraint to implementation, let alone innovation, on the AMAs. Another important constraint for field personnel has been the failure of the agencies to clearly identify the role AMAs play in federal forests and the roles that various stakeholders should take on.[50] This lack of administrative direction has also been apparent in the support provided to AMA personnel. None of the AMA coordinators has received any specialized training or technical education about adaptive management principles, and AMA responsibilities are now only one of many demands on their time.[51] Because of the lack of organizational leadership, few incentives exist for on-the-ground participants to undertake the creative approaches required within the adaptive management concept. In the absence of a clear mission for the AMAs, managers and surrounding communities have struggled to set a direction for these sites.

In the Model Forest program, the frailty of the federal claim to a role has hampered every attempt to develop and implement a national forest strategy.[52] Although the 1992 National Forest Strategy is a guide to sustainable development for the forest sector, it is vague in details about implementation. This plays out on the Model Forests as well where the task of implementation falls to the provinces. Although federal legislation provides broad grants of authority, it almost never binds the Crown to perform any particular function. However, the Model Forest Network headquarters in Ottawa makes its presence known in other ways. One example is its Model Forest Network Bulletin, *Innovations*, a nationwide publication. This monthly newsletter keeps the Model Forest Program in the public eye and provides a link between the eleven sites.

Progress on the Model Forests and the AMAs has been variable, but all have dealt and continue to struggle with social issues that stem from community involvement. Contentious relationships among stakeholders regarding membership and representation have been common. Interestingly, the designation of AMAs and MFs was conducted in exactly opposite ways. In the United States we saw a more top-down, authoritarian designation process, with little input from stakeholders, despite experience with more inclusive decision making processes. In Canada, we saw a more bottom-up and involved approach, with very explicit criteria for selection of Model Forests. Regardless, increasing participation in planning and decision-making are major goals of both programs. The inclusiveness criterion may pose more of a challenge for the Canadian efforts, as they have not had the history of stakeholder involvement in forest planning that the United States has had. Research on these efforts has been more forthcoming in the United States as well. The following chapter is a good example. It describes an assessment of the AMA experiment, based on survey research of citizen and agency participants across the ten AMAs.

12

Implementing Adaptive Management: An Evaluation of AMAs in the Pacific Northwest

Bruce Shindler

Introduction

Governments in Canada and the United States are experimenting with adaptive management as they seek to sustain forest systems. Two central components of adaptive management are monitoring system inputs and evaluating responses. Not only do these activities apply to the ongoing management of biophysical resources, they also are essential to learning about the interactions—or lack of interaction—among citizens, managers, and scientists. Throughout this book authors have noted the importance of positive community relations and the benefits of integrating citizens into decision processes. In the Pacific Northwest, an important opportunity for assessing such processes is found in the Adaptive Management Areas (AMAs). Ryan discusses the origin of these sites in the previous chapter. Ten AMAs were designated in Washington, Oregon, and northern California in 1994 as places for ecological, social, and organizational learning: "These areas should provide opportunities for land managing and regulatory agencies, other government entities, nongovernmental organizations, local groups, landowners, communities, and citizens to work together to develop innovative management approaches."[1]

This chapter summarizes research undertaken five years after formation of the AMAs to assess agency effectiveness for involving stakeholders at the ten sites. As Ryan described, both the U.S. Forest Service and the Bureau of Land Management have jurisdiction for AMA implementation. To help determine the ability of these agencies to engage the public adequately in implementation activities, citizens and federal forest agency personnel were surveyed about their experiences. The citizen surveys were designed to help managers and researchers understand what citizens think is important in their interactions with AMA personnel and to assess the nature of those interactions from the public's perspective. The agency survey followed the same format; it was intended to monitor progress from an internal point of view and to provide a means for comparing the opinions of resource professionals with those of citizens. Taken together, the data provide a report card on attempts to integrate communities and citizens into the AMA experiment and also establish a baseline for further monitoring and evaluation. Additionally, this analysis offers a method for assessing the level of agreement among citizens and agency personnel about important aspects of program implementation.

The citizen survey sampled members of the *attentive public* in communities surrounding the ten AMAs. "Attentive public" is a term used to describe individuals who have more than a passing interest in a particular topic; essentially, these are people who pay attention to a project, problem, or issue. Analysts often equate "attentiveness" with political involvement in the democratic process.[2] In this case, the sample derived from citizens who chose to become involved with their local AMA. Names and addresses were obtained from AMA mailing lists and sign-in sheets from various activities. The 418 completed surveys—a 74 percent response rate—were well dispersed across the ten sites, indicating a representative sample of people who paid attention to the AMAs. For example, 73 percent had attended an information meeting or open house; 71 percent received the AMA newsletter; 66 percent had phoned, written, or visited agency personnel to discuss an idea or problem; and 53 percent had gone on an agency field trip to a forest site. In the survey, three quarters of these individuals reported they give a great deal of attention to federal forest issues in their area.

The agency survey was completed by 105 resource professionals (reflecting an 83 percent response rate) who had various levels of responsibility at the ten AMAs. This sample derived from lists provided by AMA Coordinators of individuals who were involved on AMA planning teams. Most were resource managers from Forest Service ranger districts, others were BLM District personnel and a small group of agency scientists who had been assigned to each AMA. On balance, these were the agency members best qualified to offer an assessment of public interaction on the AMAs. For example, the survey revealed that these individuals averaged ten years at their current work station, 63 percent said their agency expected them to have frequent front-line contact with the public, and 64 percent had received formal training from their agency in public interactions. Overall, 87 percent believe that involving the public is an effective method of resource management.

Discussion of Findings

Efficacy of the AMAs

Initially, we wanted to know how people felt about the efficacy of agency efforts for implementation of their AMA, particularly those actions that involved citizens. Table 1 provides data on how respondents feel about the AMAs as places to conduct forestry research and how well the public is being incorporated into AMA activities. Although a large majority of citizens (70 percent) and agency personnel (92 percent) agree that the AMAs are appropriate places for scientific experimentation, the public's ability to be part of the adaptive management experiment appears more problematic.

Five years after inception of the program, less than half of the citizens and even fewer resource professionals thought the agencies had either identified what AMAs are intended to be or the role citizens should play in their management. Previously in 1997, researchers had cautioned the agencies over the need for clear purpose statements about AMAs

Table 1. Efficacy of AMAs

	Citizens	***Resource Professionals***	***Significance Level***
	Agreement	*Agreement*	
Scientific experimentation with forest ecosystems is appropriate on AMAs	70%	92%	<.01
Forest agencies have clearly identified for the public what AMAs are intended to be	40%	36%	NS
Agencies have clearly identified the role of citizens in AMAs	34%	29%	NS
I feel that citizens can actually participate in planning and management activities at my AMA	49%	68%	<.01
Forest Service and the BLM are open to public input and use it in making decisions	41%	73%	<.01
Efforts by local forest managers to involve the public do not have full support of national agency leaders	40%	8%	<.01

NS = not significant

and that leadership would be required to eliminate public confusion about the agencies' intent for these sites.[3] That same year an assessment of public process on the Central Cascades AMA in Oregon found that citizens believed forest projects were more successful when the public's role was defined and a desired end product was identified at the outset.[4]

In the current survey, citizens generally felt their input was discounted; less than half believed they could actually participate in planning and even fewer thought the agencies used their suggestions in making decisions. Equally distressing is that 40 percent think that local managers do not have the full support of their national leaders. Throughout the region there appears to be a growing sentiment among members of forest communities that local forest managers, in many cases individuals they have come to know and trust, are hindered from doing their jobs because of directives from Washington, D.C. or by pressure from national interests (NGO's) outside their local area.[5] In contrast, most agency members responded just the opposite to each of the last three statements about citizen participation. On reflection, there appear to be at least three potential reasons for these differences in opinion: 1) the agencies are having difficulty getting their message of open participation across to constituents, 2) the public does not believe managers will, or can, fulfill the promise of citizen participation; and 3) there is simply no compelling evidence that managers are communicating with or involving the public. In the last instance, we find a common dilemma—sometimes the problem with evaluations is that there is nothing to evaluate. In any case, there seems to be a high degree of public cynicism about cooperation on the AMAs.

AMA Implementation

Next, we began the process of assessing AMA implementation. In a framework developed for adaptive management situations, Shindler et al. identified two conceptual levels from which to monitor and evaluate citizen-agency interactions.[6] The first involves the broad goals, or desired outcomes, of productive interactions. Several syntheses of public involvement research have articulated five essential goals for such programs.[7] In general, public processes can be used to:

- improve the quality of decisions
- reach decisions that enjoy increased public support
- contribute to the building of long-term relations
- incorporate citizens ideas and knowledge in decisions
- learn, innovate, and share results with others

To determine the extent to which these broad goals were being achieved on the AMAs, a set of eight related statements were evaluated by citizens and agency members (Table 2).

Overall, findings indicate little agreement between the public and resource professionals on any item; in no case did a majority of citizens agree that goals were being met on the AMAs. Perhaps not surprisingly, agency members believed to a much greater extent that they were being successful; strong majorities felt they were showing concern for communities, building good relationships, and using citizens' ideas in decisions (all indications that "we are trying").

Responses are notable for the significant gap they reveal in the two groups' perceptions. The collective data show a difference between how agency staff feel they are treating citizens and how citizens feel they are treated by staff, reflecting a substantial and fundamental disagreement over the quality of their interactions. As a result, judgments about desired outcomes are particularly low, especially among the public. For example,

Table 2. Agency Goal Achievement at AMAs

Goal	*Citizens* *Agreement*	*Resource Professionals* *Agreement*	*Significance Level*
Showing concern for local communities and their well-being	49%	76%	.01
Contributing to good relationships with citizen	48%	67%	.01
Contributing to public knowledge by educating communities about benefits and costs of proposed plans	44%	49%	NS
Incorporating citizens' ideas and knowledge in decisions	40%	67%	.01
Increasing innovation and creativity in programs and projects	39%	53%	.05
Improving the quality of decisions by effectively involving citizens	36%	52%	.01
Building trust and cooperation with citizens	32%	48%	.01
Reaching decisions that enjoy increased public support	25%	42%	.01

NS = not significant

relatively few citizens (36 percent) believed the quality of decisions was improved through citizen involvement or that public support for decisions had increased (25 percent). Perhaps the most distressing statistic, however, is the low number of citizens (32 percent) who thought that public trust and cooperation were being built on the AMAs.

The repercussions of such disagreement in views can go beyond the normal frustrations we might expect from either side in the search for collaboration. While we have become familiar with the public's discontent, we often fail to recognize common reactions among personnel who think they are doing a good job but find out their efforts reap little success. In these situations, it is easy to point a finger at a public that "just doesn't get it," or figure that "it's their problem." These are normal and legitimate reactions by individuals who may be doing the best they can under difficult circumstances. Such situations call for a new approach and a different set of public communication tools.

The second conceptual level for evaluating citizen-agency interactions involves examining more specific attributes. Although examples of successful public involvement have been found in a wide range of situations, there is general agreement among researchers about a number of common characteristics.[8] These can be organized into four basic categories: interactions that are 1) inclusive and interactive, 2) procedurally sound, 3) innovative and flexible, and 4) outcome or results oriented.

Shindler et al. developed a framework for monitoring and evaluating these interactions in adaptive management settings using a set of core attributes for measuring whether certain objectives were being achieved.[9] These eighteen objectives (see Table 3) were used in the current survey to determine to what extent interactions on the AMAs were inclusive and interactive, procedurally sound, innovative or flexible, and results oriented. However, prior to asking citizens if these public involvement objectives were being achieved, we also asked them (on a four-point scale) *how important* each objective was. For simplicity, these results are not reported in Table 3, but for all eighteen objectives more than 80 percent of the respondents said these attributes were either important or very important to them. Many items received importance ratings over 90 percent.

Table 3 shows that in most cases citizens did not think the objectives were being met, while agency members generally had a higher opinion about the level of success. The only area where a majority of citizens believe achievement has occurred is in the *inclusive/interactive* category. It is clear that many citizens believe the agency is making an honest effort to create additional opportunities for interaction. The table also shows that most people felt welcome at meetings, that meetings are interactive and personal, and that AMA personnel are sincere, honest, and open to suggestions. Fewer believed they were shown consideration for their efforts and that public deliberation was encouraged. As mentioned, resource professionals generally gave higher ratings to their performance on these objectives, but only 50 percent could say that deliberation is encouraged. For resource managers, demonstrating good interpersonal skills is critical to the success of public engagement activities;[10] however, proficiency in this area continues to be a stumbling block for agency personnel.[11] Whether the modest ratings reported here

Table 3. Achieving Specific Public Involvement Objectives

Objective	***Citizens*** *Agreement*	***Resource Professionals*** *Agreement*	***Significance Level***
Inclusive/Interactive			
All citizens are welcome at meetings or planning sessions	63%	71%	NS
Public meetings are interactive and personal	57%	67%	NS
Agency personnel are sincere, honest and open to suggestions	53%	75%	.01
Citizen participants are shown consideration for their efforts	49%	75%	.01
Public deliberation is encouraged	47%	50%	NS
Procedurally Sound			
Decision-makers regularly attend and participate in public planning activities	47%	78%	.01
Efforts to involve citizens start early and continue through all stages of a project	42%	64%	.01
Agency information is current, reliable and easily understood	41%	47%	.05
Controversial issues receive genuine attention and a sufficient response by agency personnel	33%	69%	.01

indicate that AMA personnel have improved their outreach skills cannot be adequately assessed from this one-time study; however, the findings indicate that room for improvement exists.

There is little evidence from the three other major categories to suggest that citizens see a high degree of success in agency public involvement efforts. It is worth noting that one category—making sure agency efforts are *procedurally sound*—is part of the public process equation that does not necessarily require a high degree of interpersonal skill to accomplish. Instead, this set of objectives requires that AMA personnel take public outreach seriously and attend to procedural tasks simply because they are important and legally required. Most agency members believed they were following through on these responsibilities; however, this view was not universally shared by the public. For example,

Innovative/Flexible			
Activities foster relationship building among group members	36%	54%	.01
Efforts to involve citizens are innovative and flexible	33%	43%	NS
Agency personnel and citizens analyze information together to build a collective pool of knowledge	29%	21%	NS
When new information arises or a surprise occurs, it is usually factored into subsequent decisions	27%	72%	.01
Outcome/Results Oriented			
Local forest issues are given greater attention than national interests	43%	65%	.01
Projects/plans are carefully designed, with purposes and end products clearly identified at the outset	40%	44%	NS
Agency personnel follow through on decisions	35%	64%	.01
Citizens can see how their contributions are used in decisions	23%	31%	.05
Citizens understand how decisions are made and which information is used	22%	18%	NS

NS = not significant

most citizens did not see line officers and senior decision-makers participating in public planning activities, an important element because it provides stakeholders some assurance that leadership is being exercised and they have been heard by the individual who will ultimately render a policy decision.[12]

Similarly, citizens want to be involved early instead of being asked to come in "after the decision has already been made." They also have an expectation that agency information be current and easy to understand. Neither of these procedural functions was rated very high by citizens nor by agency members. In fairness to public outreach personnel, it may be difficult for citizens to assess each of these elements accurately. For example, it is logical that agency members are much better at recognizing when decision-makers attend meetings and at judging the reliability of their information. But the point should not be missed that many citizens often see these areas as agency shortcomings, and perception is often the reality. Opportunities exist in these cases to improve the public's understanding. However, from the low level of agreement (just 33 percent) about agency response to controversial issues, there is a clear indication that citizens feel shortchanged in the amount and type of attention given to issues they view as important. It may be there is lack of agreement among citizens and agencies on what constitutes a controversial decision. In these instances, greater sensitivity to public concerns becomes the real issue.

Overall achievement ratings in the *innovative/flexible* category were less than satisfactory, even among agency members. This is problematic, given the central role that innovation and flexibility have in adaptive management. Only about one-third of the citizens believed that relationship-building was occurring or that public involvement activities reflected innovation and flexibility. Resource professionals rated themselves somewhat higher in the first area, but tended to agree with the public regarding innovation. As before, it may be difficult for citizens to accurately judge certain objectives; for example, whether citizens and agencies have analyzed information together or if new information was factored into decisions. But the public's ratings of these items are sufficiently low to indicate these are areas of weakness. Because agency members largely agree that co-analysis of information is not occurring, it probably is a case where this type of innovation has not been introduced to any degree on the AMAs. One reason, no doubt, is the difficulty in doing so. This activity requires a substantial commitment of time as well as an ability to get beyond the common belief that the public simply does not have the background to adequately evaluate such information. As for using new information in decisions, the high level of agreement among agency members suggests this may be more of a communication gap with the public than staff's lack of follow-through on incorporating new data. Collectively, this set of responses probably reflects the public's general dissatisfaction with agency performance and responsiveness[13]—developed over many years of interactions—rather than citizens' ability to objectively rate accomplishments on the AMAs.

Ratings of objectives in the final category—*outcome/results oriented*—were just as low, especially among citizens. A majority of agency members did give themselves higher marks for paying attention to local forest issues and following through on decisions; again these are probably two areas in which agency personnel can more accurately judge their own actions. It is a concern, however, that the public does not share their views. Also of importance is the low level of agreement by both groups that projected plans are designed so purposes and products are clearly identified. This objective is critical to success for any planning process,[14] but execution seems to be missing. Another essential point is shared by both citizens and resource professionals: the public does not understand decision processes on the AMAs. This may reflect a shortage of decisions being made on these areas and thus, the public's inability to evaluate to them.

Of the eighteen objectives, the two that received the lowest ratings from both groups were about citizens seeing how their contributions were used, and understanding how decisions are made. The first is a shortcoming of the public process and falls to managers for remedy; tracking public suggestions, particularly on relatively small AMA projects, seems straightforward. The second, however, understanding how decisions are made, might be more complicated. Among citizens, this can stem from various reasons (e.g. lack of information, their own inattentiveness, poor communication), any of which can be frustrating for either public or agency participants. From a managerial standpoint, for example, some agency personnel may still be influenced by old-line, simplistic beliefs such as "the public just doesn't understand forest management," or that citizens should not have a role in the decision-making process. Either situation has deeper ramifications for reaching productive interactions. Some evidence for this tension exists elsewhere in the survey as 21 percent of the agency respondents agreed that their local publics possess insufficient knowledge to engage adequately in planning discussions about ecosystem management. In any case, decision processes that are not "transparent" and readily understood by citizens are a problem, and it is up to agency managers to address these shortcomings.

Internal Operations

Finally, it was important to understand more about agency internal operations on the AMAs. For this perspective we turned to agency personnel about their general observations of public involvement and their specific experiences regarding the type of support they receive, the success of attempted actions, and potential barriers associated with their AMA. The data have been categorized under these same headings and reported in Table 4.

Table 4. Agency Members' Assessments of AMA Public Involvement Efforts

Regarding my experiences with the AMA:	***Agree***	***Disagree***
General Observations		
Involving the public in planning and projects is worthwhile and should be part of how we do business	93%	1%
I feel comfortable working with the public	81%	5%
We have a successful public involvement program on our AMA	44%	23%
I am frustrated with our attempts to involve the public	33%	36%
Organizational Support		
My agency has clearly defined for personnel what AMAs are intended to be	40%	41%
I have received adequate training to fulfill the public contact part of my job	63%	20%
I receive adequate support from my work unit for the public involvement aspects of my job	61%	20%
I receive adequate support from administrative levels above my work unit for the public involvement aspects of my job	48%	26%

In their *general observations,* almost all personnel (93 percent) saw value in involving the public. This could mean that individuals simply feel it is the right thing to do or think it is the right thing to say; on the other hand, they may have learned over time that public involvement is a useful and necessary step toward more lasting decisions. Regardless, inclusiveness is a cornerstone of decision-making and most AMA personnel report feeling comfortable working in this public setting. On the other hand, far less (44 percent) felt their program was successful, and about one-third voiced frustration over their attempts to involve the public.

Regarding *organizational support,* only 40 percent agreed that their agency had defined what AMAs are intended to be. It is likely these opinions are linked with frustrations about attempts to involve the public, but this also could be a reflection of inadequate leadership for

Action/Achievements		
Our AMA is linked to wider community social and economic concerns	72%	11%
We have identified who our publics are and how to reach them	64%	16%
We have established demonstration projects where we can actually obtain feedback from our publics	54%	21%
I have seen new or creative ways of involving the public on our AMA	47%	24%
We have tried to find out what local people know about forestry	37%	25%
Potential Barriers		
Agency trust/credibility issues are major publics constraints among our local	81%	7%
Most local citizens don't understand the concept behind the AMA	57%	17%
The agency time frame for producing results is unrealistically short	46%	26%
I am hindered in my activities with local publics because of their perception that decisions are really made on a regional or national level	39%	37%
Our local publics don't have sufficient knowledge about ecosystem management to adequately participate in planning discussions	21%	50%

the public outreach job. In any case, a majority of personnel believe they are getting adequate training (63 percent) and support (61 percent) from their local work unit (usually ranger districts) to carry out public interactions on the AMA. Somewhat less (48 percent) see this same type of support from higher levels within the agency.

In the *action and achievement* category, findings are mixed. Substantial majorities believed their AMAs were linked to broader social concerns of the community and that agency staff had identified who their "publics" are. A smaller majority (54 percent) agreed that demonstration projects are in place to gain public feedback. However, less than half (47 percent) saw new or creative public involvement activities occurring. Additionally, only about one-third think AMA personnel have attempted to find out what local people know.

Finally, a number of *potential barriers* stand in the way of progress on the AMAs. Three problems appear critical. Almost all personnel (81 percent) agree that trust and credibility are major constraints among their local publics. This is not surprising given the rancor that surrounds most federal forest management issues. On AMAs, where some agency members have worked hard to develop positive relations with communities—in many cases well before the institution of AMAs—this is particularly distressing, and no doubt frustrating for those involved. It is likely that other barriers contribute to the lack of trustworthiness in relations; noteworthy is the relatively high number of personnel (57 percent) who thought that most local citizens do not understand the AMA concept—an interesting point given how few agency personnel (40 percent) understand it themselves. There is little doubt that the lack of clear objectives about AMA designation and implementation is a frustration for both agency and citizen participants.

A second important barrier is the belief that the time frame for demonstrating results on AMAs is unrealistically short. Previously, Stankey and Shindler acknowledged that successful implementation of adaptive management in the AMAs would take time.[15] Specifically, they noted that pressures for quick results characterize the current culture in the forest agencies, a mentality that will work to the detriment of the AMAs. Given the level of general distrust among stakeholders, sufficient time must be invested for mutual respect to develop.

A third concern is that many managers (39 percent) believed their local publics think decisions are really made by the agency at a regional or national level. Recent studies in Oregon provide growing evidence of this public sentiment,[16] a point of view that could scuttle many AMA programs. A smaller group of agency personnel (21 percent) felt local citizens did not have sufficient knowledge to participate in planning for ecosystem management. Although this sentiment represents a relatively small minority opinion, such opinions still reflect a "we know best" attitude that can be particularly detrimental to public planning processes.

Conclusion

The most striking features of these data are the perceptual gap that exists between citizens and resource professionals, and the demonstrated need for improved interactions among participants on the AMAs. In the first case, a principal issue raised by this series of ratings is not whether citizens or agency personnel are "right," but why their perceptions differ so widely. Answers to this question are likely to involve obvious explanations; for example, because of their day-to-day involvement and level of personal commitment, agency members may feel quite strongly that they are achieving many aspects of the public outreach job, whereas citizens may not share this same perspective. But some explanations are likely to involve more complex ideas such as differences in the scope of projects that various AMAs have attempted, the degree of trust that exists in these communities, and the quality of leadership evident among agencies and citizens groups. For example, it probably is much easier for citizens and the agency to reach agreement about a Jobs-in-the-Woods Program that provides local employment than it is to reach consensus on how much to harvest in a timber sale. In other situations, the level of trust among participants may be so eroded that no amount of information or encouragement will alter the immediate outcomes.

In addition, the context in which any of these judgments are made is extremely important. Each public process is situationally dependent on a number of factors, many of which are predominantly local. More in-depth qualitative analysis can help reveal the influence of contextual factors on these interactions. In any case, the most promising finding here may be the level of importance given to public involvement activities. There is fairly strong agreement among both citizens and resource professionals that effective, high-quality interactions are essential to AMA success. Ultimately, however, demonstrable results will be necessary before any real long-term gains accrue.

There is also substantial evidence from these findings that improvements are necessary if the AMA experiment is to continue. Many of these ideas are reflected in the notion of *civic science* put forth by Lee in his observations of adaptive management.[17] He argued that the challenge for agencies in effectively managing large ecosystems is to build community relationships that incorporate both science and politics. At its most basic level, adaptive management must be a public activity, open not only to the participants who must exercise responsibility but also to those who value and depend upon these resources.

The first area for improvement is the extent to which citizens are being included in planning and decision processes. The data suggest that, at least for the attentive public, the agencies are making gains in this area—particularly in the quality of personal interactions with citizens. The public responds best to sincerity, honesty, and genuine effort. However, it would be a mistake to take the level of public agreement in these data tables (simple majorities at best) to mean that planning processes are highly inclusive or that the public participation part of the job is complete. Even among those individuals who pay attention and are actively engaged, many citizens still are not convinced that public deliberation is encouraged.

The second general area for improvement involves procedural functions. These elements often are easier to implement than other more complicated components of public involvement. Some immediate gains can be made in this area by recognizing the importance of these tasks and making them a priority. For example, current and reliable project information is often available; providing timely documents in a clear, understandable format is usually achievable on most forest units. Also, making a commitment to engage citizens early in project discussions seems reasonable as long as agency personnel themselves have a clear idea about what they hope to achieve. And although attendance at public planning sessions may in some cases be an added responsibility, visible participation by decision-makers is an important symbol of organizational commitment. This is a clear sign to citizens that the meeting is worthwhile and that their input is likely to be taken seriously.

Third is the degree to which public involvement activities on the AMAs are any different from previous attempts and will result in outcomes that are recognizable by the public. An expectation was created in the planning documents, and in many cases by the individual AMA planning teams, that the agencies will be more flexible and more creative in getting projects accomplished. Thus far, as this survey shows, few people could describe what has actually occurred as innovative. Innovation usually involves some risk, but the adaptive management philosophy—as practiced on the AMAs—has not supported a risk-taking environment.[18] The ability to fail and to learn from one's actions is a principal component of adaptive management. Lee notes that learning from errors and factoring this new information into subsequent attempts is what makes adaptive management truly adaptive.[19] Yet, the findings here suggest that little failure has occurred—probably because few new activities have been implemented—and thus, little learning has been achieved. The upshot of agency efforts thus far is the lack of progress for getting things accomplished "out on the ground" where citizens can see, feel, and react to the results.

The fourth area is probably the major stumbling block for successful public interactions, particularly if responses from resource professionals are any indication. This involves the internal operations of the management agencies and their inability to come to any substantial agreement on what the AMAs are supposed to be. The low level of organizational support for personnel in adaptive management functions is directly related to the lack of results observed by the public and the barriers identified by AMA personnel. Little real progress can be made with citizens—especially those who think local managers are unable to make decisions on their own—until internal problems and politics are resolved in substantive measure. This is an agency-wide dilemma that is not likely to be settled on the individual AMAs.

Part Four

Challenges for the Future

There is little doubt we are facing rapid change in technology, a new dynamic in the social, political and economic interactions of forestry, and a host of emerging issues that previously did not exist in forestry. This section examines these ideas in some detail; however, the results here are more speculative than conclusive as many of the issues and trends are quite new. The first two sets of authors attempt to bring into focus the huge issues of certification, biotechnology, international trade, and climate change. We deem certification to be such an important and pervasive topic that it receives a chapter to itself. McDermott and Hoberg take on this challenge and provide a succinct history of certification, describing major developments in both Canada and the United States up to the fall of 2002.

In the following chapter, Alavalapati, Das, and Wilkerson review trends in biotechnology development, trade liberalization, and policies related to climate change. In this discussion they review the economic literature that attempts to model anticipated or potential changes to forests caused by these trends. We felt that no book examining the broad scope of forestry in North America in the early twenty-first century could ignore these issues. They will continue to challenge us in our search for sustainable forest management systems.

The final two chapters look more specifically at the institutions that will be required to further the sustainable forestry agenda. Chapters in Part 3 reviewed current experiments with institutional reform. Stankey, McCool and Clark (chapter 15) articulate why such new institutions are necessary. They discuss the fundamental nature and character of modern forestry problems, and argue that tinkering on the margins of existing governance and regulatory systems will not be sufficient to solve such problems. Rather, they suggest we need new institutions altogether. Their argument is a compelling one, and this same theme is featured in the final chapter.

Chapter 16 is intended to draw on references made by authors throughout the book, and focuses on the theme of institutional reform and the creation of new institutions. As the title suggests, it examines the question of whether we have arrived at our destination—that is to say, does either country have a sustainable forest management system in place. The answer is a qualified no, but Beckley, Shindler, and Finley discuss what will be required to move us further down the metaphorical paths that have been used to frame this book.

The "journey" is important and its use here is more than a metaphor. Adaptive management, which many consider to be an integral component of sustainable ecosystem-based management, suggests there is no destination, only the journey of continuous experimentation and adjustment. The philosophy of adaptive management recognizes that we live in a dynamic and complex world. The best we can do is to experiment, learn from these efforts, and try to keep pace with change in both the natural and social environment. Attention to change and the flexibility to adapt are the hallmarks of sustainable forest management.

13

From State to Market: Forest Certification in the United States and Canada

Constance L. McDermott and George Hoberg

Introduction

Over the past decade, a significant trend has emerged in the governance of forest management in Canada and the United States. Until recently the state has served as the final arbiter responsible for ensuring that forest practices are consistent with the public interest. Recently, however, private organizations have become increasingly influential in ensuring that forests are sustainably managed.[1] Forest certification, a system of verifying that forest practices meet environmental, social and economic objectives, began as an effort by international and local environmental groups to use the market place to raise the level of social and environmental performance of forest companies. As such, it represents a new institutional form initiated by stakeholders and operating in the market and civil spheres, rather than the state sphere. Soon after its inception, conflicting ideas about appropriate forest management, as well as different conceptions of appropriate certification decision-making processes, resulted in the emergence of separate environmentalist and industry-backed forest certification schemes. This chapter will outline the development of those different schemes under differing regional contexts in the United States and Canada, and highlight how many of the challenges facing certification are challenges of governance, involving diverse interest group strategies at multiple decision-making scales.

Certification has been only one of many strategies employed by environmental groups over the last few decades to influence forest industry practices, from the local to the international level. By the 1980s, international timber boycotts were attracting public attention to environmentalist concerns over tropical deforestation, loss of biodiversity and the overall environmental and social sustainability of forest management practices. In response, governments and large-scale industry turned to the authority of traditional institutional mechanisms, including intergovernmental conferences and world trade forums, to craft environmental resolutions. These international forums have included the United Nations Conference on the Environment and Development, assemblies of the International Tropical Timber Organization, the Intergovernmental Working Group on Global Forests, the Montreal Process; the Helsinki Process; the Tarapoto Proposal, etc.[2] Such higher level processes, however, often failed to appease those not included in the decision-making circles.[3] Furthermore, they have generated broad principles of forest management that are not easily translatable to on-the-ground practices.[4]

About a decade after international alarm was raised over tropical deforestation, domestic conflicts over North American forest practices, including the harvest of old growth forests on the Northwest coast and the use of biocides along the Eastern corridor, gained political prominence. Environmental group strategies to tackle domestic issues ranged from traditional government lobbying to blockades and court injunctions. Despite at least temporary success in gaining public attention, however, these methods have been criticized for failing to influence the central sources of environmental problems. Such strategies have been viewed as blunt instruments, sometimes disproportionately harmful to rural communities and indigenous peoples, and ultimately ineffective in promoting good management practices.[5] They are in essence, reactive rather than proactive strategies.

Forest certification was envisioned by some of its founding supporters as a voluntary, incentive-based alternative to boycotts and more confrontational strategies.[6] Certification would allow companies to attach a "green" label to wood products coming from forests which met agreed-upon criteria for sustainable forest management. Despite this promise of positive incentives, however, the use of certification to provide market differentiation did not receive initial support from most government and trade forums, at either the international level or within the United States and Canada.[7] Instead, both national and international government and industry groups responded with the development of competing certification systems that were less threatening to existing models of forest management.

In this chapter, we outline the local, regional, national and international development of four major certification systems currently active in Canada and the United States. These four systems include the environmentalist-backed Forest Stewardship Council (FSC); the international standards organization (ISO); the Canadian Standards Association (CSA); and the United States-based Sustainable Forestry Initiative (SFI). We place particular attention on the Forest Stewardship Council, as the system supported by environmental groups and the catalyst for the development of competing industry-led schemes. After a general overview of the four major systems, we follow with some regional case studies that illustrate differences in the evolution of certification between regions in Canada and the US. In conclusion, we discuss the differing balance of interests and decision-making scales that have shaped this phenomenon in the two countries and the effects this has had on the definition and implementation of certifiable forestry.

From Local to International: The Convergence of Certifiers and the Creation of the Forest Stewardship Council (FSC)

In the early 1990s, several non-profit organizations in the United States and Canada, as well as a for-profit United States firm, began conducting independent forest certification assessments. These early North American certifications involved mostly small-scale woodlots and a few moderate-sized forest companies. The early certifiers were scattered mostly along the west and east coasts, from California to British Columbia and from New England to the Canadian Maritimes. They ranged from locally based organizations supporting small-scale eco-forestry, to enterprises promoting corporate responsibility consistent with larger-scale environmental product markets. By the mid-1990s, the United States -based certifiers consolidated into two organizations, the non-profit SmartWood Program of the Rainforest Alliance, and the for-profit Scientific Certification Systems.

Meanwhile, leading international environmental organizations, including the Worldwide Fund for Nature and the Natural Resource Defense Council, were involved in initiating an international accreditation and standard-setting body, the Forest Stewardship Council. The Forest Stewardship Council (FSC) was founded in Toronto, Canada, in 1993 and established its headquarters in Oaxaca, Mexico, in 1994. The FSC accredits certifiers who meet the requirements of the FSC's global framework for forest certification.[8] By 1995, SmartWood and SCS became accredited under the FSC, resulting in the further consolidation of United States certifiers under a single international label.

The FSC was designed as an alternative to governmental and industrial decision-making forums. It is a membership organization, in which membership is based on adherence to the organization's goals and values. In an effort to balance the interests considered in decision-making, the FSC is organized into three Chambers, representing environmental, social and economic interests respectively. Select representatives for the three Chambers, through a system of equal votes, have established Ten Principles and Criteria of "environmentally and socially responsible forest management" as the foundation for all certification bearing the FSC label.[9] The FSC has also developed a Chain of Custody certification process, which tracks certified forest products from the certified forestry operation to the end consumer.

When SmartWood and SCS received accreditation, their existing certifications became retroactively certified under the FSC. These two United States -based certifiers have thus far been the only certifiers active in the United States, and have continued certifying at accelerating rates.

Table 1. List of FSC-Accredited Certification Bodies

Country of Origin	*Approved Area of Operation*	*Forest Management*	*Chain of Custody*	*Organization*
Canada	Canada	✓	✓	Silva Forest Foundation
Germany	World-wide	✓	✓	GFA Terra Systems
Italy	World-wide	No	✓	ICILA
South Africa	South Africa	No	✓	South African Bureau for Standards
Switzerland	World-wide	✓	✓	Institut für Marktökologie IMO
Netherlands	World-wide	✓	✓	SKAL
United Kingdom	World-wide	No	✓	BM TRADA Certification
	World-wide	✓	✓	SGS Forestry QUALIFOR Programme
	World-wide	✓	✓	Soil Association
United States	World-wide	✓	✓	Rainforest Alliance SmartWood Program
	World-wide	✓	✓	Scientific Certification Systems

Source: FSC Documents, September 21, 2001 at http://www.fscoax.org/html/available_documents.html

In Canada, however, an increasing number of certifiers from around the world have been actively seeking forest industry clients. By 2001, a total of eight certifiers worldwide had the necessary accreditation to operate within North America (see Table 1). Despite the interest of many of these certifiers in the Canadian forest products industry, FSC-accredited certification has proceeded more slowly in Canada than the United States. As will be discussed later in this chapter, the entrance of multi-national certifiers lacking histories of local and regional environmental group support was accompanied by heightened conflicts over on-the-ground certification activities.

Around the same time that the first certification organizations were accredited, FSC initiated the development of national and regional "working groups" charged with the creation of region-specific indicators and verifiers to be added to the FSC International Principles and Criteria. Out of this initiative, the FSC-United States national working group was formed in 1995, which completed a set of national certification standards in 2001.[10] These national standards were designed to serve as a baseline for nine active regional working groups: the Southwest, Southeast, Mississippi Valley, Ozark Ouachita, Northeast, Appalachia, Pacific Coast, Rocky Mountains, and Great Lakes Regions. These regional groups were then responsible for modifying and elaborating on the national standards as necessary to address differences in regional context.[11]

FSC-Canada was established in 1996, and currently includes four regional working groups: the Maritimes, Great Lakes/St. Lawrence, British Columbia, and Boreal Regions. FSC-Canada has also created a fourth Chamber for its membership and standard-setting processes, known as the "Indigenous Peoples Chamber." This Chamber is allowed equal vote with the Environmental, Social and Economic Chambers. FSC Canada has thus far chosen not to produce a national baseline standard, thereby devolving more decision-making responsibilities to the regional level. In both the United States and Canada, establishing FSC regional standards has proven a very challenging task. By the fall of 2002, only three North American regional standards had been endorsed by the FSC International, those of the Maritimes in Canada and the Rocky Mountains and the Lake States in the United States. Several of the other three regional standards in Canada, and seven others in the United States, are well along in process and many will likely have been endorsed by FSC International by the time this volume goes to press.

Meanwhile, FSC-accredited certifiers have continued to conduct certification assessments on the basis of the FSC Principles and Criteria and their own field guidelines developed over the past decade. However, the creation of regional standards processes has resulted in new regional-level governance structures and an accompanying expectation that forest certification should be controlled at this level of decision-making. As will be discussed later in this chapter, this expectation became particularly strong in Canada, where few forest certification assessments of any kind had been completed prior to the initiation of the regional standards processes.

While certifiers and FSC standard-setting working groups were busy defining and assessing the types of forestry supported by environmentalists and other interest groups, many environmental organizations were concurrently pursuing other channels of market-focused activism. Perhaps the most pivotal campaign issue affecting the international trade of North American wood products was the protection of the old growth rain forests along the West Coast, from California to British Columbia. A series of boycott campaigns, including the two-year-long Rainforest Action Network (RAN) campaign against the U.S. home building retailer, Home Depot, contributed to commitments by numerous major U.S. retailers to stop buying wood from "old growth" forests.[12]

While some European retailers have supported the FSC since its inception, it was not until after these intensive boycott campaigns that Home Depot and other major retailers in the United States developed policies giving preference to forest products certified under the FSC.[13] Reflecting the growing interest of wood product buyers for certified wood, the North American Certified Forest Products Council was formed in 1998, with a policy that singled out the FSC as the only credible forest certification system in existence.[14] The interest of retailers, however, was not matched by forest managers on the supply side. In the United States, Canada, and elsewhere around the world, forest companies were putting their support behind alternative certification schemes.[15]

The International Industry Model: ISO

In Canada, 60 percent of the value of forest products is sold in the export market.[16] Given the international focus of the Canadian forest industry, many businesses have looked to an already existing international standards institution, ISO, as the appropriate body for developing environmental standards acceptable in the global marketplace. The ISO has been creating industry standards since 1947, and carries the approval of the World Trade Organization. The ISO, first formed for the purpose of creating technical standards for consumer products, has since secured the participation of nearly one hundred national standards organizations. ISO standard-setting is achieved through consensus-based technical committees and subcommittees.[17]

By the time of the United Nations Conference on Environment and Development (UNCED) in Rio in 1992, ISO was participating in international dialogue over the evaluation of environmental management.[18] The result was the ISO 14000 series, including ISO 14001 environmental management system certification. Rather than attempting to resolve international conflicts over the definition of appropriate environmental standards, ISO 14001 allows companies to set their own performance requirements. ISO certification is then based on the effectiveness of the management "systems" companies put in place to meet their own environmental goals.

ISO's lack of environmental performance standards and the absence of an environmental labeling system drew strong criticism from environmental organizations in North America and elsewhere.[19] Perhaps just as important, ISO was criticized for its decision-making procedures. A report released by the Worldwide Fund for Nature, an organization which had participated early on in ISO 14001 standards development, stated that ISO was not equipped "to deal with environmental issues and their many stakeholders."[20] Certainly the development and implementation of ISO processes did not include the same participants, such as small and medium sized industries and public interest groups, as were actively involved in the FSC.

Despite a lack of environmentalist support, ISO environmental management certification has proceeded fairly rapidly in Canada. As of June 2002, over 107 million hectares of Canada's forest lands had been certified under ISO 14001.[21] In contrast, ISO has made very little advance across forest lands in the United States. As of 2001, six ISO 14001 certificates had been issued to forest companies in the United States, as compared with 129 in Canada.[22] United States forest industry interest is growing, however, as illustrated by the major forestland owner Weyerhaeuser's recent announcement of the ISO certification of its forest properties in Oregon and Washington states.[23] In neither country, however, has ISO 14001 served to appease environmentalists' demands for a reform of forest management practices.

National Industry Initiatives: SFI and CSA

In addition to, or in lieu of, ISO 14001, many forest companies in North America and elsewhere have supported the development of national-level certification schemes, perhaps with the expectation that domestic programs will be more sensitive to their interests.[24] In Canada, the vehicle chosen to administer certification is the Canadian Standards Association (CSA), which has developed Sustainable Forest Management (SFM) certification.[25] In the United States, the American Forest and Paper Association (AF&PA), a consortium of timber and pulp companies, has developed a certification system known as the Sustainable Forestry Initiative (SFI). Initially, both CSA and ISO allowed first-party registration, whereby companies could receive certification on the basis of their own internal audits. Both the CSA and SFI programs are compatible with the ISO approach, and include elements of ISO systems evaluation as well as environmental performance indicators. Like ISO, neither of these national schemes originally included a product label.

While these systems were initiated at a national level, they link themselves to international processes, as well. Both the CSA and SFI programs claim foundations in commonly cited international agreements, including the Brundtland Commission Report released in 1987 by the United Nations Conference on the Human Environment, and the 1993 intergovernmental Montreal Process, which addressed sustainable management in non-European temperate and boreal forests.[26] The amount of attention these two programs have paid to international processes, and the ways in which their decision-making structures developed, however, reflect key differences between the two countries.

Forestry decision-making in Canada is shaped not only by its dependence on international wood product markets, but also by the government ownership of 94 percent of its forest lands.[27] The choice of housing an SFM forest certification system within the CSA, an established national standards organization, is consistent with a high level of government landownership. It is also consistent with a concern for international acceptance of CSA standards. In fact, the CSA made an unsuccessful attempt in 1995 to have its SFM series adopted by ISO as an international standard for forest management.[28]

The CSA SFM standards are based on the criteria for sustainable forest management developed by the Canadian Council of Forest

Ministers (CCFM). These standards are supplemented by more specific forest management values, goals, objectives, and local indicators, which forest companies must develop in consultation with local-level public advisory groups. While requirements for public participation are strong at this level, it has been argued that the emphasis on local community input orchestrated by the forest company is not conducive to the same degree of regional-level environmental group participation as occurs with the FSC. Furthermore, deferring objective-setting to the community level is likely to produce disparities in the interpretation of the CCFM standard among companies pursuing CSA certification.

While CSA SFM differs from FSC in that it is housed within a national standards organization, it focuses on management systems and it requires locally based public advisory groups, CSA has gradually been incorporating new elements that bring it closer to the FSC. CSA SFM certification currently requires third party assessment, involving CSA-accredited certifiers. As of July 2001, CSA took yet another step closer to the FSC by introducing product labeling and chain of custody verification systems. This has enabled CSA-certified forest companies to sell wood bearing a certified forest product label.[29] Despite its third-party auditing and chain of custody procedures, however, the CSA system has failed to gain environmental group support. At the same time, industry's lack of direct control over the CSA process, together with CSA's difficult public participation requirement, have been cited as factors inhibiting industry participation in CSA certification.[30]

In the United States, a large domestic wood products market and predominantly private land ownership pattern have been cited as factors contributing to high levels of industry support for the AF&PA SFI certification program.[31] The solidarity with which industries have participated in the AF&PA and associated SFI program is, in fact, quite striking. In 1995, when the SFI program was first implemented, Association members were responsible for 95 percent of US paper production and 65 percent of solid wood production, and owned 90 percent of the industrial forest land base.[32] By 1996, the SFI secured access to this large wood products market by declaring SFI certification a prerequisite of AF&PA membership.

SFI first evolved as a system for company self-evaluation (i.e. "first-party" certification), using standards produced by AF&PA members and forestry professionals. Under competition from the FSC and pressure from interest groups outside of industry, the system has continually added more features in common with the FSC model. As of 2001, companies now have the option of third-party certification as well as the use of a product label. Interest group participation in SFI decision-making has gradually expanded to include a wider range of interests at different decision-making scales.

The development of SFI standards and procedures is handled at the national level. Although its central executive body, the Sustainable Forestry Board, incorporates a form of multiple stakeholder participation, most major environmental organizations have chosen not to be involved.[33] Like the CSA system in Canada, SFI also makes use of a national accreditation body, the American National Standards Institute (ANSI), which is responsible for accrediting SFI certifiers. At a more localized level, State Implementation Committees have been established in thirty-four states. These committees are responsible for addressing a range of implementation issues, including the development of state-level certification assessment criteria and handling of public complaints over the actions of local certified companies.[34]

Meanwhile, by the year 2000, SFI had further expanded its client-base through collaboration with the American Forest Foundation, an organization of non-industrial private forestland owners. The American Forest Foundation revamped its already existing non-industrial private landowner certification system, originally established in 1941, to meet the new management expectations associated with SFI certification. This resulted in a mutual recognition agreement between these two certification systems,[35] which further consolidated the participation of United States forestry producers under a single industry-led scheme.

SFI's continual enhancement of its program and procedures does not appear to have created many barriers to certification on the ground.

Table 2. Summary of Certification Systems Used to Assess Forestry in the United States and Canada

Acronym	*Certification System*	*Origin*
FSC	Forest Stewardship Council	International
ISO 14001	International Organization for Standardization, Environmental Management Systems	International
CSA SFM	Canadian Standards Association, Sustainable Forest Management	Canada
AF&PA SFI	American Forest and Paper Association, Sustainable Forestry Initiative	United States

As of the spring of 2002, 43 million hectares were SFI certified in North America, 31 million hectares under the third-party system.[36] The SFI philosophy has been described as "a rising tide raises all boats." In other words, certification should be easily accessible to all, with a focus on "continual improvements."[37] Although such an approach may appeal to many in the forest industry, it has been rejected by most environmental organizations. Thus the SFI, together with ISO and CSA, have failed to gain acceptance among critical drivers of green product markets.

Regional Developments in Forest Certification in the United States and Canada

Meanwhile, FSC-accredited certification has proceeded much more slowly than its competitor programs, accompanied by numerous controversies and conflicts among FSC supporters themselves. Nevertheless, in both the United States and Canada, the FSC has maintained the endorsement of the environmental groups responsible for the initiation of the forest certification movement. Beyond the consistency of environmentalist support, however, there are considerable differences between countries and regions, regarding the structure and level of the certification decision-making processes that have evolved. The following case studies of the United States Pacific Coast, the

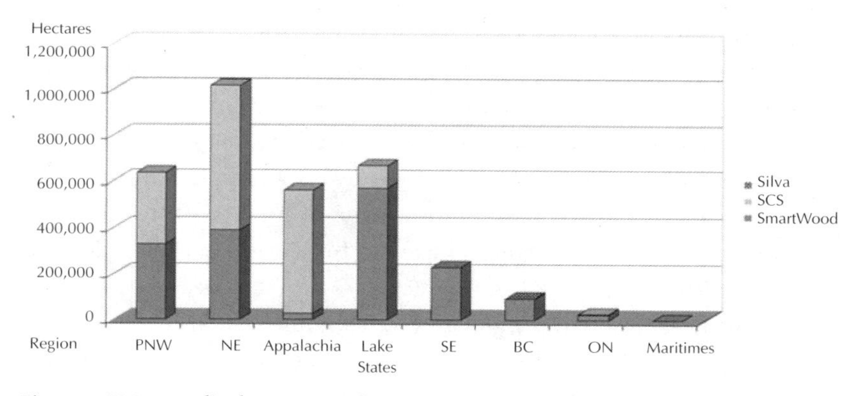

Figure 1. FSC-Accredited Forest Certification by Region and by Certifier Total for US: approximately 3.1 million hectares. Total for Canada: approximately 120,000 hectares.

Sources: FSC, Forests Certified by FSC-Accredited Certification Bodies as of October 12, 2001 at http://www.fscoax.org/html/available_documents.html; SmartWood, List of Certified operations at http://www.smartwood.org; SCS, The Forest Conservation Program, Forest Lands Certified as "Well-Managed" at http://www.scs1.com/forestry.

Canadian Maritimes, and British Columbia, illustrate some of these regional differences.

The FSC in the United States

The Pacific Coast

Forest management in the United States has long been a source of public controversy. Perhaps the most heated and widespread debate in recent years has centered on the protection of the old growth rainforests of the Pacific Coast. Adding to the public nature of debate in the region, roughly 50 percent of forested lands in Washington, Oregon, and California are owned by the federal government. Many of the remaining remnants of old growth are located on these federally owned lands. The greater number of legal avenues for environmental protest in the United States than in Canada,[38] particularly in regards to public lands, has enabled environmental groups to win court injunctions and spur federal action plans that limit old growth logging on federal properties.[39] Following upon these successes, in 1996 the Sierra Club U.S. initiated a "Zero Cut" campaign aimed at ending all commercial logging on U.S. federal lands.[40]

In response to the demands of some of its constituents, the FSC-U.S. created a Federal Lands Committee to make recommendations for FSC policy regarding national forest management. Meanwhile, no U.S. Forest Service or Bureau of Land Management properties have been assessed for certification under the FSC, although an assessment has been conducted on an U.S. Army Base. Thus the region's most controversial issue, that of federal land management, has been relegated to the national level and effectively postponed. Nevertheless, significant controversies over non-federal lands, including the issue of growth protection and restoration on private property, have continued to stall FSC regional standards development.

The Pacific Coast working group (FSC-PC) initiated the development of regional certification standards for Washington, Oregon and California in 1996. The process was facilitated in its early stages by the Pacific Forest Trust, a non-profit organization involved in the creation of land trusts for forest protection and sustainable development. The most active participants in the process have included FSC-accredited certifiers, environmental groups, and moderate-sized industrial forest companies and wood processors. Large U.S. forestry firms have not directly participated in Pacific Coast regional standard-setting. Instead, many of them have opted for involvement in the industry-led SFI initiative.

The slow pace of standards development, and the lack of industry participation, however, has not stalled certification on the ground. SmartWood and SCS, the two FSC-accredited certifiers active in the United States, have been operating in the Pacific Coast since the early 1990s, pre-dating the initiation of FSC certification. SmartWood entered the region through affiliation with three local non-profit organizations, who already had strong ties with local environmental groups.[41] Early forest certifications conducted by these SmartWood network members involved small-scale, non-industrial forestry operations. SCS granted its first forest certification in the Pacific Coast to a moderate-sized industrial client, Collins Pine Company's Almanor Forest in Oregon. Since Almanor Forest was located well outside of coastal rainforest habitat, it was not a source of major controversy.

When SmartWood and SCS received FSC accreditation in 1995, their early certifications were absorbed into the FSC system without major fanfare. Having already gained some measure of regional acceptance, these certifiers have since continued to certify forestry operations on the Pacific Coast at increasing rates. As of 2001, around 635,000 hectares of land had received FSC-accredited certification within Washington, Oregon, and California (see Figure 1). Recent certifications include some relatively large industry holdings, such as the Mendocino Redwood Company in California and Roseburg Forest Products in Oregon. Thus, the development of FSC certification in the Pacific Coast region has been shaped to a significant degree by the on-the-ground implementation of certifiers, despite conflicts and lack of industry buy-in at the level of regional standard-setting processes. The FSC in Canada, however, has followed substantially different paths.

The FSC in Canada

The Maritimes

The first FSC-accredited certifications in Canada occurred in 1997 and 1998, within two years after the launch of FSC regional standard-setting processes. These included the SCS certification of J. D. Irving, Ltd. in New Brunswick, an industrial forest operation and supplier for the large U.S. retailer Home Depot. The Irving certification was followed by

SmartWood's certification of Haliburton Forest and Wildlife Reserve, Ltd. in Ontario. Both the Irving and Haliburton properties, unlike the majority of Canadian forest lands, are privately owned. Private ownership, however, did not remove forest managers from public controversy. The certification of the more industrially focused Irving raised an outcry among both local and international environmental groups, championed by the Sierra Club of Canada.

By the time the first two certifications were conducted, the Maritimes standards process was immersed in debates over regional scale environmental and social controversies. Maritime standards development was handled by the Maritimes Regional Committee, facilitated by the non-profit sustainable development organization, the Falls Brook Centre. The standard-setting process was consensus-based, without a highly formalized decision-making structure. J. D. Irving participated in this regional process as the only representative of larger scale industrial forest interests.[42] Irving's certification in the midst of the regional standards development brought to the fore a number of heated controversies including the use of clearcutting, the conversion of natural forests to plantation, the provision of protected areas, the use of biocides (a particularly contentious issue in an area prone to periodic spruce budworm epidemics), a public relations debacle with a monastery (in a different province from the certified land), lack of transparency in the certification process, and dissatisfaction with the standards used. The Sierra Club of Canada responded with an appeal to the FSC to revoke the FSC certificate.[43]

Further contributing to concern over Irving's certification, issues of aboriginal rights had also become prominent in the Maritime region, meaning that FSC-accredited certification was under pressure to ensure that First Nations[44] were adequately involved in shaping certification standards.[45] The alarms raised by Irving's certification, as well as more general concerns over precedent and interest group representation, may well have led to the rapid acceleration of the Maritime regional standard-setting process. At year's end in 1999, the Maritime standards became the first set of North American regional standards to be endorsed by the FSC.[46]

In the midst of the push to complete regional standards, J. D. Irving withdrew from the Maritime Regional Committee, and returned its FSC-accredited forest management certificate. Soon afterwards, completed standards were submitted to the FSC and J. D. Irving instigated its own appeal. The company claimed that there was inadequate stakeholder representation in standard-setting and, furthermore, that the standards were disproportionately restrictive relative to those being developed in other ecologically similar regions (e.g. the Northeast region in the United States). The FSC responded by placing conditions on acceptance of the Maritime standards, including a requirement that FSC-Canada take steps towards harmonizing standards within Canada, and between Canada and the United States. The conditions also called for harmonization with Sweden, the only temperate-zone country that had yet received full FSC endorsement of its standards. The Maritime Standards Committee was instructed to formalize their decision-making procedures and aim for better representation of interests within the province. More work was also required on the development of standards governing pesticide use.[47] Thus conflict in the FSC-Maritimes standard-setting process, led to international pressure to formalize decision-making processes at the regional level. At the same time that the Maritimes Standards Committee was being asked to harmonize its standards internationally, one of the most controversial environmental issues, the use of biocides, was devolved back to the regional level for further resolution.

While the longer-term effects of increasing formalization remain to be seen, the regional focus of decision-making in the Maritimes appears to have initially slowed the progress of FSC certifications on the ground. With the departure of Irving, the total forest area certified under the FSC in the region was approximately 20,000 hectares (see Figure 1). Meanwhile, across the border, Irving received FSC certification for its Allagash Woodlands operations in Maine. The environmental community also appealed this certification.[48] Nevertheless FSC-accredited certification in the New England area continued to far outpace the Maritimes, even though the Northeastern regional standards had yet to receive FSC endorsement.

British Columbia

The early days of forest certification in British Columbia in some ways parallel those of the Pacific Coast. The first assessment in British Columbia was completed in the mid 1990s, by the Silva Forest Foundation, a British Columbia-based non-profit environmental organization strongly supported by environmental groups. This certification, involving a small-scale woodlot and a trusted local certifier,[49] preceded FSC-accredited certification in the region. Conducted in close communication with British Columbia environmental groups, this early Silva certification helped to establish environmentalist expectations for the type of forestry that certification should promote.

It was not long, however, before the debate over forest certification in British Columbia escalated to an international level. The rise of FSC-accredited certification in British Columbia corresponded with a major environmental campaign over the protection of old growth on its central coast, an area dubbed as the "Great Bear Rainforest". Leading environmental groups in British Columbia, including Greenpeace and the Sierra Club among many others, incorporated a push for FSC certification into the Great Bear Rainforest campaign. This involved supporting European and American boycotts of old growth and/or any wood from British Columbia that was not certified under the FSC. In sum, FSC-BC was caught up in a regional war between large-scale environmental groups and major forest industries staged in the highly competitive international marketplace.

European retailers were the first to respond to environmentalist demands. In the early 1990s major Do-It-Yourself retailers in Great Britain, including B&Q, formed the "WWF 95"certified buyers group and publicly announced their commitment to purchase FSC-certified wood products. In addition, some European buyers responded to boycott pressure by canceling shipments of wood from British Columbia coastal companies. North American retailers were slower to respond to environmentalist pressure, perhaps due in part to their closer relations and greater dependency on North American forest industry suppliers.[50] Nevertheless, by the end of the 1990's, major North American retailers including Home Depot and Lowe's, were also committing to the ban of "old growth" wood from their stores and/or to stocking only wood certified under the FSC.[51] These North American market signals for certified products came at a time of Asian economic crisis that hit the British Columbia coastal forest industry particularly hard. Making matters still more tense, quotas set under the 1996 US and Canada Softwood Lumber Agreement had already severely limited the British Columbia coastal industry's access to U.S. markets.

In 1998, the same year that Irving publicly announced its New Brunswick certification, British Columbia's fourth largest wood products company, Western Forest Products, announced its intention to become certified under the FSC system.[52] Western hired SGS Qualifor, an international for-profit certifier based in England and accredited by the FSC, to conduct the certification assessment. Just as with Irving in the Maritimes, British Columbia environmental groups responded with alarm. The large for-profit certifier SGS appeared in marked contrast to the British Columbia-based Silva Forest Foundation, the organization backed by many of the province's environmental groups. Furthermore, Western Forest Products differed vastly from the kind of small-scale eco-forestry model promoted by Silva. A number of British Columbia environmentalists were adamantly opposed to the precedents that would be set if SGS was allowed to go through with the certification assessment. Further adding to these concerns, another multi-national consulting firm, KPMG, was applying for FSC accreditation. With the entrance of such established, large-scale consulting interests (many with long-standing ties to the forest industry), pressure was mounting for actions to ensure that certification under the FSC would continue in a manner acceptable to the majority of its environmental group supporters.

First Nations also had major concerns about the SGS certification assessment of Western's forestry operations. In British Columbia few treaties have been signed, and thus land title and the rights to harvest forest products remain unsettled across nearly the entire province.[53] British Columbia First Nations were becoming increasingly active in the FSC, with the intention of ensuring that certification adequately address their land and resource rights. First Nations involved in the development of the regional standards were not eager to see the regional process by-passed through the certification assessment of a major forest company.

Given the level of conflict developing between interest groups in British Columbia, the FSC-BC regional working group put forward a request to the FSC and FSC-accredited certifiers that no certifications be completed until the first draft of the FSC-BC standards was released. This first draft was completed in May of 1998.[54] However, the release of the unenforceable draft standards was not enough to allay fears that the FSC would be "bought out" through the certification of status quo industrial forestry.

The experiences of J. D. Irving and Western Forest Products appears to have effectively discouraged other British Columbia forest industries from proceeding with FSC certification until British Columbia regional standards were approved. Meanwhile, many companies were undergoing certification under ISO 14001, and a few pursued CSA certification as well. A division of one major forest company, Weyerhaeuser's British Columbia Coastal Division (formerly owned by MacMillan Bloedel), displayed a particularly complex consideration of options. Weyerhaeuser's Coastal Division pursued ISO and CSA certification and then later applied for FSC-accredited Chain of Custody certification for its mills. Taking advantage of international agreements aimed at the mutual recognition of different certification systems, the Coastal Division has been able to sell wood to Germany from its CSA-certified operations under the German Kerhout label. Meanwhile, other Canadian companies sought certification under the SFI, catering to the certification scheme popular among forest industries in the United States.

While pursuing competing certification schemes, British Columbia companies also took actions to keep their options open with the FSC. Industries, along with environmental and indigenous groups, became increasingly focused on the completion of the British Columbia regional standards. Western Forest Products placed a representative on the FSC-BC Steering Committee, as did another moderate-sized firm, Lignum. A number of other British Columbia industrial forestry companies applied for membership within the FSC and/or provided funding for standards development.

The promise of the FSC exerting a major influence over the forest industry in the province placed considerable pressure on the regional standards to resolve a host of controversial issues. The scope of these problems was enormous, ranging from the conservation of coastal old growth, to reform of government land tenure, to the sharing of rights and benefits with British Columbia First Nations. Reflecting the heavy demands on the regional standard-setting process, FSC-BC began further to formalize its structures and procedures, which increasingly resembled elements of state governance. Members were allowed to nominate and vote for the regional Steering Committee, which is the organization's executive branch. A technical "Standards Team" was then placed at arm's length from the Steering Committee and charged with developing standards for approval by the Steering Committee. A Vancouver-based environmental consulting firm, Dovetail Consulting, was hired to facilitate the standards development.[55]

Perhaps as a result of all the above-mentioned factors, British Columbians comprised 20 percent of the membership of the FSC by the fall of 2001,[56] making them by far the most active regional group world-wide. At the same time, tension over what constitutes certifiable forest practices led to few forestry operations receiving certification. By 2001, five forestry operations were certified within the province, totaling roughly 96,000 hectares (see Figure 1). None of these certified operations resemble large-scale industrial forestry. They include several small woodlots, a small logging operation, and a moderate-sized company, Iisaak Forest Resources, a joint venture between the Nuu-Chah Nulth First Nations and Weyerhaeuser committed to low intensity, eco-system-based management. Meanwhile the FSC-BC working group set its sights on finalizing the British Columbia regional standards. A third draft of the British Columbia standards was submitted in April 2002 for approval by FSC-Canada, and ultimately endorsement by FSC at the international level.[57]

The Canadian Boreal Forests

By the time that the Maritimes and British Columbia standards were close to completion, FSC-Canada was placing increasing attention on the Boreal region, spanning from the northeastern to the northwestern borders of the country. The incorporation of numerous provinces into an overarching standard for Canada's northern boreal forests has proven a challenging governance task, involving the careful balance of regional concerns with demands for national consistency. As of the fall of 2002, a draft of the boreal standard was nearly ready for public review and comment. No on-the-ground forest management certifications had yet been completed in the region, though several firms, including some large, industrial-scale firms had expressed their intention to seek FSC certification once the boreal standard has been endorsed.

The United States and Canada: A Comparative Summary

By the fall of 2002, over 3 million hectares had been certified under the FSC in the United States, while only some 973,000 hectares received FSC-accredited certification in Canada. In contrast, by the first quarter of 2002, ISO had certified around 107 million hectares of forest land in Canada[58] involving 129 certificates versus only 6 certificates issued in the United States in 2001.[59] Meanwhile, the SFI system had third-party certified 77 million hectares in the United States and Canada,[60] while CSA, as of 2001, had completed the certification of 14.7 million hectares in Canada.[61] Looking solely at the number of hectares certified, however, is inadequate for determining the influence of certification on green markets, its impact on forestry practices, or the balance of interest group control over certification decisions.

The explanation for the faster advance of FSC-accredited certification in the United States will depend on the level of analysis used. Regional and national scale perspectives on forestry in the two countries highlight economic and political factors. One such factor is the predominance of private lands in the United States, in contrast to the concentration of government land ownership in Canada. Factors that might be considered likely to facilitate certification on private lands are less public scrutiny and controversy, as well as less bureaucratic control, resulting in greater management flexibility. An explanation for the lesser industry participation in standards processes in the United States, on the other hand, could be the large domestic wood products market, the greater solidarity among domestic producers, and the concurrent growth of the industry-led SFI initiative.[62] Regardless of the reasons for differences between the United States and Canada, of particular importance is how these differences do or do not affect the defining goals of certification itself and the balance of control over decision-making processes. A ground-level analysis of certifications thus far completed may prove surprising.

Despite the challenges of public scrutiny and regulatory bureaucracy associated with government lands, over 40 percent of the more than 3 million hectares in the United States certified under the FSC are owned by municipal, county, or state governments. This indicates that public landownership has not precluded certification. Also running counter to some macro-level predictions, more large-scale, industrial forest industries have been certified in the United States than in Canada, indicating that domestic markets and domestic industry pressures have not prohibited industry involvement. At the same time in the United States, the continuing practice of certifying forest management prior to the completion of regional standard-setting processes does not appear to have worked counter to the interests of local or regional environmental groups supportive of small-scale, non-industrial forestry. The median size of forest areas in the United States currently certified under the FSC is roughly 2,800 hectares.

Certifiers have in fact taken a number of steps to encourage the participation of small-size forest management operations. Two such steps include the development of group certification and the certification of forest managers. Under both of these programs, a collection of forest owners and/or managers can apply for certification as a single entity, based on collective agreements to meet FSC requirements. If group certification is taken into account, the median size of certified forest management *properties* would in fact be smaller than the 2,800 hectare median size of *forest management areas* certified under the FSC. The small size of many certified operations argues against the assertion by some theorists that economies of scale have weighted the FSC system in favor of larger land-owners. [63] While any fixed costs of a certification assessment may put smaller operators at a disadvantage, other factors, such as the nature of the certification standards and the priorities and constraints of certifiers in their choice of clients, may help to facilitate the certification of small-scale and non-industrial forestry operations.

Looking yet more closely at the dynamics of scale, there are significant differences between certifiers in terms of the distribution of the size of operations. The median size for SmartWood certificates is about 1,600 hectares. For SCS it is roughly 33,000 hectares. Even when lumping the two certifiers together, however, only 9 percent of the total number of forest management certificates in the United States have been awarded to forestry operations exceeding 100,000 hectares in size.[64] Half of these larger operations, furthermore, are government-owned and produce relatively low volumes of timber. Thus considering the number of certificates issued, together with the area certified, provides critical insights into the dynamics of on-the-ground FSC-accredited certification (See Figure 1 and Table 3).

Table 3. Number of Forestry Operations Certified under FSC by Region and by Certifier

	PC	*NE*	*Appalachia*	*Lake States*	*SE*	*BC*	*ON*	*Maritimes*
SmartWood	41	12	4	11	4	2	3	1
SCS	12	3	2	1	0	0	1	0
Silva*	N/A	N/A	N/A	N/A	0	3	0	0
Total: 100								

*Silva is accredited to certify in Canada only.

Sources: FSC, Forests Certified by FSC-Accredited Certification Bodies as of October 12, 2001 at http://www.fscoax.org/html/available_documents.html; SmartWood, List of Certified operations at http://www.smartwood.org; SCS, The Forest Conservation Program Forest Lands Certified as "Well-Managed" at http://www.scs1.com/forestry.

As discussed earlier, the forest certification movement can be viewed as a strategy of environmental groups to bypass traditional industry and government decision-making processes governing forest management. Within this strategy, as long as controversy over certification has remained focused more on individual assessments and certifiers, i.e. at more local levels, and the certifiers have established local reputations for certifying the types of forestry acceptable to environmental group supporters, it has been possible to proceed with assessments without becoming as embroiled in larger debates at regional and higher scales. In contrast, the relatively late arrival of FSC-accredited certification to Canada, and the sudden interest of both major industries and numerous multi-national certifiers, has placed the struggle for control over certification decisions at regional and international levels. This has put environmental organizations involved in FSC regional standard-setting under pressure to formalize decision-making processes and standardize procedures in order to counterbalance the participation of traditionally powerful forestry interests.

Meanwhile, industry-backed certification schemes have focused on the rapid implementation of certification on the ground, while placing less emphasis on environmentalist support. Which certification scheme ultimately carries the day, or whether the various competing systems will be able to co-exist for the long-term, remains to be seen. It appears clear, however, that this new institutional form is not a passing fad, as its critics claimed less than a decade ago, when forest certification was little more than an idea.

14

Emerging Issues of Globalization: Implications for Forest Use in the United States and Canada

Janaki R. R. Alavalapati, Gouranga G. Das, and Cynthia Wilkerson

Introduction

Historically, both U.S. and Canadian forests were managed with the objective of optimizing revenue from timber production. Now the concept of sustainable forest management (SFM), which supports a balance among social, economic, and ecological benefits, has become the guiding principle for both countries. Forests today, more than ever, must provide a host of environmental services ranging from conservation of soil and water, biological diversity and carbon sequestration, to outdoor recreation. As a result, the societal definition of "costs" and "benefits" of forestry operations has become more holistic. Benefits of forest ecosystems have expanded, beyond the traditional market values of timber products, to include non-timber forest products (e.g. mushrooms and pharmaceuticals), as well as non-market values such as existence, bequest, cultural and spiritual values (see McFarlane et al., chapter 7, and Beckley and Bonnell, chapter 9).

In response, several changes in the use and management of forests are being made by forest agencies of these two countries (see Ryan, Duinker, et al., chapter 3). The government of British Columbia, for example, announced its intent to double the protected forest area to 12 percent of the province by the year 2000, and nearly accomplished that goal.[1] British Columbia's Forest Practice Code and associated regulations imposed a series of constraints on timber harvesting, including limiting the size of clear-cuts, and adding connectivity requirements. These changes in laws and regulations have caused a 35 percent reduction in the area available for commercial harvesting in British Columbia.[2] In the United States, many environmental laws such as the Clean Air Act of 1970, the Clean Water Act of 1972, and the Endangered Species Act of 1973 are influencing forestry operations on both private and public forestlands. In 1989, for example, the listing of the northern spotted owl as an endangered species resulted in a significant reduction of timber harvest in the Pacific Northwest, thereby causing a shift to greater timber production in the U.S. South. As a result of this and other environmental regulations, timber harvest from U.S. public lands fell from 12 billion board feet in 1989 to 3.7 billion board feet in 1995; a reduction of about 6 percent of the total U.S. timber production.[3] More recently, a house bill H.R. 1494 entitled "National Forest Protection and Restoration Act," proposing an elimination of commercial logging on all U.S. national forests, was introduced into the U.S. Congress. If or a similar bill were passed, such a policy would have significant consequences for regional shifts in timber production and corresponding changes in the use of forests in the United States and Canada.

This chapter presents an overview of three emerging issues that we believe will have significant impact on the future use and management of forests in the United States and Canada. These issues are global climate change and policies to address it, trade liberalization and bilateral trade issues between Canada and the United States, and finally the potential positive and negative impacts of biotechnology in forestry. In addition, we review output from economic models that chart the potential effects of changes related to these issues.

First, in order to reduce greenhouse gas concentrations in the atmosphere, both the United States and Canada have strongly supported the inclusion of forest carbon sequestration activities as a means to achieve emission reduction targets.[4] Natural forests may be preserved, or cleared and replanted, to realize carbon sequestration benefits. Forest management activities could be altered to sequester more carbon, and new tree plantations could be raised to sequester additional carbon by providing incentives to landowners. Alternatively, governments could facilitate carbon markets wherein landowners and polluting companies participate, respectively, by selling and buying carbon credits. These changes may have economy-wide implications for forestland use, forest commodity prices, outputs, and households' welfare.

Second, forest product companies in the United States and Canada will vigorously pursue the elimination of trade barriers for all forest products, to increase their access to European, Japanese, and other developing countries' markets.[5] The United States, for example, at the World Trade Organization (WTO) meeting in Seattle, urged other members of the WTO to move up the deadline for tariff elimination on pulp and paper products from the year 2004 to 2000. While these two countries seek to remove trade barriers for forest products in the global market, they have continued to maintain separate bilateral forest product trade restrictions. An unusual coalition of U.S. timber companies and environmental groups has successfully lobbied the Bush Administration to impose new tariffs on Canadian softwood lumber imports to the United States.

Finally, forest preservation sentiments will grow both in the United States and Canada in the face of rising demand for forest products. In the United States, besides national policy proposals such as "National Forest Protection and Restoration Act", many state governments have undertaken massive forestland acquisition programs for the purpose of preservation.[6] In Canada, several provinces have implemented major programs and policy changes to increase the amount of protected areas in their provinces as well. In the face of growing sentiment for preservation of natural forests and rising demand for forest products, Sedjo and Botkin and Dekker-Robertson note that the adoption of biotechnologies and increasing output from plantations through intensive management practices may be potential ways of addressing this paradoxical situation. While these tools may help increase protected areas, they are themselves a subject of environmental controversy.[7]

The rest of the chapter is organized as follows. The effects of markets for carbon on timberland owners' behavior and the creation of carbon plantations in response to climate change and the Kyoto Protocol are treated in the next section. In the following section, a conceptual framework illustrating the impact of the Canada-United States softwood lumber trade dispute is presented. We articulate the rationale for the adoption of biotechnologies in the U.S. Forest Sector and discuss its consequences for the United States, Canada, and the rest of the world in the third section. The final section concludes with a summary and discussion. In our overview of these issues, we pay special attention to trends, tradeoffs, and welfare implications in both the United States and Canada.

Forest Carbon Sequestration and Implications for Forest Management

The 1997 Kyoto Climate Protocol and the subsequent Intergovernmental Panel on Climate Change meetings recognized forest carbon sequestration as an acceptable way of meeting carbon emission targets. The Kyoto Protocol allows nations to claim as a credit any carbon sequestered through afforestation and reforestation since 1990, while carbon lost as a result of deforestation is a debit.[8] In spite of the withdrawal of support by the United States, the details of the Kyoto Protocol were worked out in November 2001 at Marrakesh, Morocco including forest carbon sequestration as an acceptable way of meeting carbon emission targets. Research has also shown that the cost of sequestering carbon through forestry practices could be lower than many alternatives. Therefore, forest carbon sequestration has attracted a lot of attention in the United States and Canada. Shongen and Mendelsohn found that global warming is expected to expand timber supplies and benefit U.S. timber markets.[9]

Tree plantations to sequester carbon could be encouraged through direct government intervention, by the use of a forest subsidy paid to the landowner to grow trees, or by providing credits based on the amount of carbon sequestered by a forest. These approaches may change the current land base in the forest sector. Carbon subsidies and carbon credit payments could impact forest planting decisions, investments and forest management decisions, rotation lengths,[10] and land prices. For example, Stainback and Alavalapati found that forestlands would

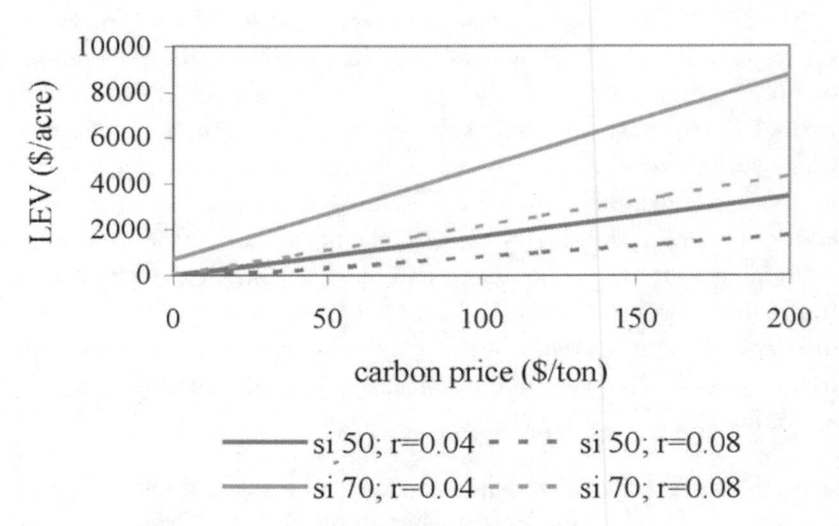

*Figure 1. Land Values under Slash Pine in Response to Carbon Prices**
Si = site index; r= interest rate
**Source: Stainbeck and Alavalapati, "Economic analysis."*

be more profitable if payments for carbon sequestration are considered (see Figure 1 for details).[11]

As a result, more land could be devoted to forestry as opposed to other land uses such as agriculture and urban development. Also, an increase in the optimal rotation age in response to carbon payments may cause an increase in the proportion of sawtimber production.[12] If value-added opportunities associated with sawtimber are higher than pulpwood, societal welfare could improve. The above results may not be applicable to old growth forests. Harmon et al.[13] concluded that harvesting old growth forests in the Pacific Northwest would contribute to an overall increase in atmospheric carbon dioxide. As such the impact of carbon sequestration economics may affect U.S. and Canadian forests differently. For example, in the U.S. South, where private forestlands predominate, carbon payments may stimulate landowners to bring more land into forestry.

Alavalapati and Wong[14] examine the impact of creating 50 million hectares of carbon plantations in North America and Europe using a dynamic general equilibrium model. They simulate the impact of establishing 1 million ha of plantations in the U.S. Pacific Northwest

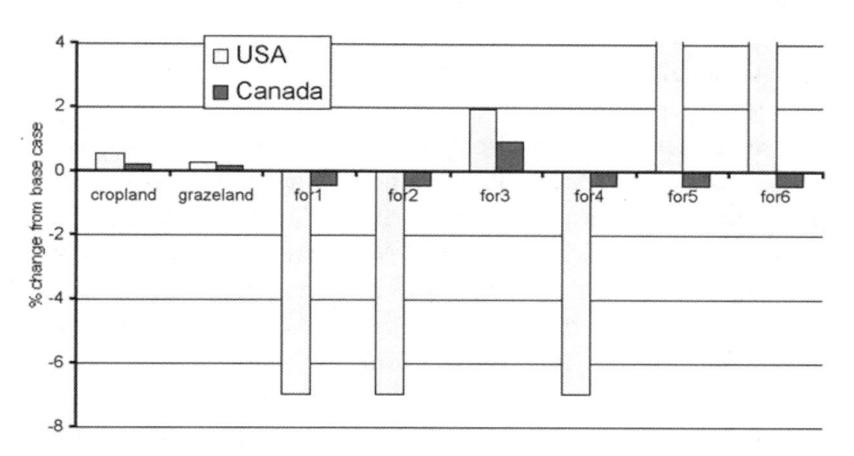

*Figure 2. Land Use Changes in the United States and Canada in Response to Proposed Carbon Plantations**

Notes: The policy scenario involved the expansion of 16 million hectares of carbon plantations in the U.S. and 5 mil ha Canada. Land is segregated into 6 classes depending on its growing season length. Carbon plantations are simulated on land classes 3, 4 and 6 in the U.S. and on land class 3 in Canada. Leakage effects are 2.15 mil ha in the U.S. and 0.35 mil ha in Canada.

**Source: Alavalapati and Wong, Forest Carbon Sequestration Policies.*

and 30 million ha in the U.S. South, 5 million ha in Canada's temperate deciduous forests, 2 million ha in the European Union; 6 million ha in the former Soviet Union; and 6 million ha in other European countries.[15] According to the model, the anticipated effect of creating 50 million ha of plantations is to induce a shift in land re-allocation among the crop, livestock and forest sectors due to the fixed land base (see Figure 2).

Establishing carbon plantations in the Pacific Northwest and U.S. South will induce corresponding decreases in cropland in the same land classes. However, crop production is shown to expand in other land classes in order to maintain adequate food supply. The area of grazing land is shown to decline in response to the change. Although the total area of forests increased, forestland in other regions of the United States would decrease according to the model. This is a trade-off from the policy shock; as the expected influx in supply of timber from carbon plantations drives down prices and forestland rents, a shift to other land uses occurs. As a result the policy shock is shown to decrease household income by 0.06 percent, wages by 0.07 percent, and the aggregate welfare by 728.59 US$ million in the United States. Thus, the results of Alavalapati and Wong[16] raise an important question; "is the loss of welfare an acceptable price to pay for averting or delaying the global impacts of climate change?"

Alavalapati and Wong[17] also simulate the impact of a U.S. domestic climate plan by providing carbon subsidies to the forest sector in the United States. One key result from this study suggests that Canadian forest producers are likely to lose competitiveness in the scenario of domestic carbon policies in the United States. Given that the U.S. and Canadian forest industries are highly linked, the impacts of forest carbon policies on Canada are dominated by the actions of the United States, her largest timber trade partner. This suggests that Canada will have to take pre-emptive steps to improve their competitiveness in global timber markets if the United States undertakes any type of forest carbon policy.

In Canada, many provincial governments are becoming vocal critics of the federal government's stated intention to ratify the Kyoto Protocol, even if it means doing so unilaterally. Environment Canada is running a national advertising campaign on the negative effects of global climate change, so the commitment of the federal government to the issue appears strong. In the United States, President Bush has clearly articulated his opposition to the Protocol, but that does not mean the issue will go away. Whatever is decided, in either country, this issue will continue to have a major effect on the forest sector, even if the only effect is to create considerable uncertainty.

Trade Liberalization and the Bilateral Trade Dispute between Canada and the United States

During the 1999 World Trade Organization Summit, the United States strongly argued for trade liberalization and the elimination of tariffs on all forest products. Many forest product companies in the United States and elsewhere have supported the idea of trade liberalization with an expectation that it would lead to increased access to new markets across the globe. On the other hand, many environmental groups expressed concern over trade liberalization in the expectation that freer trade will promote importation of wood products from tropical countries and expand tropical deforestation.

Wu et al.[18] have examined the potential impact of a worldwide 33 percent tariff reduction on forest products using a multiregional and multisectoral trade model. This was the reduction suggested in the Uruguay Round's trade discussions. Their results show that under a 33 percent tariff reduction, the price of wood products and pulp and paper products would decrease in the United States and Canada. In both countries the output of wood products would decrease while the output of pulp and paper products would increase. Although the drop in price of wood products might stimulate domestic consumption, additional wood imports from other regions would likely meet this demand. With respect to pulp and paper products, the decrease in price in response to tariffs reductions will increase the demand in the United States and the rest of the world. Since Canada and the United States are the major suppliers of pulp and paper products, output would rise both in Canada and the United States.

With respect to the North American Free Trade Agreement (NAFTA), the limited log market and existing low tariffs in NAFTA countries may not generate large effects on forest products trade in North America.[19] The reduction in tariffs is shown to enhance pulp and paper exports from the United States, Canada, and the European Union. On the other hand, tariff reductions will increase wood product exports from Asian and Latin American countries.[20] In general, in response to trade liberalization of forest products, the above studies predict that there will be an increase in wood product trade in developing countries. In terms of pulp and paper products, Canada, the United States, and the

European Union would remain as major players in the global pulp and paper markets.

In a regional context, the United States and Canada are significant partners in forest products trade. Over 70 percent of Canada's forest product exports find their destination in the United States, so the U.S. market is clearly critical for Canadian producers. On the other hand, imports of forest products from Canada account for over 60 percent of total U.S. forest product imports indicating that Canadian forest products are critical in meeting U.S. demand. This suggests that these two nations have market power to influence bilateral trade in forest products. Figure 3 explains how trade restrictions impact output, exports, and prices and sheds light on the on-going bilateral trade dispute relating to softwood lumber.

In the absence of trade, Qc is the quantity produced/consumed at a market price, Pc, in Canada while Qu is the quantity produced/ consumed at a market price, Pu, in the United States. In the presence of freer trade, the excess supply, ES, from Canada and excess demand, ED, from the United States results in a bilateral price, Pw. As a result, Canadian exports Qe quantity to the United States and consumes only, Qct. On the other hand, U.S. consumption will increase from Qu to Qut, and domestic production will drop from Qu to Qu' because of imports from Canada. In the presence of a tariff on Canadian imports, the excess supply curve shifts up thereby resulting in a new bilateral price, Pwt. As can be noticed from the figure, this new price will decrease Canadian exports from Qe to Qe' and expand production in the United States. This is the reason why U.S. forest producers lobby hard for tariffs on Canadian softwood lumber. They allege that Canadian forest firms have unfair advantage through lower stumpage rates and lower environmental restrictions.

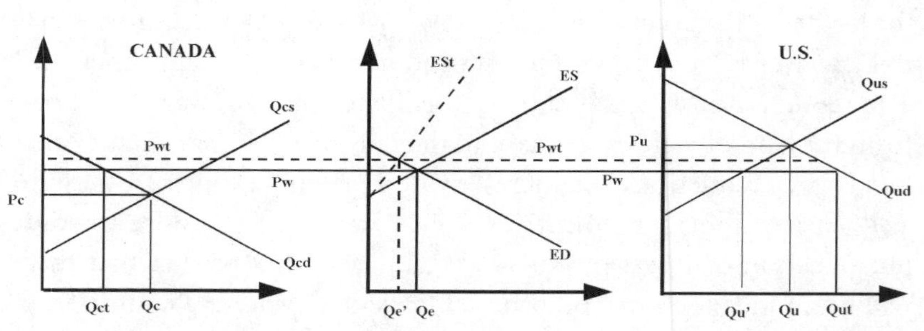

Figure 3. An Illustration of Trade Restrictions between Canada and the United States

The dispute, which started in 1982, had gone through a series of bilateral agreements (see Box 1 for historical details of the dispute). On April 1, 2001, Jimmy Carter, the thirty-ninth president of the United States, wrote an article that Canadian "subsidized" imports would hurt timber producers in the United States (see Appendix A for details).

It is unlikely that this dispute will end as long as Canadian imports affect the profitability of U.S. producers.[21] The impacts of this dispute are quite significant on the welfare of both nations. Alavalapati et al.[22] conducted a general equilibrium analysis to examine the impact of a 1 percent increase in tax on Canadian lumber and wood products exports into the United States. They found that forest products' output would drop in Canada and increase in the corresponding sectors of the United States and other regions of the world. General equilibrium effects suggest an increase in output of other sectors in Canada, and a decrease in the corresponding sectors of other regions. Although forest firms in the United States would benefit from trade restrictions, the authors found a decrease in U.S. welfare and an increase in Canadian welfare in response to a 1 percent tax on Canadian lumber and wood products exports while global welfare is would to drop by 5.78 $US million.

In another study, Zhang[23] investigated the welfare implications of the 1996 United States-Canada softwood lumber agreement in a partial equilibrium setup. The analysis shows that a welfare transfer occurred from U.S. consumers to producers in both the United States and Canada. Thus, on the whole net loss to the United States was about 4.7 $US billion (estimated costs to U.S. consumers are 12.5 $billion) whereas Canadian producers' net gain was 3.1 $US billion. As consumers are expected to bear the costs of the lumber dispute, in the form of higher forest product prices, they oppose the arguments of U.S. forest companies. In fact, the U.S. National Home Builders Association supports Canada on this bilateral trade issue. On the other hand, in the interest of reducing timber harvests in Canada, many environmental organizations support U.S. forest companies in their lobbying efforts. With more political clout in the U.S. Congress, forest products firms in the United States are likely to intensify their efforts to bring a new set of regulations on Canadian lumber imports. In other words, further lobbying efforts by the lumber producers in both the United States and Canada could generate effective rent-seeking activities.

Text continues on page 262

Box 1 History of Bilateral Agreements on Softwood Lumber between Canada and the United States, 1982-2002

Lumber I Oct. 1982	The U.S. industry first petitioned against Canadian softwood lumber imports under the U.S. countervailing duty law, alleging that Canadian stumpage prices constitute a subsidy.
May 1983	After a full investigation, the U.S. Department of Commerce concluded that stumpage did not confer a subsidy.
Lumber II May 1986	U.S. industry again petitioned for countervailing duties. The U.S. Department of Commerce (USDC) found that Canadian stumpage systems conferred a weighted average subsidy to its lumber producers by approximately 15%.
Dec. 1986	Canada negotiated with the U.S. further and entered into a Memorandum of Understanding (MOU) to resolve the dispute.
1986 MOU	Pursuant to the MOU, Canada agreed to collect a 15% export charge on all exports of lumber from Canada. The MOU also provided that provincial governments could reduce or eliminate the 15% export charge by implementing so-called "replacement measures", defined as increased stumpage or other provincial charges on softwood lumber production. Both B.C. and Quebec chose to adopt replacement measures over the export charge.
Lumber III Oct. 1991	While keeping all measures in place, Canada wanted to exercise its contractual right to terminate the MOU. The USDC self-initiated a new countervailing duty investigation by imposing a temporary bonding requirement on imports.
May 1992	The USDC issued its final affirmative determination on subsidization. It found that the forest management programs in B.C., Alberta, Ontario and Quebec, and the log export controls imposed by B.C. conferred countervail subsidies.
July 1992	The USDA's subsidy determination was followed by the issuance of a final affirmative determination on injury by the U.S. International Trade Commission (ITC). As a result, final countervailing duties of 6.51% were imposed on lumber imports from Saskatchewan, Manitoba, the Yukon Territory and the Northwest Territories.

Aug. 1992	Canada appealed both determinations on subsidy and injury to bi-national panels established under the Free Trade Agreement (FTA). The panel remanded the decision twice back to the USDC because of lack of sufficient evidence or legal basis. After the second remand, the USDC accepted the finding that stumpage and log export restrictions were not countervail subsidies and terminated the countervailing duty order.
Aug. 1994	The United States trade representatives requested the establishment of an Extraordinary Challenge Committee (ECC) under the FTA alleging conflict of interest on the part of the two Canadian panelists. The majority of the ECC supported Canada.
Dec. 1994	USDC terminated the countervailing duty and ordered liquidation of all entries. All countervailing duties (CVDs) tentatively paid by Canadian exporters (~$800 million) were refunded. Canada agreed to participate in further negotiations.
May 1996	Canada and the U.S. finalized an agreement on softwood lumber covering the five-year period to March 31, 2001. Under this agreement, the first 14.7 billion board feet of softwood lumber from British Columbia, Alberta, Ontario, and Quebec enter the US duty free. Over this quantity, a tax at the rate of $50 per thousand board feet up to the first 650 million board feet and a $100 tax per thousand board feet beyond will be applied.
Mar. 2001	The Canada-U.S. Softwood Lumber Agreement expired.
Mar. 2002	The U.S. Department of Commerce (DOC) calculated a single country-wide subsidy rate of 19.34% to be applied to all producers and exporters of softwood lumber products covered by CVD investigation. CVD duties are applied on a final mill basis. The DOC determined that certain softwood lumber products from Canada are being sold, or are likely to be sold, in the U.S. at less than fair value. DOC calculated an 'all-other' basket dumping margin of 9.67% and separate individual company margins.

At this writing, the U.S. trade restrictions have been in effect for about one year. In an unprecedented move, all four Canadian opposition parties agreed that the ruling Liberal Party needs to do more to help beleaguered forestry workers, firms and dependent communities. The Canadian government has received a favorable ruling from the WTO and is awaiting a ruling (that would have more impact) from NAFTA. In the meantime, the government has proposed a package of loan guarantees to keep Canadian forestry firms at risk from going out of business until these anticipated favorable rulings reverse the current duties. History has shown, however, that this issue goes away only temporarily, if at all. It has cast a serious, negative pall on U.S. and Canadian relations, and will likely continue to do so for the foreseeable future.

Impact of Forest Biotechnological Innovations

Growing concern for the environment and rising demand for outdoor recreation and other ecological services are prompting government agencies to limit timber production on public lands and increase forest set asides. For example, on April 4, 2001 the government of British Columbia announced an agreement of forestry companies, First Nations, and environmental organizations to protect 2,316 sq miles, encompassing twenty watersheds from logging. Another 2,000 sq miles representing sixty-eight valleys, has been designated as 'deferred' until further study can be conducted in the subsequent twelve to twenty-four months to establish their conservation priority status. In the United States, the bill H.R. 1494 entitled "National Forest Protection and Restoration Act", also known as the McKinnery/Leach Bill, proposed an elimination of commercial logging on federal public lands in the United States. Such a policy would reduce U.S. timber supply by approximately 5 percent. With no signs of a reduction in the demand for wood products,[24] forest setasides and timber harvesting restrictions will widen the gap between the supply and demand for forest products.

Adopting biotechnological innovations[25] and intensive fiber management through tree plantations are potential ways to meet the rising demand for forest products while simultaneously satisfying demand for forest setasides.[26] Some argue that this approach would help conserve tropical forests by reducing the need for imported tropical timber. Specifically, biotechnology could contribute to improvements in the quality of fiber and increased opportunities for product differentiation, while directly producing environmental benefits such as increased diversity.[27] Inventions such as herbicide and insect tolerant seedlings, pest and disease resistant seedlings, and improved fiber properties (to name a few) have already resulted in significant productivity gains in the forest sector. However, a variety of institutional and economic factors are inhibiting the promotion of forest biotechnologies. For example, policy-related uncertainties, the public good nature of forest lands and products, the longer time frame needed to create new tree varieties, and tenuous land ownership structures are causing investment levels to be lower than those considered socially optimal.[28]

Despite these problems, during the last few decades there has been considerable improvement in three areas of forestry biotechnology, namely, improvement of trees; biopesticides for forest management; and propagation, conservation and restoration.[29] For example, biotechnology development, through tissue culture and somatic embryogenesis, molecular genetics, genetic engineering, monoclonal antibody techniques and manipulation of important biological processes at the DNA level, offers innovations to tackle the longer time problem. Manipulation of tree form and structure, increased photosynthetic efficiency, development of herbicide resistance seedlings, manipulation of physical, chemical and structural features of the cell wall, are important silvicultural and biological processes that contribute to the economic productivity of commercial forestland.[30]

Sedjo[31] documents four major types of genetic research pertinent to forestry. They include: 1) breeding techniques and hybridization (growth-enhancing hybrid poplars in the U.S. South, for example); 2) application of molecular biology to enhance efficiency of traditional breeding and also to identify and modify genes for better biochemical performance; 3) cloning techniques, which are conduits for replicating the superior traits via gene transfer; and 4) tree improvement programs for desirable attributes such as disease and pest resistance, growth performance, quality of wood fiber and tree form, and good fiber characteristics related to convenience of processing. Managing intensively to produce such desired traits would produce significant cost-savings. The potential economic benefits of biotechnological innovations are also well documented in the literature. For example, Sedjo[32] estimates that traits such as fiber length and uniformity would save \$10 per m^3 in pulp mills. Herbicide resistance benefits are estimated to be around $35 per acre and $160 per acre, respectively, for fast-growing softwoods and hardwoods. The Bt pest resistance benefits from insect tolerance are estimated to be $375 million annually.

In situations where countries are linked through international trade, the benefits of biotechnology would extend beyond national boundaries. Das and Alavalapati[33] examine how the trade-related technology flows from the United States could influence the trade and welfare of various regions. In particular, they use a multi-regional, multi-sectoral model[34] to analyze the potential impact of biotechnology in the logging sector and its potential impact on other sectors and regions in Canada, the United States, and the rest of the world. The model consists of seven regions— United States, Canada, Western Europe, South America, South East Asia, Japan, and Rest-of-the-world—and seven sectors—Agriculture, Manufacturing, Natural resources, Logging, Lumber and wood products, Paper products and publishing, and Services. Parry, Das and Alavalapati expect a 0.6 percent multi-factor productivity increase (i.e., equal magnitude of productivity improvement in all primary factors of production so that the technical change has no factor-bias) in the U.S. logging sector in response to biotechnology and estimate its impacts on output, prices, global trade, and welfare.[35]

Tables 1 & 2 show the production implications and economic welfare impact of biotechnological advancement. At the sectoral level, Das and Alavalapati[36] observe that agricultural and forest production, and exports from the United States, Canada, and Western Europe will increase in response to biotechnology advancement. Regional income, exports, and welfare will improve as a result of technical progress and its transfer to other regions via intermediate inputs.

The above results suggest that the technological change in the U.S. logging sector and its transfer into other regions via trade linkages will improve total factor productivity, thereby improving regional income and welfare. The United States and Canada would muster greater benefits relative to other regions as a result of biotechnology change. Conversely, South East Asia and South America would experience relatively fewer transmitted benefits of biotechnological advancement and thus do not show much growth in output and productivity according to the model. The wood products and pulp and paper sectors, which use the logging sector's output intensively, register higher growth in output. Thus, on the whole, forestry biotechnology has a potential to meet the rising demand via higher production and trade volume.

Table 1. Impact on Production and Exports of 0.6% Biotechnology Change in the U.S. Logging Sector*

	U.S. Production	*Exports*	*Canada Production*	*Exports*	*W. Europe Production*	*Exports*
Agri. Products	2.20	3.58	2.43	2.89	0.84	2.26
Forestry Products	2.80	2.34	1.60	0.47	0.28	0.15
Lumber	3.02	1.43	1.30	0.63	1.29	0.10
Pulp and Paper	2.55	2.66	2.37	2.09	1.40	1.52

All the figures are in percentage changes
* Source: Das and Alavalapati, "Trade mediated biotechnology transfer."

Table 2. Impacts of 0.6% Biotechnology Shock in the U.S. Logging Sector on Income, Exports, and Welfare in Selected Regions*

	Regional Income	*Exports*	*Welfare (in US$ million)*
United States	2.57	1.63	147558.12
Canada	2.95	2.82	13716.05
Western Europe	2.69	1.1	174115.70

All the figures except for welfare are in percentage changes.
*Source: Das and Alavalapati, "Trade mediated biotechnology."

Many environmental groups, however, oppose the use of transgenic crops in agriculture as well as in forestry. Their concern is that genetically modified organisms (GMOs) can have adverse effects on human and ecosystem health. Furthermore, they believe that the use of herbicides, super-resistant pesticides or weeds, and fertilization, increases the risk of allergenic or toxic reactions. Critics of biotechnology development believe that GMOs will affect global biodiversity and global weather patterns. Earlier this year the Biosafety Protocol (an extension of the 1992 Convention of Biodiversity) was adopted to refuse trans-border shipments of GMOs. Increasing health and safety concerns may thwart future developments in biotechnology. In addition, more regulation and/or clear lack of public support in this area may slow the growth of biotechnology applications and adoption. Although GM trees are not as widely used as compared to GM agriculture crops, the opposition is still in line with the arguments put against GM crops. Several incidents in which genetically modified trees were vandalized have occurred in both the United States and Canada. According to Kaiser, the ecological risk of using the Bt proteins (such as pollen spread) and GM trees lies in the mode of usage.[37] It seems that there is no disagreement about the potential of biotechnologies in furthering productivity. What is at stake is ensuring the long-term safety and garnering popular support for GMOs. There are many indications that public support for GMOs is waning faster than it is growing.

Conclusions

Growing concern for the environment and rising demand for wood products and outdoor recreation are giving rise to a myriad of changes in the United States and Canada. Many of these changes are macro in nature with implications to the region of origin, as well as other regions across the globe. In this chapter, we review three emerging issues—forest carbon sequestration, trade liberalization and bilateral trade disputes, and forest biotechnology—that we believe will affect the use and management of forests in the United States and Canada. We have not attempted to provide a comprehensive treatment of these issues, but rather suggest that these are some of the forest issues to watch in the future. These issues influence the comparative advantages of timber production and processing and thus provide a major stimulus for foreign direct investment (FDI) in the form of purchasing or establishing production, warehouse, marketing, or consulting capacity.[38] For example, FDI from U.S. forestry firms quadrupled from 1982 to 1998.[39] FDI represents 38 percent of total forest industries in Malaysia, with the majority of the companies coming from Japan, the United States, Singapore and Taiwan.[40] FDI projects to be completed in Argentina over the next decade amount to US$2 billion with major participation by multinational companies from the United Kingdom, the United States, New Zealand, Germany and Chile. In the face of growing volume of FDI flows and increasing environmental regulations, capital flows from Canadian and U.S. forest sectors to different locations might occur. FDI flows and burgeoning volume of international trade impact consumption and production of timber products. Although both the United States and Canada are key players in international forest products trade, the situation is slowly changing with the increased participation of other countries. For example, between 1988 and 1999, Indonesia expanded its paper production capacity from 1.2 to 8.3 million tons per year and its pulp production capacity from 0.6 to 4.9 million tons per year. During the past decade, Indonesian paper exports doubled and pulp exports increased tenfold.[41] Furniture exports grew by over 37 percent annually in Brazil during the 1990s.[42]

Another factor that might influence the future of forest sectors in the United States and Canada are mergers and acquisitions of major forest products firms. Through mergers companies are getting bigger and more powerful and national boundaries are being steadily eroded.[43] Over thirty major international acquisitions/ mergers occurred in the forest products industry between 1995-2000.[44] Now two forestry companies (Stora Enso in Finland and SCA in Sweden) rank among the hundred largest transnational companies by foreign assets.[45] More than 160 joint ventures took place between office furniture makers in China and firms based in the European Union in the mid-1990s.[46] These trends indicate that factors affecting globalization will have significant implications for the future use and management of forests in the United States and Canada. The increasing power and scope of these international forest products interests may challenge the ability of provincial, state and even national governments to set policies according to the values of their own citizens. Some would argue that this situation already exists.

Finally, in order to analyze the future impacts of the issues discussed in this paper in an objective and unbiased way, we offer several suggestions. First, attention must be paid to both intended and unintended consequences. Often, we run into situations where unintended effects of changes will outweigh the anticipated effects. For example, the uncertainty and risks associated with biotechnology may outweigh the anticipated benefits.

Second, analyses must consider both short-run and long-term impacts and market and non-market values. Some changes may be associated with more benefits up front and huge costs in the future. The converse may be true for other policies. Opponents of biotechnological innovations perceive that immediate benefits, in the form of increased productivity may not offset the costs associated with human health and environmental changes in the long-term. Furthermore, many analyses pay attention only to market outputs and inputs (timber and fertilizer for example) and leave non-market outputs and inputs (clean air and water, biodiversity etc.) unexamined. Failure to incorporate non-market values in the analyses may either underestimate or overestimate the impacts of these changes.

Third, we need to be aware that the forest sector is linked to other sectors within a region, and that regions are linked to one another through trade. Examining the effects of policy changes from a partial equilibrium framework assumes that changes in the forest sector do not affect other sectors in the economy. This may not be true. In response to changes in the forest sector, a variety of trade-off or multiplier effects can be expected in other sectors. We noticed that the creation of tree plantations for sequestering carbon, for example, would cause a reduction of agricultural land use in the United States. With respect to regional trade linkages, it was shown that the benefits of biotechnology changes in the U.S. logging sector will spill over to other regions through traded intermediate inputs. Failure to capture intersectoral and interregional linkages may lead to biased estimates, and policies based on these estimates will be erroneous.

Finally, we must recognize that sustainable forest management is a moving target and is a function of societal preferences. Demographic, socioeconomic, and technological changes will continue to influence societal preferences, thus influencing the principles of sustainable forest management. What is cherished today may not be appreciated tomorrow. Our belief and value systems influence the way we use resources, and thereby influence policy development. Much of the discussion presented in this chapter is based on an anthropocentric philosophy, where resource use is justified to further human welfare. However, concern for the environment and wider recognition of intrinsic values for nature may have serious impacts on future policy development.

Social scientists and policy makers should continue to track developments related to the forestry issues examined in this chapter. In addition, they should continue to monitor changes in social values related to forests in particular and our relationship with nature more generally. Sustainable forestry involves a myriad of complicated trade-offs. One role of social science in the sustainability debate is to outline clearly what such trade-offs are likely to be and what potential consequences they entail. We have attempted to do just that in this chapter through economic models that demonstrate the projected effects of policy and behavioral change. However, the social and political landscape with respect to trade in forest products, managing forests for carbon, and biotechnology is constantly shifting, and therefore needs continued study and monitoring by the social science and forestry communities.

Appendix A

A Flawed Timber Market

by Jimmy Carter, the thirty-ninth president of the United States

Along with all the other former presidents, I was a strong supporter of the North America Free Trade Agreement when it was initiated in 1994. Free trade among the United States, Mexico and Canada has, in general, been good for the people and the economies of all three nations. However, we are now facing a crisis in the marketing of lumber that could be devastating to 10 million American landowners, 20,000 sawmill owners and more than 700,000 workers, and also to the environment. This problem has aroused the concern of labor, industry and environmentalists. There are many facets to this complicated issue, but they can be summarized in relatively simple terms.

In Canada, the national and provincial governments own 95 percent of the timberland. In the United States, private investors own the overwhelming portion of woodlands. In Georgia, for instance, 70 percent of all forestland belongs to about 600,000 private nonindustrial owners, most of whom are also involved in farming. Timber companies like Weyerhaeuser, Georgia-Pacific and International Paper own another 20 percent, while the remaining 10 percent is in public ownership, as in parks and military bases.

Rosalynn and I are typical family landowners. On our relatively small woodland tracts, some of which our family has owned for seven generations, we maintain a proper mix of hardwood and softwood trees for optimum wildlife habitat, and we market our timber selectively when it reaches full maturity. We cut relatively small areas at a time and replant as quickly as possible after harvesting. Within 10 years, we begin periodic thinning, always providing the best conditions for optimum growth of the next generation of full-grown trees.

When we sell some mature trees, we obtain bids from sawmill owners, who are under contract to cut under strict conditions that protect the permanent value and productivity of the farm. It is an almost universal practice of families like ours to protect the land from erosion and to replant another crop immediately after harvest. Our sawmills must pay full market price for standing timber, saw and dress the lumber as efficiently as possible, and sell it on the retail market.

Canada has no equivalent free market for the overwhelming portion of its timber. Provincial governments grant an annual allowable cut to sawmill owners at whatever low price is necessary to maintain full employment in the timber industry. These sawmills usually pay a fraction of the price that American sawmill owners pay, creating a great disparity that is beginning to wreak havoc with the timber industry in the United States, from the farm family that owns some woodland to the small or large sawmill owners who cannot compete on the retail market with the heavily subsidized lumber being imported from Canada.

These disparities between the American and Canadian timber industries have existed for more than 25 years. In 1996, the United States and Canada signed a five-year pact, which expires this month, that tried to limit the problem by restricting Canadian exports of lumber into the United States. But quotas are not the answer. What we need is a permanent agreement that ensures free trade but ends the artificial price restrictions that the Canadian government has put on timber. This will allow both Canadian and American lumber interests to compete on equal footing.

Without a dependable timber market in the United States, many landowners cannot afford to invest in reforestation and forest maintenance, and the consequence will be land that is barren or converted to other uses. The cost to society is great - less carbon dioxide sequestered in the trees, a loss of air and water filtration, less green space and wildlife, and more soil erosion and urbanization.

Our family has other personal income and can survive even if our nation's timber industry is crippled, but hundreds of thousands of American families depend on a fair and stable market for their livelihood. Their interests must be protected.

Source: *The New York Times,* March 24, 2001.

15

Building Innovative Institutions for Ecosystem Management: Integrating Analysis and Inspiration

George H. Stankey, Stephen F. McCool, and Roger N. Clark

Introduction

> "Right now we're doing a lousy job of 'it': ecosystem management. Do I find that disappointing? No, I don't, because it's an incredibly new thought, new concept."[1]

> "I thought multiple use was difficult to implement, but ecosystem management is impossible—what am I supposed to do?"[2]

In 1992, U.S. Forest Service Chief Dale Robertson announced adoption of a policy of ecosystem management. This policy would utilize an ecological approach in achieving multiple use management to blend the needs of people and environmental values in such a way as to ensure the diversity, health, productivity, and sustainability of the National Forests. He defined ecosystem management as the skillful, integrated use of ecological knowledge at various scales to produce desired resource values, products, services, and conditions in ways that sustain the diversity and productivity of ecosystems.

Other federal land management agencies followed suit. For example, the U.S. Fish and Wildlife Service (FWS) argued that while an ecosystem approach would not change its underlying mission—to conserve, protect, and enhance fish and wildlife and their habitats for the continuing benefit of the American people—it would change how that mission was pursued.[3] This would involve 1) increasing agency effectiveness in conserving fish and wildlife, 2) improving cross-program coordination within the FWS, and 3) increasing the quality and quantity of partnerships with external stakeholders. Grumbine[4] applauded the emphasis on ecosystems by land management agencies, particularly recognizing the importance of partnerships among citizens, managers, and scientists. However, he lamented the apparent lack of awareness of the radical implications associated with creating such partnerships. Along with others,[5] he cited the lack of effective institutions as a major constraint to successful implementation of ecosystem management.

The search for new institutions to address complex ecosystem-based problems reflects an interest in innovative governance across all sectors and a common desire to improve the human condition.[6] There is widespread concern with expanding access to, and involvement in, public decision-making processes and for improving the state of knowledge that informs public policy decisions. This is especially germane given the increasing complexity of issues facing modern society. Pierce and Lovrich[7] observe that the interface between pressures for expanded public participation in decision-making processes and the technical complexity of many issues confronts society with a quandary: "how can the democratic ideal of public control be made consistent with the realities of a society dominated by technically complex policy questions?" As society devotes efforts and resources to understand such complexity better, it will need to make an equal commitment to developing effective and equitable responses founded on accepted definitions of democratic governance. Moreover, as such definitions evolve, governance institutions must demonstrate a parallel capacity to adapt and respond.

The problematic nature of the institutions needed to elevate ecosystem management from rhetoric to a substantive change in how resource management is undertaken, forms the focus of this chapter. We first discuss the concept of institutions, briefly review historical developments that have altered the context within which natural resource management takes place, and analyze how these changes have given rise both to a dissatisfaction with existing management strategies and the search for a new way of doing business. We argue that increasing scientific complexity and political ambiguity combine to form a new class of problems confronting natural resource management that are largely resistant to resolution by traditional institutional organizations and processes. They also highlight the importance of problem framing[8] and achieving an appropriate match between problem and institution. Although the nature of ecosystem management makes it unlikely that any single "best" institutional model exists, our discussion seeks to introduce and describe some key concepts and frameworks that will help in the search for effective solutions.

Institutions and Natural Resource Management

In the broadest sense, institutions embrace the rules, structures, and standards—formal (e.g., tax laws) and informal (e.g., a handshake)—that society employs to achieve desired ends. More formal definitions emphasize a "pattern of expected action of individuals or groups enforced by social sanctions" and "normative patterns embedded in and enforced by laws and mores."[9] Given the breadth of the definitions and the scope of changes confronting institutions, Cortner et al.[10] offer the simple definition that "It is through institutions that humans search for the means to solve social problems." However, there is little agreement at the conceptual level as to what the term means, how such entities might be studied, or the steps required to change them.[11]

A brief review of American history reveals an ongoing search for means to solve natural resource management problems. For example, Dana and Fairfax[12] and Cortner and Moote[13] describe how changing public values, statutory developments, and political realities have reshaped the institutional landscape. Recent years have witnessed diverse proposals calling for reform of natural resource management institutions in the United States. Clarke and McCool[14] proposed amalgamation of the Bureau of Land Management (BLM), Bureau of Reclamation (BoR), and FWS into a Western Ecosystems Management Agency while Wood[15] proposed combining the Forest Service and BLM. O'Toole[16] has called for sweeping reforms of the Forest Service and the various legislative mandates that direct its actions. Nelson[17] has called for divesting much of the current National Park System to state and local governments and non-profit organizations.

Such proposals are grounded in an implicit assumption that changing the formal institutional structures constitutes an adequate means of achieving desired change. However, the array of informal norms, customs, and practices encompassed by a broad definition of institutions might be equally, or even more important to achieving change.[18] Thus, in the search for effective means through which an ecosystem management approach is implemented, there needs to be equal attention to formal structures and processes (e.g., administrative organizations, policies, laws) *and* to the array of informal processes and norms that operate within and between formal entities.

Despite substantial agreement about the need for institutional reform to implement ecosystem management, there is less agreement on what needs to be reformed, and what would facilitate or constrain such reform. Current processes often implicitly assume a single decision-maker, addressing a well-defined problem with perfect information and unlimited time to develop a strategy. However, real world conditions typically are the opposite, with multiple actors attempting to arrive at decisions involving poorly defined problems, with imperfect information, and facing tight deadlines.[19] Such situations also often involve structural distortions in political power, resulting in equity impacts that often are overlooked or ignored.

If current institutions are inadequate to practice ecosystem management—however problematic its definition—two questions present themselves. First, what are the particular characteristics of the problem situation that make existing institutions inadequate? Second, what might more effective or responsive institutions look like? To answer such questions, it is helpful to review briefly the historical context within which current institutions were created and the nature of the changing social and political context that now challenges their efficacy.

From Multiple Use to Ecosystem Management

Over the last several decades, increasingly acrimonious challenges have confronted natural resource management, especially forestry, in the United States (as well as globally). On the National Forests in the United States, the concept of multiple use dominated for many years. The doctrine reflected an effort to achieve a balance among complementary and competitive uses; the 1960 Multiple Use-Sustained Yield Act directed that management be "harmonious and coordinated . . . without impairment of the productivity of the land, with consideration being given to the relative values the various resources, and not necessarily the combination of uses that will give the greatest dollar return or the greatest unit output."[20]

However, the 1960s saw growing criticisms of the concept.[21] At their core was concern that despite the rhetoric of multiple use, in reality, a philosophy of dominant use prevailed, focused on commodity production.[22] There was also criticism of the levels of discretion accorded agency decision-makers; Wondolleck[23] cites a federal judge's comment that the Multiple Use-Sustained Yield Act "breathes discretion at every pore."

In 1970, in response to mounting public pressures, a University of Montana review committee was charged with assessing forest policies and practices on Montana's Bitterroot National Forest. The report concluded, "multiple use management, in fact, does not exist as the governing principle on the Bitterroot National Forest."[24] The Committee's chair and Dean of Forestry Arnold Bolle quoted Professor William Duerr from the New York State College of Forestry: "The subculture of forestry is out of step with American culture."[25]

During the 1970s, Congressional intervention in this debate heightened with passage of the Resource Planning Act (RPA) in 1974 and its subsequent amendment in 1976, the National Forest Management Act (NFMA). NFMA in particular began to reshape the forest management policy debate, instigating a national planning effort. The decade of the 1980s witnessed an extraordinary investment in development of individual plans for the 155 National Forests, but with nightmarish results; while Forest Service officials estimated the process would generate perhaps two to three hundred appeals, the first seventy-five plans filed generated six hundred appeals.[26] Because the planning process remained primarily a technical enterprise, lacking adequate, substantive, and representative participation by the public whose values, interests, and uses would be affected by those plans, the political constituency needed to support the agency's vision failed to materialize.

If the 1980s represented an era of forest planning, the 1990s heralded an era of scientific assessments. Triggered by concerns about the status of the northern spotted owl and its old-growth forest habitat in the Pacific Northwest, Congress charged a scientific assessment team to develop a scientifically credible management plan. Their report[27] heralded a series of science assessments across Canada and the U.S. to build management policies and processes grounded in the best-state-of-scientific knowledge.[28] Ecosystem-based management emerged as a central element in many of these efforts.

An ecosystem approach typically contained three key elements:[29] (1) a larger spatial and longer temporal scale than previously appreciated; (2) a focus on the relationships among biotic and abiotic systems, rather than individual species; and (3) a focus transcending traditional physical, administrative, and disciplinary boundaries. In summary, "ecosystem management requires collaboration and cooperation among a greater number of agencies and stakeholders involved in or affected by ecosystem management. . . ."[30]

Two schools of thought emerged as ecosystem management gained currency in the natural resource management debate. One held that ecosystem management represented nothing new, and that the term would only confuse the public.[31] An alternative perspective argued that ecosystem management would require such a change in thinking and behavior as to qualify it as a paradigm shift.[32] Lackey[33] concurred that implementation of an ecosystem-based approach would require a systemic shift in world view, but one, he concluded, unlikely in the current institutional and legal environment. Based upon a literature review, Moote et al.[34] synthesized five core principles distinguishing ecosystem management from its traditional predecessors:

- Socially defined goals and management objectives
- Integrated, holistic science
- Broad spatial and temporal scales
- Collaborative decision building
- Adaptable institutions

Although the temptation is to draw an either-or contrast between traditional resource management planning processes and ecosystem management-based approaches, any planning process reflects an adaptation to learning, to new knowledge, to changing public values and political contexts, etc., and thus represents a blend of old practices and new thinking. Nevertheless, it is possible to discern how this new thinking has evolved. Treating these planning models as archetypes, Bengston[35] depicts the key differences between traditional and ecosystem management approaches (see Box 1).

The increasing inability of natural resource management authorities to respond effectively to emerging issues and challenges[36] has stimulated a search for an alternative planning model, with increasing interest in ecosystem management as the elemental framework for such an alternative. Yet, the potential value of an ecosystem approach remains muted by the lack of effective institutions through which it could be implemented. Despite a growing consensus that current institutions lack sufficient capacity to achieve the objectives of an ecosystem-based approach, the specific nature of institutions that might replace them remain arguable.

Scientific Uncertainty, Political Ambiguity, and Decision-making

What are the distinguishing characteristics of the scientific and political context within which an institution intent on implementing an ecosystem management approach must operate? Thompson and Tuden[37] argue that the decisions any institution makes are influenced

Box 1. Traditional Resource Management Compared to Ecosystem Management

	Traditional management emphasis	*Ecosystem management emphasis*
Philosophical base	Utilitarian	An Aldo Leopold ethic
Objectives	Maximize commodity production	Maintain the ecosystem as a whole, provide for sustainable commodity production
	Maximize net present value	Maintain future options
Constraints	Yield of outputs must be less than or equal to their periodic growth	Long-term ecosystem sustainability Maintain esthetics Management practices are socially acceptable
Role of science	Views management as applied science	Views management as combining science and social values
Value	Instrumental Single or dominant value	Instrumental and intrinsic Plural values
Major themes	Focus on outputs	Focus on inputs and processes
	Management consistent with industrial production processes	Management mimics natural processes
	Single outputs (e.g., timber primacy)	All species important
	Resource scarcity (e.g., timber famine)	Biodiversity loss
	Mechanistic, reductionist view of resource system	Systems view of world
	Scale: typically site- and stand-level	Ecosystem and landscape
	Planning unit governed by political or ownership boundaries	Planning units ecosystem-based
	Economic efficiency	Cost-effectiveness, social acceptability and equity

(Adopted from Bengston 1994, p. 517)*

*These comparisons do not represent an either/or situation; rather, these attributes help reveal the shift from what was the traditional emphasis to what is the emphasis in ecosystem management. In short, ecosystem management does not necessarily reject attributes of traditional planning, but augments them with new perspectives.

by the status of two elements—causation and preference for outcomes—and whether there is agreement or disagreement within the institution about those elements. That is, do we understand cause-and-effect? Do we understand what will happen if some action/policy is taken or how past actions explain the present state? Such questions concern the level of scientific understanding that characterizes some phenomenon. Similarly, is there agreement or disagreement over which outcomes are preferred and how are those preferences distributed? Thus, when a decision is made, there can be differences over means (causation), ends (preferences), or both.[38]

Thompson and Tuden add two companion ideas to their matrix: each decision situation requires a distinctive decision strategy, and, for each strategy, a distinctive organizational structure is best suited for making a choice (Figure 1).

In each cell, the first line denotes the type of decision process (DP) most appropriate for the conditions depicted for causation and preference for that cell; the second line identifies the most appropriate institutional structure (I) for implementing that strategy. For example, in Cell A, where agreement exists on both causation and preferred outcomes, a computational approach is appropriate. Problems are technical or mechanical and decisions regarding them are routinized and relatively straightforward. Under such conditions, the most effective institutional structure to facilitate decision-making is the bureaucracy, embodying specialists operating in a formal, centralized, and hierarchical manner.

Cell B depicts a situation where disagreement exists over causation, but consensus exists on the preferred outcomes. Because of both

Beliefs about Causation	Preferences about Outcomes: Agree	Preferences about Outcomes: Disagree
Agree	DP: Computation I: Bureaucracy A	DP: Bargaining I: Representative C
Disagree	DP: Judgment I: Collegium B	DP: Learning and consensus building I: ? D

Figure 1. Decision-making under Varying Conditions of Agreement about Causation and Preference. DP = Decision process. I = Institution.
After Thompson and Tuden, "Strategies, structures, and processes."

differential problem perception as well as differential interpretations of the facts, the collective wisdom of experts is required to understand the problem. For example, in implementation of the Northwest Forest Plan, although there was substantial agreement among scientists and managers on the goal of restoring old-growth forest conditions, there was disagreement on how such conditions could best be achieved. Through a collaborative effort among a group of experienced scientists, a strategy was developed that would improve understanding of the cause-and-effect relations between silvicultural treatments and old-growth conditions. This example illustrates the preferred decision-making strategy of majority judgment, exercised through the institutional structure of the collegium.

Decisions in Cell C face agreement on causation, but disagreement on preferences. When there are conflicts over the values and uses of the resource, some (winners) will achieve their preferences while others (losers) will not, making it important that decision-makers are aware of the equity implications of any eventual decision. Because this process involves bargaining and compromise among competing values and interests, it must be supported by a decision-making structure that represents, as much as possible, the full range of values and interests at stake. It must also support venues for the dialogue necessary for such negotiations. Thus, representative structures, such as state legislatures, are key for operating under these conditions.

In Cell D, we find a situation in which disagreement exists over both causation and preferences. Thompson and Tuden characterize the Cell D world as one of anomie and normlessness, where former goals and values have lost their meaning and relevance. "(T)his is a most dangerous situation for any group to be in; certainly by definition and probably in fact the group in this situation is nearing disintegration."[39] However dangerous Cell D conditions might be, they nonetheless typify many of the problems resource managers face today. Typically, the technical complexity of resource management challenges is high, characterized by uncertainty as well as a lack of consensus about problem definition and causation,[40] thereby confounding informed choice.[41] The high levels of scientific uncertainty are matched by equally high levels of political ambiguity;[42] in a pluralistic society in which the search for preferred outcomes is characterized by intense political rivalries and strategic posturing, the capacity to reach accord is problematic.

The disjunction between the Cell D context within which choices must be made and the current processes and structures that shape those choices, is sharp. That is, the computational processes and bureaucratic structures that dominate traditional planning (appropriate to Cell A problems) remain the principal means by which choices are made, despite a lack of agreement on either causation or preferences. This mismatch between problem and remedy contributes to the sense of institutional inadequacy widely reported, yet it also holds implications for how ecosystem management might best be implemented.[43]

There are two consequences of such mismatches. First, the failure to resolve the planning issue means that significant opportunity costs are incurred; scarce resources are wasted, and planning often must be repeated to accommodate the uncertainty and conflict ignored the first time. Second, such failures compound societal dissatisfaction with, and distrust of, the agency charged with resource stewardship, resulting in decreased institutional capacity to respond to future issues. It also produces political pressures to limit organizational discretion, replacing it with legislated and/or standardized procedures that often prove equally ineffective in dealing with the problems that derive from complex, ambiguous systems. So what processes and structures are appropriate to Cell D conditions? Thompson and Tuden[44] find little experience or theory to answer such a question. Rather, they call for creation of a "structure for inspiration"; i.e., an institution in which a charismatic leader can "rally unity out of diversity."[45]

In an essay on the role of complexity in policy analysis, Roe[46] takes sharp exception to this conclusion. ". . . '[I]nspiration' . . . sticks in the throat . . . . Analysis was once all the vogue, but increasingly inspiration is there to challenge it."[47] To the contrary, he argues that analysis remains the only viable option for survival in a world of complexity and ambiguity. "Issues of extreme uncertainty and complexity can be analyzed quite effectively without falling back to inspiration alone;"[48] he adds "the analytic methods required for these sometimes desperate situations are, however, not those taught in most of our methods courses and seminars. Yet we proceed ahead today as if the old methods will get us across this complex public policy terrain. When those peter out, we seek inspiration, a.k.a luck, leadership, intuition, or the high octane of political will."[49]

Roe's caveats about the utility of analysis are crucial. There is a growing conviction that traditional methods and principles of analysis are inadequate or even inappropriate in resolving Cell D problems.[50] Rittel and Webber[51] describe such problems as "wicked", using the term in a meaning akin to malignant or tricky, as opposed to malicious, while Ackoff[52] describes them as "messy" because of their interconnectedness. Such problems are never solved in the sense of finding an unequivocal, enduring answer because the social and political context constantly redefines the problem as values and preferences change. These problems share certain characteristics, such as reflecting a symptom of a higher order problem, involving situations in which trial and error learning has a limited role, and are subject only to solutions that are more or less useful.[53] Moreover, there is a concern that conventional analysis—technically sophisticated, involving complex algorithms—often is substituted for more reflective, value-based, and deliberative debate. Undoubtedly, analysis aids such debate, but only if used appropriately. Socolow[54] concludes "we should not be surprised to learn that the disciplined analyses brought to bear on a current societal dispute hardly ever do justice to the values at stake. Terribly little is asked of analysis." The key here is that although Cell D problems possess a component subject to analysis, "analysis cannot find the correct solutions."[55] Still, disciplined analysis can provide important keys to resolving socially problematic situations *once* we understand what things people care about.[56]

Roe also provides insight regarding how underlying systems operate, within the structural context identified by Thompson and Tuden. For example, building on work by Perrow,[57] Roe argues that systems vary in terms of the extent to which they are either tightly or loosely coupled. Tightly coupled systems are highly time-dependent, invariant in terms of the sequences underlying them (i.e., "B" cannot occur until "A" happens), inflexible in how objectives are achieved (i.e., there is one "right" way), and lack the capacity to tolerate delays or the unexpected. Loosely coupled conditions are generally the opposite; delays are possible and common, sequences are not invariant, there are many ways to achieve objectives, and there is a capacity to tolerate delays without imperiling the system.

Systems are also either complexly or linearly interactive. Complexly interactive systems involve the unfamiliar, unplanned, and expected, while linearly interactive involve the expected and familiar, and even when the unexpected does occur, it is both apparent and comprehensible. The Thompson-Tuden framework and the critiques offered by Roe and Perrow provide insight as to the nature of the decision environment facing society today and of the kinds of institutional structures and processes required to operate effectively in this environment. The following points summarize the situation.

First, the decision environment is characterized by high levels of scientific uncertainty and political pluralism; moreover, it is a domain of unpredictability, unfamiliarity, and often low levels of comprehension. Causes do not necessarily lead to predicted effects because the systems are not linked in a tight, deterministic fashion, but are loose and probabilistic. The inevitable surprises that result often lead to acrimonious debate not only over what happened, but also what caused it (why) and potential corrective actions (what would resolve it).

Second, the work of Thompson-Tuden and Roe reveal why the failure to recognize the decision context within which one is operating is dangerous. For example, when management actions are guided by an assumption that biophysical systems are tightly coupled and linearly interactive, when in fact, they are not, adverse consequences can result. For example, despite being managed in accordance with the prevailing theory of maximum sustained yield, Peru's anchovy fisheries declined 75 percent in two years.[58] The decline was not simply a result of overfishing, but more a result of flawed assumptions about how the underlying system operated. Rather than one governed by stability, order, and constancy (characteristics of tightly coupled systems), the system was loosely coupled and complexly interactive. The result of the mismatch between assumptions and reality led to strategies that jeopardized the long-term survival of the very species they were intended to protect.

Third, the two frameworks define the challenges with which effective institutional structures and processes must contend. Traditional reliance upon centralized, hierarchical control, or an assumption of order and predictability, results in institutions being ill equipped to deal effectively with complex, idiosyncratic, and ambiguous problems. For example, in reviewing seven institutional structures/processes through which society exercises choice (e.g., market mechanisms, law, administrative policies), Dryzek concludes "any 'winner' among the types of social choice would . . . be little more than the best of a poor bunch."[59]

Roe[60] stresses three points in assessing the institutional structures needed to operate in an ecosystem context. First, in the Thompson-Tuden framework, while Cell A thinking and action works well for Cell

A problems, the ability of such processes and institutions to contribute to effective action in other cells is limited. Second, because the central elements of Cell C and D problems are their uncertainty and complexity, surprise is inevitable. Lee[61] argues that the first element of prudent behavior in the face of uncertainty is to recognize this inevitability; it is also important to acknowledge that surprises are the source of learning. Inevitable surprise stands in contrast to the traditional hallmarks of the bureaucratic world—the ability to anticipate events and outcomes. It is also not a consequence of a lack of ability, competence, or foresight. Rather, in the face of uncertainty and the conditions that characterize Cells C and D, learning often only derives from trial-and-error experiences. This places a premium on organizational processes and structures that promote resilience and adaptability[62] and encourage learning coupled with action.[63] Third, in a world of complexity and uncertainty, the most effective analytical processes are those that promote triangulation—"that is, use of multiple methods, procedures, and/or theories to converge on what should or can be done for the complex issue in question."[64] In short, new and evolving institutions must be able to embrace situations found in each of the cells—those known and anticipated as well as those that are surprises.

These frameworks offer insight as to the kinds of institutional structures and processes that might be employed to achieve the objectives typically associated with an ecosystem-based management approach. This includes the ability to work at larger spatial and longer temporal scales, across multiple tenures, and to accommodate multiple uses and values, all under circumstances of uncertainty, ambiguity, and high levels of complexity. The challenge of accommodating such attributes will require more than simply improving the operational effectiveness of current systems (e.g., better EIS processes, more data). Basic change to underlying technical and behavioral systems and the institutions that support these systems[65] also likely will be required.

Both frameworks also present a number of questions. Thompson and Tuden appeal for a "structure for inspiration" capable of creating "unity out of diversity."[66] However, neither the requisites for such conditions nor the relative contribution of alternative institutional structures and processes are apparent. For example, how might the relative strengths of the choice processes or the institutional structures that fit other decision contexts apply in the world of wicked problems and what new processes and structures must be identified to augment them?

Roe recommends triangulation as a means of enriching and extending analyses of ecosystem management problems. For triangulation to fulfill its intended role, "the instruments used . . . should thus be as radically different (indeed, orthogonal) as possible."[67] This implies both a capacity and a willingness on the part of organizations to expose themselves to alternative perspectives, to process and assess them appropriately, and to fashion actions, involving potentially profound changes, consistent with the results of the analysis process. However, the record of organizational willingness to undertake such critical introspective assessments is not impressive; as Michael[68] observes "contrary to popular belief, most people under most circumstances are not all that eager to learn," often driven by organizational resistance to changing the status quo.[69]

Re-conceptualizing Ecosystem Management: Four Foundation Attributes

We previously contrasted the predominant distinctions between traditional and ecosystem management. We now build on that analysis, with the intention of expanding understanding of the necessary attributes of effective ecosystem management processes and structures. In tables 1 to 4, the specific attributes and the characteristics of those attributes are depicted for both the traditional planning and ecosystem approaches.[70]

First, the philosophical basis (Table 1) upon which the planning system is grounded requires attention. An ecosystem-based approach to management will require new ways of thinking, of framing problems, of defining roles and responsibilities, and of thinking about the relationship between humankind and nature.

Table 1. Philosophical Perspectives on Resource Management

	Traditional Emphasis	*Ecosystem Management Adds*
Nature of problems	Technical	Social
Nature of knowledge	Objective, knowable	Socially constructed
Source of knowledge	Experts	Society
Role of people vis-à-vis environment	External	Internal
Predictability	Certainty	Uncertainty

As a caveat to interpreting this and the following three tables, we stress that the qualities associated with traditional and ecosystem planning should not be seen as reflecting an "either-or" choice. Rather, the qualities reflect the central tendencies that permeate the underlying conceptions that practitioners use to resolve these issues. Creating effective ecosystem management institutions will not involve abandoning or rejecting the foundations of sound traditional (i.e., rational-comprehensive) planning. Rather, the challenge is identifying the key elements required to augment and expand these qualities. It is not a matter of an ecosystem approach *as opposed to* a traditional approach; we are not pitting the new against the old.

The role of knowledge in Table 1 illustrates this issue. In traditional planning, knowledge tends to be defined in positivist terms; objective, knowable, and subject to measures of validation and reliability. It is seen as independent of the values and judgments held by the observer; it is "real." Such knowledge is a powerful means for understanding the world. However, it is not the only means through which such understanding occurs. Knowledge can also be understood as a product of the experiences, values, and perspectives of the observer. Moreover, knowledge is provisional and tentative; i.e., it reflects the best state of knowledge at a given moment, but evolves as new understandings emerge and new theories replace the old. Knowledge results from "a social process that remains unfinished and open to the future."[71]

The legitimate sources of knowledge also differ in these contrasting management conceptions. Traditional planning conceives of knowledge as the domain of the technical specialist, because it derives from positivist inquiry—valid, reliable—and because only such experts are qualified to conduct such inquiry. An alternative conception sees knowledge emerging, not only from scientific inquiry, but also from "the experience of acting in the world" that generates an "articulate, particularistic, and embodied form . . . called personal or experiential knowledge."[72] Such knowledge derives directly from the labor process (i.e., engagement in a specialized activity) and is "finely tuned to the concrete exigencies, needs, and requirements of local conditions."[73] Because wicked problems tend to be idiosyncratic and resistant to conventional technical-rational solutions, improving understanding of the kinds of knowledge needed to address them is called for. It also calls for a philosophical openness to the question of where such information resides (e.g., among local stakeholders, subsistence users) as well as the

Table 2. Analytical Perspectives for Resource Management

	Traditional Emphasis	*Ecosystem Management Adds*
Problem definition	Reductionist	Holistic
Spatial scale	Sites/stands	Landscapes
Temporal scale	Short-term	Long-term
Resource orientation	Functional	Interdisciplinary
Value breadth	Instrumental	Symbolic
Value definition	Economic marketplace	Social and political marketplace

operational challenges of facilitating their access to decision-making processes.

Table 2 identifies elements of the analytical procedures and processes undertaken to implement natural resource management programs. Many ecosystem management definitions focus on the importance of landscape-level, rather than site- or stand-level analyses. There is a growing appreciation for holistic and interdisciplinary studies, departing from the traditional emphasis on reductionist, functional studies. Finally, interest in ecosystem-based approaches has focused increasing attention on the question of time, given that the relevant time scales for many ecological processes (including humans) are measured in decades, if not centuries.

The analytical orientation of ecosystem management must also be sensitive to a wider range of public uses and values. For example, spiritual or subsistence values possess limited expression in the economic marketplace. However, the failure to acknowledge the importance of such non-instrumental values in resource management planning because of the lack of an economic market has proven politically perilous. There is a need to augment traditional market-based analyses with values whose expression is revealed primarily in the political marketplace.[74]

The attributes described in Table 2 reveal limitations facing efforts to implement ecosystem-based management. Efforts to focus attention on landscape level analyses or long-term studies often are constrained by a lack of adequate theories, appropriate methodologies, and empirical experience. Understanding ecological processes at the landscape level, for instance, is not simply a matter of aggregating a suite of site- or stand-level studies. Many phenomena operating at the landscape scale or over long time frames are emergent properties revealed only at these analysis scales. Moreover, managing ecosystems at larger or longer scales might require modifications in management of sub-units of these systems, such as at the site level. As spatial scale increases, so does the likelihood that multiple jurisdictions, with competing objectives and differing priorities, become involved.

As conventionally framed, ecosystem management calls for major reforms in the ways in which problems are constructed and decisions debated and implemented. Table 3 portrays some of the dimensions for which new approaches to decision-making will be required. Many decision-making attributes, particularly related to ecosystem-based approaches, involve not only operational changes, but systemic shifts at the philosophical level as well. The elements shown in the right-hand column of Table 3 reflect a shift from the dominant social reform model of planning to a notion of social learning.[75]

To understand complex ecosystems, as well as gain the political understanding and support necessary to implement ecosystem management strategies, the decision-making process must be both technically informed and politically inclusive. Table 3 suggests that decision-making processes in ecosystem management need to be increasingly open, inclusive, and adaptive (to strengthen the rigor of knowledge underlying any decision), and the decisions themselves must

Table 3. Elements of Decision-making in Natural Resource Management

	Traditional Emphasis	*Ecosystem Management Adds*
Planning Tradition	Social reform	Social learning
Decision Context	Cloistered	Open
Computational Processing	Routinized	Flexible
Power	Concentrated	Diffused
Strategy	Linear	Iterative and adaptive
Decision-flow	Top-down	Bottom-up
Decision-makers	One	Multiple
Relation to People	Reactive	Interactive
Leadership	Hierarchical	Shared

Table 4. Elements of Natural Resource Management Institutional Structure

	Traditional Emphasis	*Ecosystem Management Adds*
Organization	Compartmentalized Hierarchical	Integrative Loosely coupled
Locus of control	Centralized	Local
Motivators	Penalties	Incentives
Jurisdiction	Within boundaries	Cross-boundary
Disciplinary composition	Homogeneous	Heterogeneous

flow more from the ground-up and through interactive processes in order to gain credibility and legitimacy.

A central assumption of traditional management decision-making, deriving from depression-era government, is that the organization represents the public interest. In this schema, technically trained experts represent the public interest because of their abilities to bring value-free, objective analysis, planning and decision-making processes to bear on problems. However, the growing diversity of society and the expanding range of expectations for the values sought from ecosystems, challenges this assumption. Although planners can represent the interests of managing agencies, their ability to represent a wider suite of values is problematic.

Table 4 focuses on the specific attributes that characterize natural resource management institutions. As previously suggested, it is in this institutional landscape that significant challenges to successful implementation of ecosystem management are found.[76]

A central characteristic of ecosystems is their integrative nature. The need for integrative institutions is driven by two interrelated factors. First, complex links exist within and between the bio-physical and socioeconomic systems with which natural resource management must contend. Second the array of goods and services sought by society mean that the decisions to supply (or deny) some particular set of outputs hold consequences and implications for other goods and services.[77] Simply put, everything is connected, and decisions affecting one component of the system will affect other components. Such consequences hold clear implications for those advocating ecosystem management, because any decision taken with regard to an ecosystem also carries with it equity impacts, as well as effectiveness and efficiency effects.

Integrative structures imply an interactive, strategic, and process-oriented approach to problem-solving. Despite the rhetoric of integrated resource management noted in the literature and in organizational programs over the last twenty-five years, empirical evidence of achievements remain scarce, probably for the same reasons that underlie the skepticism about ecosystem management; is there sufficient institutional capacity and political will to make the difficult choices required for integrated management schemes?

There are two basic organizational structures to consider—the hierarchical organization (Figure 2) and a loosely coupled organization (Figure 3).[78] This should not be seen as an either-or choice; elements of both can take us where we want to go in ecosystem management. However, hierarchical organizations (sometimes referred to as "stove pipe" organizations, given their vertical structure) often create boundaries and barriers that inhibit rather than facilitate integrative efforts. Another problem associated with hierarchical organizations is that they usually have rigid concepts of teams and how they function. When such teams are composed of similar interests, then cross-interest efforts are difficult to establish and sustain. As a result, such organizations can constrain and limit local options for responding creatively to emerging ideas.

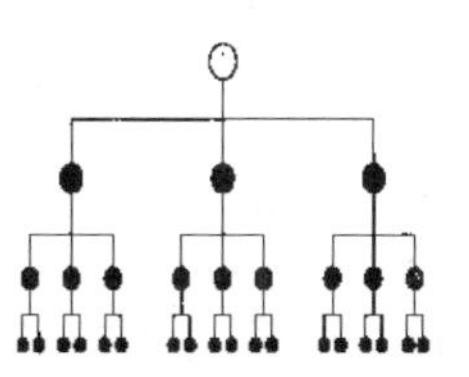

Figure 2. Traditional Hierarchical Organization

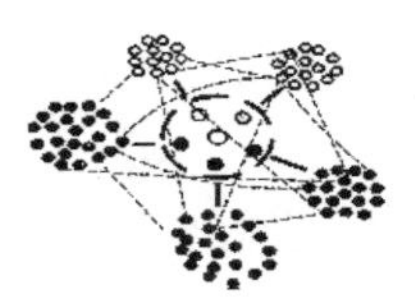

Figure 3. Loosely Coupled Organization

Alternatively, a loosely coupled organizational model considers how people form around the needs of particular tasks or topics (Figure 3). Such collaborative groups can be more or less formal and exist for relatively short or long periods, depending on the problem.

As is the case for the hierarchical organization, there are pros and cons to the loosely coupled model. It is flexible and responsive and generally effective in resolving complex problems. But because it can stand in contrast to the larger organization's structure, this can lead to internal conflicts that limit its effectiveness. Loosely coupled organizations are hard to manage from the top because of their non-traditional arrangement and, as a consequence, their ultimate success is particularly tied to effective leadership and management to ensure that organizational, team, and individual goals are met.

In developing an organizational structure that facilitates learning-based, integrative, and ecosystem-based management thinking and behavior, what can be said about these alternative organizational models? Senge[79] writes: "It is . . . clear that rigid authoritarian hierarchies thwart learning, failing both to harness the spirit, enthusiasm, and knowledge of people throughout the organization. . . . Yet, alternatives to authoritarian hierarchies are less than clear. . . . While traditional organizations . . . control people's behavior, learning organizations invest in improving the quality of thinking, the capacity for reflection and team learning, and the ability to develop shared visions and shared understandings of complex . . . issues . . . these capabilities . . . allow learning organizations to be both more locally controlled *and* more well coordinated than their hierarchical predecessors."

One reason underlying the lack of capacity and/or will to implement integrative structural and management changes is related to its effect on the distribution and locus of decision-making power. Most organizations are reluctant to share, let alone relinquish, power and as a result, efforts to implement reforms involving such relinquishment will not come about easily[80] unless there are incentives to do so.

As noted earlier, the interest in an ecosystem approach has heightened awareness of the uncertainties that characterize biophysical and socioeconomic systems as well as the implications and consequences of management actions on those systems. For institutions to operate effectively under such conditions, there needs to be an environment that encourages and supports engaging risk and uncertainty overtly, consciously, and deliberately.[81]

The factors and attributes presented in Tables 1to 4 provide the beginnings of what the structures and processes of an ecosystem management approach might look like; in a sense they represent the architectural design criteria for new structures. The emphasis here is on the plural; there are multiple pathways and alternatives for designing and implementing effective ecosystem management-based institutions. Indeed, it is important that the search for ecosystem management institutions not be constrained by a decision criterion of "what's the best or right way?"

Principles to Guide Creation of Ecosystem Management Institutions

The factors in Tables 1 to 4 can be analyzed to help identify key underlying principles. However, efforts have been made to identify such principles more explicitly in an *a priori* manner. For example, Costanza et al.[82] identify six core principles essential for sustainable governance:

1) responsibility (access to environmental resources carries attendant responsibilities);
2) scale-matching (institutions should match the scale of the environmental problem);
3) precaution (in the face of uncertainty about potentially irreversible impacts, err on the side of caution);
4) adaptive management (decision-makers acknowledge uncertainty, and continuously gather and integrate information);
5) full cost allocation (all internal and external costs and benefits of the use of environmental resources should be identified and appropriately allocated); and
6) participation (all affected stakeholders are engaged in the decision-making process concerning environmental resources).

These principles are similar to a list proposed by Paehlke and Torgerson[83] for effective environmental administration. These include:

1) non-compartmentalized (avoid the bureaucratic tendency to separate and distinguish various functions and responsibilities);
2) open (focus on developing unbounded, transparent processes that facilitate access by interested groups);
3) decentralized (focus on processes that facilitate engagement by those possessing local knowledge and that link decisions to the geographic and cultural context);
4) anti-technocratic (this argues for more open communication, and receptivity to, the ideas and knowledge held by non-technical individuals); and
5) flexible (sensitive and responsive to differences that resist recognition and classification).

Many of the qualities contained within these two lists overlap or reinforce one another. For example, both cite the need for a broader, more inclusive approach to management (open, participatory). Both acknowledge the need for a management structure capable of enhancing learning and applying it to local circumstances and issues (flexibility and adaptive management). Both hint at the need for linking scale and problem solution, rejecting the utility (and common reliance upon) standardized rules and guidelines. This is an important issue; institutional structures and processes that work well at one scale will not necessarily do so at others, so linking scale, problem, and institution is a critical challenge.

Yet there are also inconsistencies between and within these suggested principles. For example, Costanza et al.[84] argue for the idea of precaution (in the face of uncertainty, err on the side of caution). Yet perhaps the most significant defining characteristic of many ecosystem management problems is the inherent uncertainty that surrounds them, and thus precaution does not necessarily mean a lack of positive action. Moreover, the capacity to learn in ways that reduce uncertainty is constrained by a strategy that errs on the side of caution as opposed to experimentation.[85] The precautionary principle also contrasts to the basic precept of an adaptive approach; adaptation evolves through the learning that results from thoughtful implementation of management policies and the systematic feedback of outcomes from those experiments into subsequent policies. It is difficult to practice precaution and adaptive management simultaneously.

Leadership and Culture— Do They Trump (Other) Institutional Changes?

As institutions move towards an ecosystem approach to resource management, they must consider more than simply changing structures, policies, or rules. In particular the issues of leadership and organizational culture present significant challenges and equally significant opportunities.

Leadership is usually thought of as being embodied in specific people or positions, particularly those at the top of organizational hierarchies. But because ecosystem management includes bottom-up, as well as top-down processes and outcomes, there will be an increasing need for effective leaders with the appropriate skills at all levels of organizations. And because ecosystem management embraces engagement of a wide variety of interest outside the formal organizations, there is a need for external leadership as well. We need to move from a worldview of leadership as people in *positions* to one that sees leadership as *processes.*[86]

Typically, leaders in relatively rigid hierarchical organizations operate under a command-and-control paradigm.[87] This was effective when the issues were simpler and when there was a clear mission. But ecosystem management requires other forms of leadership and power sharing.[88] Collaboration, integration, and mutual learning require leadership processes different from those emphasized in traditional organizations. As Dentico[89] suggests, "People do leadership together, because leadership is a relationship which celebrates diversity and thrives on collective involvement." The bottom line is that one can have all the necessary ingredients in place at the institutional level, but if the type of leaders and forms of leadership are inconsistent with the needs for ecosystem management, success will remain problematic.

Culture also can make desired changes difficult to achieve. In simple terms, culture is a combination and expression of shared norms, values, and beliefs. These are deeply imbedded in people and organizations and influence the roles people play and how incentives and disincentives are used (or not). Kaufman's studies of the U.S. Forest Service, for example, described both the institutional culture and how individuals in it were socialized.[90] Shared norms and values were particularly significant influences. Most resource management organizations historically have been comprised of like-minded, similarly trained professionals and the norms and beliefs held by these individuals were highly resistant to change and transformation, at least in the short term.[91] As a result, when changes to organizational structures, policies, programs, and practices are made, they might not have the desired effect if contrary to the dominant organizational culture. Denial and resistance often lead to both overt and subtle ways to stifle undesired solutions to problems on which there is not agreement. For example, although there has been widespread rhetorical ascription among U.S. natural resource agencies to ecosystem management, there is less evidence of substantive changes in terms of organizational structures, processes, and behavior. Making changes is relatively easy; adopting new policies, assigning new job titles, renaming organizational units, etc. However, it takes time for individuals to work through the social-psychological transition involved in changing how we think and behave.[92] Successful transition requires skills and processes often not found in contemporary organizations. This is difficult in most situations, but even more so when cultural issues underlie fundamental shifts in direction.

This implies that tinkering with organizational structures, policies, or practices is only part of what is called for to make the systemic shifts required for successful implementation of ecosystem management. Attention also must be paid to transformations in leadership and culture before the changes will have the desired effect. All of this will take time and patience. But patience is hard to come by in a society fixated on results-oriented, quick fixes to what are systemic and complex problems.

Conclusions

The foregoing discussion reveals considerable agreement among scientists, planners and the public that many aspects of traditional institutions are no longer adequate to the task of managing ecosystems, particularly when the issues confronted are technically complex and imbedded in multiple, oft-changing interests. However, there is much less agreement regarding what needs to be done. Unfortunately, as we strive to meet the diversifying demands of a growing human population with an ever more limited resource base, it is likely that problems will become increasingly complex and contentious, widening the discrepancy between organizational capacity and effective problem resolution. The need for aggressive and innovative institutional structures and processes to cope with this situation will become more, not less, important.

The potential of innovative institutions to promote ecosystem management is great. Not only could they lead to more effective and appropriate decisions, but they could also promote increased levels of social equity and provide a premium for the learning and management essential to operating in a world of uncertainty and ambiguity. There remains a place for traditional approaches to decision-making, but we need to tailor decision processes to the problem being addressed. There is a particular need for attention to framing problems so that the most appropriate institutions and processes are brought to bear.

Natural resource managers widely recognize the shortcomings of applying the "rational-comprehensive model", but in the face of institutional, political and personal barriers, have been reluctant to shift to other approaches. Shifting to new models—whatever they might be—carries the attendant risks of uncertainty not only with respect to the substance of the decision, but how the decision is made. To some extent, new institutional structures represent a changing political reality. Natural resource agencies, long characterized as attending to and maintaining their own agendas, no longer have the political power to implement the legally mandated plans for ecosystems. In a sense, while they have retained the legal mandate to make decisions, they have lost the consent of the governed. The differentiation between the power to plan—legally defined—and the power to implement—defined politically—means that agencies must account for new constituencies with evolving interests in their decision-making. If they do not, they cannot survive in their present form.[93]

Our attempt here has been to define the nature of the problems confronting natural resource agencies and how they might respond through new institutional architectures and processes. Such new designs must have the support of a diverse public which ultimately benefits or pays the costs of inappropriate or ineffective decision-making processes. In a very real sense, the search for new designs reflects broader society pursuits for democracy—an elusive, evolving goal requiring new forms of leadership.[94] More than voting is needed to ensure the survival of democracy. Attributes such as volunteerism, self responsibility, and learning characterize not only the political realm, but the arena of ecosystem-based management. Consequently, the search for systems of governance for ecosystem-based management might well benefit from an assessment of experiences in other sectors, such as education and health. The challenge of ecosystem-based management lies in developing the leadership needed to address the complex, wicked and messy problems of a world of increasing diversity, expanding natural resources demand, and evolving values and attitudes toward the environment.

16

"Are We There Yet?" Assessing Our Progress Along Two Paths to Sustainability

Thomas M. Beckley, Bruce Shindler, and Carmel Finley

Introduction

The title of this book suggests that Canada and the United States are on a journey. Concern over sustainable forest management and ecosystem management derive from convictions that we need to address ecological problems we have created in our forests. Creating systems to nurture the ecological attributes of a more sustainable forest is essentially a social and political endeavor. The notion of sustainable forest management as a process is important. This approach recognizes the continuous and evolving nature of traveling the path to sustainability. It is an approach that allows us to cultivate our understanding of forests through the acquisition, dissemination, and evaluation of knowledge, and provides a framework for generating and deciding among alternatives. The process must reconcile scientific inputs with social and economic values. We believe social scientists who study resource management and policy are well situated to contribute to this journey.

In this concluding chapter we offer our assessment on how much progress has been made down the twin paths of sustainable forestry and ecosystem management. There will always be some debate about whether as a society we have arrived at the final destination, but we prefer the continuous journey analogy. Thus it is more important to establish whether we are heading in the right direction, and if so, to assess our degree of progress.

There is considerable variety of opinion among the authors in this volume regarding the extent of progress toward sustainable forest policies in Canada and the United States. That diversity of opinion reflects perceptions in the broader forestry community regarding the relative ease or difficulty in charting the course toward sustainability. Duinker, Bull and Shindler (chapter 3) are differentially enthusiastic about our prospects for the future as they review the dramatic changes of the last decade in our two countries. Weber and Herzog (chapter 10) clearly believe that grassroots ecosystem management groups represent a breakthrough, and are strong advocates that grassroots management should be adopted more widely.

Others are more skeptical. They suggest that our twin paths will be long and filled with obstacles. Ryan (chapter 11) reviews experiments with new institutions in each country, but concludes that Adaptive Management Areas and Model Forests represent tentative first steps down those paths. Both programs have a narrow geographical mandate and lack real power to implement change. In the case of the AMAs, there is little financial support from sponsoring agencies; while provincial governments have never fully adopted the MFs as part of their management framework. Any long-term gains from these sites will require a more substantial commitment from their host institutions.

Stankey and colleagues (chapter 15) make the most forceful and detailed case that natural resource management in the new millenium requires new institutional tools. They argue that new and conflicting demands from our forests, and new sorts of problems (described by them and others as "wicked" or "messy") require completely new institutional forms and structures. This message is the recurring theme throughout this book—that old institutional forms characterized by centralized, bureaucratic, top-down, expert-driven management that privileges science content and favors technological solutions will not lead us to sustainable forest management. Although our authors disagree in their assessments of how far any current institutional changes have taken us down the two paths to sustainability, they all—even the optimists—agree that such change is needed. Many side with Stankey et al.'s contention that entirely new institutions are required.

The old institutions have failed to assess adequately and to incorporate the values of our post-industrial society into forest management and planning. These new demands (discussed in Part Two and summarized in Figure 1 below) include such things as carbon sequestration, biodiversity, amenity values, non-timber forest products, as well as our traditional (and increasing) demand for forest products. The chapters in the second part demonstrate the public's strong preferences for a style of management that places ecological integrity and amenity values over economic growth and wealth.[1]

It is somewhat ironic that the social scientists represented in this volume and elsewhere increasingly present results from public surveys and other public involvement exercises suggesting that what people want, above all other forest values, is healthy, intact, biologically diverse ecosystems. At the same time, some of the most renowned ecologists (e.g., Hamish Kimmins from the University of British Columbia) and forest scientists (e.g., Jack Ward Thomas at the University of Montana, former Chief of the U.S. Forest Service) suggest that ultimately forestry is about people and choices. As such, the problem really is about devising better processes for involving the public to define management objectives for public land.[2] Both the ecologists and foresters seem to be telling us that what is important in forestry today is process, democracy, and identifying social values, and then incorporating those values into management objectives. On the other hand, social scientists are relaying the message from the public that forest management is really all about ecological integrity, or at least that ecological integrity should be the highest priority for management. With members of these traditional disciplinary divides championing each other's causes, perhaps we are making progress toward integration.

Changes in the Social, Economic, and Political Context of Forest Management

Virtually all the authors in this volume agree that we are currently in a time of transition with respect to forest management in North America. When paradigms are in transition, it often takes awhile for actions and institutions to catch up with the change in thinking. Kimmins suggests we are currently experiencing "future shock" in forestry.[3] He describes future shock as a situation where the expectations of society have surpassed the ability of our institutions to meet those expectations. Future shock is also in evidence when individuals are unable to adapt to the accelerating rate of change they encounter. The former is the predicament in which forestry institutions are mired, and the latter is the predicament in which many forestry professionals find themselves today.

What are those new expectations? Figure 1, taken from earlier work by Beckley,[4] illustrates the range of values demanded by the North American public for the twenty-first century. Any element under this figure could be expanded—it is merely meant to be illustrative—and the list of values is growing. The problem, of course, is that all these expectations must be met from the same limited resource.

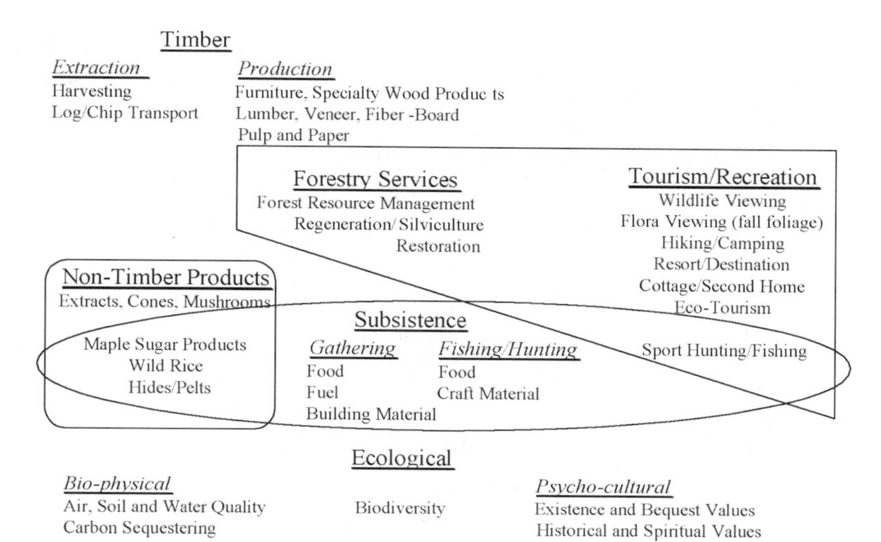

Figure 1. Human Values for Uses of Forests (North America)

Under the sustained-yield management model, or multiple-use, or integrated forest management, we usually considered timber and allied products first (the upper left corner in Figure 1), and then considered "everything else." Forest managers attempted to supply "everything else" in sufficient quantities to appease the various stakeholder groups, but in most cases all these other values were viewed by resource managers as constraints on extracting as much fiber as possible in as cheap and efficient a manner as possible. The demand for forest products has not lessened. Various sources can show that it has increased.[5] However, in North America, it has not increased as fast as demand for other forest values. The result is a relative shift in our priorities for forests. Typical of North Americans, we do not want any less of what we are already getting (pulp, newsprint, dimensional lumber, etc.) and now we want a whole lot more. Professional foresters have been doing everything in their power to supply these new and varied demands, but we may well be exhausting the capacity of the forest to supply everything society wants of it. If conflict is any measure of scarcity, then it would appear that many forest values are in short supply.

While we are beginning to reach the limits of what the forest can supply in North America, the situation is even more alarming in the rest of the world. In relative terms, the Earth continues to shrink at a dramatic rate. Earth was home to some two billion people in 1945, and it had taken 10,000 generations of human procreation to reach that level. It is estimated that by 2032, there will be nine billion people on the planet, so we will quadruple the population in just four subsequent generations.[6] As population increases, the planet shrinks in its capacity to provide a high material standard of living for all. Indeed, several studies have concluded that we would need several planets the size of Earth (and equally endowed with natural resource wealth) in order for the rest of the world to enjoy the lifestyle we lead in North America.[7]

The world is smaller not just because we are more crowded on its surface, but also because of revolutions in technology (primarily communications and transportation). In addition, global trade is commonplace and on the rise. With the collapse of communism, trade restrictions have fallen away and we now find ourselves in a highly volatile world economy. In such an environment, national sovereignty is constantly chafing against the demands of the world marketplace. The values, attitudes and preferences of customers are extremely important for producers. Canada, as a producer nation, is highly sensitive to its international image with respect to forest practices. But this begs the question—should Canada's forests be managed according to the values of Canadians, or according to the values of customers of Canadian forest products? The United States is somewhat less susceptible to this trend, but nevertheless, its isolationist tendencies will certainly be challenged in the years to come (Alavalapati et al., chapter 14).

The demand for new goods and services adds layers of complexity to the management mix compared with the time when foresters were simply managing for timber, and all other values were considered constraints. Forestry was always more political under the old paradigm than we often care to admit. However, in those days, the politics was more of the "backroom" variety, involving mutually beneficial collaboration between industry and government. This is not to suggest a conspiracy between government and business. Rather, in the heyday of the sustained-yield era, government and industry worked together with a single objective to "get the wood out." The public had confidence that the cadre of professional foresters, armed with the tools of scientific forestry, could deliver the goods—sustained yields of timber and recreational values (including wildlife). That confidence has eroded, and today the public views the forest as vulnerable and shrinking. As such, the stakes over its disposition increases. Today, forestry conflicts are common. Conflicts are featured prominently in the media, and now the practice of forestry seems more contentious than ever.

Many people now consider the risks involved in forest management to be higher. The high stakes at the forefront of forest policy include, for example, climate change, biodiversity, habitat loss, protected areas, and the introduction of genetically engineered species into the wild. Because many of these issues can involve irreversible change and potentially catastrophic consequences (or at the least, people perceive them to), they tend to be highly charged, emotional issues. One problem about managing in this context is that the greater the risk involved—or the greater the uncertainty about risk potential—the less likely the public is to support management initiatives, especially those they are unfamiliar with. Our management agencies must be more forthcoming about the difficult decisions and the choices involved. Scenarios involving risk and uncertainty only reinforce the need for forums to discuss openly and to interpret forest information, practices, and conditions.

Similar to the earlier discussion by Stankey and colleagues, Funtowicz and Ravetz[8] describe situations that involve high stakes and system uncertainty as ones that require a fundamentally different sort of solution. For years most resource professionals dealt with applied problems (signified by the area "A" in Figure 2) that stemmed from sustained-yield forestry. The stakes were often low and the forest system was perceived as a more simple entity, and thus more easily managed. Once we acknowledged that system uncertainty had increased, forest professionals and the public relied on solutions that would come from science, and the expertise of professional consultants who applied

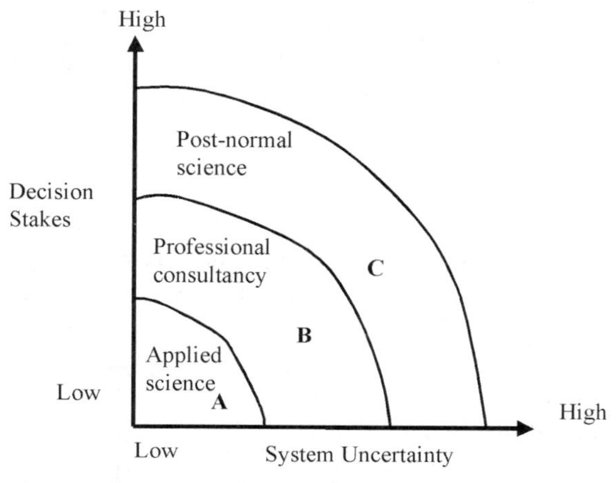

Figure 2. Three Types of Problem-solving Strategies (adopted from Funtowicz and Ravetz, 1992)

specialized knowledge (area "B" in Figure 2). Decision-makers and the public had faith in technical solutions. Ecologists and biologists now are indicating that natural forest systems are more complex than we thought; to paraphrase Frank Egler, "Nature is not only more complex than we think, it is more complex than we *can* think."[9] Today, most members of the forestry community recognize that the stakes are higher as well. Clearly, many forestry problems and issues now fall into the domain of post-normal science (area "C" in Figure 2). Because forests are complex systems and encompass a broad range of social values, we are finding it difficult to predict the indirect, systemic effects that management actions may create.

Some of the current efforts in forest management could be classified as experiments in post-normal science and policy making. In the following section we assess the extent to which a number of these endeavors are addressing the new demands on forest systems. Our initial assessment is that we are just beginning to understand this post-normal world and still have a long trek ahead of us. While our two paths seem to be taking us in the right direction, much more institutional innovation will be required.

Are We There Yet?—Progress Along Two Paths

In spite of our cautionary declarations, a number of interesting developments in forest management suggest Canada and the United States are making forward progress down their paths toward sustainability. This section examines some of the key applications that we view as positive developments where the major institutional players (agencies and organizations) are stepping up to meet the demands placed on our forest systems by a broader, deeper, and more concerned public.

Certification

Certification is probably the most dramatic recent change in the way we practice forestry today. To review, McDermott and Hoberg (chapter 13) define certification as

> . . . a market-based instrument of forestry reform that allows companies that meet criteria for sustainable forest management to carry a "green-label" on their wood products.[10]

Certification appeared suddenly in the forestry sector in the mid-1990s and has expanded rapidly since. In addition to certifying forest products, certification schemes may certify forestry practitioners as sustainable managers and designate specific land bases as operating under sustainable management practices. Virtually all major forest products companies in Canada as well as many in the United States are positioning themselves to be able to obtain at least a minimum level of certification, if indeed they do not already have it. Those hoping to stay in front of this wave are seeking certification under standards and "green labels" with more rigid requirements. The certifiers are themselves in a competitive game to become the preferred certifier among the major industry players. And while certification was once thought to have limited applicability, it has become more mainstream.

In many respects, certification is a function of a failure on the part of government to convince the public that they are minding the store. Certification, in part, has come about because people have lost faith in government stewardship and oversight, particularly of public forests. However, certification is also becoming popular on private forestland (as is the case in the United States). From society's perspective, certification provides a mechanism to ensure that public goods from forests are not abused, whether those public goods derive from public or private land. Most policy in the environment and resource management realm is created to address market failures. These include externalities such as pollution, or equity issues regarding fair distribution of access to public resources. Certification is better characterized as a market solution to a policy failure. Through certification, environmentalists and others are trying to use the carrot of secure market share to induce producers of commodities to practice sustainable forestry. While this is a market rather than a policy solution, it still represents an important new institutional form that has tremendous potential to promote and enhance sustainable forest practices at the field level. One of the advances represented by certification is that various schemes explicitly address high conservation value forests, public involvement, community benefits, and the like that are "demand" issues and have been poorly supplied in the past by either industry or government.

While certification has tremendous potential to transform forest practices and contribute to sustainable management, it is too early to judge whether it will realize that potential. Certification has received more attention in Canada than in the United States, in large part due to

Canada's export dependence. Nevertheless, the preference in the United States for market-based, economic instruments that induce good behavior over policy instruments that punish bad behavior, probably means that certification has a bright future in the United States as well. The competition between certification schemes is intense.[11] Some go very far in their requirements for new ways of doing business. Others do little more than certify status quo management as sustainable. Whether the consumers eventually understand the difference between various green labels enough to differentiate between them may ultimately determine whether certification has an effective and long-lasting impact on sustainable forest management.

Institutional Experimentation—Model Forests and Adaptive Management Areas

After years of resistance, government and industry have recognized that the "rules" of forestry have changed and that forests, especially publicly owned forests, can no longer be managed in a sociopolitical vacuum. In many cases, however, the institutional framework that governs public land management is antiquated because it was designed to serve the scientific, sustained-yield paradigm. Many of the constraints come, not at the field level, but at the planning or administrative level. A classic example is the Annual Allowable Cut (AAC) calculated on Canadian Crown land licenses. The AAC is determined through various models, or in some cases by administrative fiat (in British Columbia), and licensees are then told how much wood they may harvest over a five-year period. However, the corporations that hold these licenses are limited in their mandate. Their licenses only pertain to timber, not to other forest resources. If they opt to take a conservative approach (to protect or enhance non-timber values) and consistently harvest below the AAC for their designated land base, their provincial government will usually "reward" such cautious behavior by drawing back some of the land and reallocating it to another company that will harvest the wood. Such institutional legacies are constraints to progress in sustainable management and add momentum to calls for new institutional approaches.

Recent attempts have been made to transcend the constraints of the sustained-yield institutional structures. Two discussed here (Ryan in chapter 11, Shindler in chapter 12) involve Canada's Model Forest Program and the Adaptive Management Areas in the United States. Thus far, both have been extremely limited. The Model Forests are limited geographically, but more importantly they are constrained by their mandate. They are creations of the federal government (Natural Resources Canada), but in Canada the provinces exercise authority over natural resources. As such, the Model Forests have only advisory power, not decision-making authority. The case of the AMAs is worse. They have been more subject to political shifts as they were the brainchild of the Clinton administration. But even during Clinton's tenure in the White House, they were underfunded and poorly supported beyond the lowest levels of the agencies (Forest Service and BLM).

Where AMAs and MFs have been successful is in building new partnerships on the ground between traditional adversaries. At most locations, these institutional experiments have made honest attempts to create collaborative processes. Frequently this has resulted in more meaningful dialogue between stakeholders—including industry and environmental groups, First Nations people, recreation user groups, and government agencies at multiple levels. These efforts attempt to combine scientific information with knowledge from local stakeholders and values from local communities. Conceptually they may attempt to merge these ideas with social values from larger scales, but this is a more difficult match. In the end, they are practicing post-normal science. Bridging the gap between stakeholders and different sources of knowledge and information—in essence, creating a place for mutual learning—has been a substantial achievement. Yet this will not be enough if the management and geographical mandates of these experiments remain as small as they currently are. In a positive move, the Model Forest program was recently renewed (funded) for another five years. Part of its mandate for the next five year period is to move "beyond the boundaries," to take the experience of the model forests and apply them more broadly across the forest landscape.[12] In contrast, the AMAs in their current form appear to be history; they have not received funding since the 2001 fiscal year, and whatever administrative support they had is waning. However, some (and other designated sites) may see new life as "charter forests" under the Bush administration. Their mandate and how well this will play out is uncertain.

In summary, Model Forests and AMAs have been simultaneously wonderful, frustrating, enriching, eye-opening experiences for many who have been involved with them. But their broader impact has been minimal. Still, the governments that have sponsored these efforts and the partners that have participated in them should be applauded. To achieve sustainability on a broad scale we need more institutional experiments such as these. They have the potential to provide meaningful settings where thoughtful consideration of possible futures can occur. These are the kind of places where forest managers can regain a leadership role in their communities and re-establish trust among a broad base of stakeholders.[13] As these sites experience success, however this may be defined, we need to give them greater management authority. As Stankey et al. suggest in chapter 15, this would mean ceding real power, which governments are always loath to do. But we will not get very far down the path of sustainability by just minor tinkering with the old system.

Institutional Experimentation—The Grassroots

Adaptive Management Areas and Model Forests were initially established as top-down, bureaucratic efforts to implement institutional reform. While many view the terms "top-down" and "bureaucratic" as pejorative, we are merely using them as descriptors. Government efforts, by definition, are bureaucratic and top-down, so this type of reform is better than none at all. However, there has also been a dramatic increase in institution building and capacity building from what many describe as the bottom up. Concerned local citizens are organizing and taking action at the grassroots level. Weber and Herzog described several U.S. examples of this movement in chapter 10.

Elsewhere in the United States, watershed councils have (almost) overnight become a new community-level institution. Many western states are providing seed funds to start-up groups as a way to involve local citizens in conservation programs. The common thread for all groups is that they are interested in accomplishing projects on the ground. Astute resource professionals from state and federal agencies are seeing the value in cooperating with watershed councils. While not always marriages made in heaven, these associations can lead to substantial gains in finding more sustainable solutions. At the very least they get more people involved in the discussion. As advocates for this form of local governance, Weber and Herzog detail a number of additional attributes.

In Canada the grassroots movement has been less spontaneous and input from the public has been much more managed. Stedman and Parkins describe typical settings and sentiments in their treatment of

Public Advisory Groups (chapter 8). Canadian legislation regarding the public's role is more vague than in the United States, where laws guarantee the public a role in planning and decision-making. This has also meant that fewer Canadians have utilized the judicial system to protect their access to the planning process. However, if provincial governments create a minimum requirement for public involvement—and provide standing for forest users to innovate and experiment alongside managers—then we are likely to see a dramatic increase in citizen activism. As we described from the outset, Canadians are communal by nature; given greater opportunity for participation, results could surpass what is occurring in communities across the border.

Institutional Reform

While we are arguing forcefully for profound institutional change and the creation of new institutions altogether, it is important to continue to encourage incremental institutional reform as well. Governments are supposed to reflect the will of the people, in policies, programs, and priorities (e.g., how they spend our tax dollars). Governments often lag behind public opinion, but over time official policies for natural resource use and management eventually catch up. The types of reform and policies outlined by Duinker et al. and Stankey et al. become critically important. Some view documents like the National Forest Strategy in Canada as a "feel good" exercise, or a mechanism to convince the international community that sustainability is a reality in Canada. While there is some truth to such a characterization, it is important that emerging forest values are being recognized and incorporated more quickly into meaningful programs on the ground.

Our experience indicates that incremental reform is often best achieved in more localized settings. As Shindler and others[14] have noted, incremental reform also benefits from strong leadership. Places where recognizable progress has been made are in settings where agency leaders support and legitimize the efforts of field personnel to conduct business on a multi-partner basis. This usually involves enabling motivated personnel to develop effective public communication strategies, while at the same time encouraging them to experiment with more innovative practices for sustaining forests. This may mean providing administrative flexibility that allows these individuals the freedom to run with good ideas. Often there is a need for leaders to clarify roles and responsibilities. Today, the roles agency personnel are being asked to play are much different from those in the past, when citizen participation was minimal and technical expertise was foremost. Our institutional personnel need assurance that a broader scope of public contact is essential and that any reasonable program in sustainable forestry will be a long-term endeavor. Not only must agencies more clearly define their relationship with the public, but they must also identify the role they are willing to let citizens play. These actions are best when they play out publicly and collectively. Incremental reform occurs in places where people can come together to learn about the uncertainties of ecosystem management, to discuss and understand the risks involved, and to weigh the tradeoffs and consequences.

A Seat at the Table for Social Science

Another institutional change that has accompanied the new paradigm in forestry is the recognition that social science has the potential to contribute to sustainable forest management.[15] This potential is not universally acknowledged within the forestry community, and for years establishing the beachhead was a challenge.[16] Occasionally we still get the odd question from the forestry "old guard" about how sociology can help build a better stream crossing.[17] However, such misunderstanding about the nature of social science contributions in forest management is dissipating. Progressive colleagues from ecology and forestry not only welcome social science contributions in forestry, but they see them as necessary. Hamish Kimmins, a prominent ecologist at University of British Columbia, recently wrote,

> Forestry is about people—their needs and desires—and not fundamentally about biophysical issues such as biodiversity and specific ecological conditions. The reason why these and other issues are of pivotal importance in forestry is that we now understand that they are important to sustaining the values and environmental services people want from forests.[18]

As stakes increase, and complexity increases, and the diversity of demands on forests increases, people are becoming more accepting of the fact that social science can help with tools (such as public opinion research), analysis (of new institutions), and the monitoring and evaluation of process issues such as facilitation, mediation, and conflict resolution. Social scientists can help identify forest values as well as their distribution across social groups (classes, gender, age cohorts, stakeholder groups, geographical locales). Documentation of emerging social values, through various means, is the main contribution of chapters in Part Two of this volume.

Post-normal science requires that this sort of information be combined with more traditional science-based information such as timber inventories, home ranges or habitat suitability indices. Our tool kit for integrating these diverse types of information is developing rapidly. In western Newfoundland, non-timber values maps that show hunting, trapping, berry-picking areas, cabin locations, important viewsheds, and domestic firewood cutting areas are being integrated with traditional resource data through geographic information systems to provide an integrated database that will facilitate post-normal science in planning processes. We already have the tools to do much of this work. If it is not happening in a given locale, it is likely due to a lack of political will, or imagination, not to technological limitations.

The door has been opened for social science to make a positive and long-lasting contribution to the pursuit of sustainable resource management. In order to take this chance we need to do several things. First, we need to draw on our full range of resources to make more creative and positive contributions. Social scientists in natural resources have often taken the role of critic of status quo institutions and management. To make a contribution to post-normal science, we need to build bridges instead of tearing them down. We need to use the tools and language of foresters to do this. In particular, we need to help citizens and stakeholders articulate their forest values in forestry terms. Erdle and Sullivan issue this challenge.

> The set of important forest values is becoming increasingly broad, and in many jurisdictions now relate to timber, wildlife habitat, aesthetics, biological diversity, water regimes, ecological health, and recreation. The point here is not to argue what these values should be, but to suggest that once they have been identified for the forest in question, through whatever local mechanisms exist, it is critical that they then be defined in terms of the forest condition, using an appropriate characterization of stand composition, stand structure, and forest pattern.[19]

We believe that social scientists, particularly those with credentials and experience in forestry, are particularly well suited to take up this challenge. Rather than simply continuing to relay the message that the public does not like clear cuts, we need to help provide forums where the public and forest managers together can articulate things that matter most in their particular communities. This might be some desired future forest age class or species composition to enhance recreation experiences in a forest. Or it might be about how much wood to cut to ensure a sustainable flow of harvestable timber over time.

An Agenda for Institutional Change

Although the public continues to express preferences for forest management that maintains the ecological integrity of forests, the institutions in both countries that hold management authority over our forests were created to fuel economic growth through responsible resource extraction. As concerns over environmental degradation occurred, governments *added* departments and agencies to regulate environmental externalities. Thus, the institutional legacy that we find ourselves with today is one where many jurisdictions (federal, state and provincial) maintain parallel departments or ministries with competing mandates toward the environment. In the U.S. this is best exemplified by the Environmental Protection Agency (EPA)—i.e., the watchdog agency—and the two primary institutions (Forest Service and Bureau of Land Management) whose mandate, at least in part, is to generate wealth from the publicly owned forest estate. In Canada a similar divide exists at the national level with ministerial divisions between Environment Canada and Natural Resources Canada. But more important, the same divide is in evidence at the provincial level where most natural resource management authority lies.[20]

The need for institutional reform is but one of many challenges that we face on the twin paths to sustainable forestry and ecosystem-based management. But because our two countries are so dependent on these institutions—from a legislative, scientific, experience, and leadership standpoint—they represent the most critical focal point for change. Below we outline five agenda targets as priorities for our collective attention in the near-term. They derive from the substantive discussion in the previous chapters, but also reflect much of the rich, ongoing deliberation among resource professionals, policy-makers, and scientists concerned with creating a more sustainable brand of forestry. We use the term *agenda* purposefully; this is not intended to be an exhaustive discussion of these points. Rather we offer them as targets for those who follow—the integrated groups of legislators, professionals, and citizens who will be required jointly to determine the institutional changes necessary to create an enduring program for sustaining forest values.

First, we must organize major reform of government institutions to eliminate many of the barriers between the agencies responsible for delivering on the multiple-use mandate (i.e., economic development and recreation services) and those charged with protecting the environment. The perpetuation of economic development and environmental watchdog agencies within government creates a wasteful, competitive, contradictory alignment. There is the need to create more holistic management institutions that take ecosystem health as their primary mandate. Economic development through resource use will still be important, but ecological integrity must come first. Under this scenario, those with regulatory responsibilities must be active participants in management experiments.[21] For example, they can help focus learning objectives on critical questions (e.g., about species survival, watershed restoration, or air quality) to help us understand the risks and uncertainties of experimental options. Thus, we are more likely to find alternative strategies that support both endangered species objectives (for example) and larger societal goals of sustaining forest ecosystems.

Second, we must implement reform within government institutions so that planning and management activities for multiple resource values are more integrated. Even within departments of forestry or natural resources at the federal, state, or provincial level it is common to have a timber branch, a wildlife branch, a recreation branch, and so on. There are plenty of functional reasons why integrating work and work units is more efficient and effective.[22] Information is dispersed; the relevant knowledge to solve complex resource problems is distributed among many groups and individuals. Issues revolve around a shared resource in which dynamic systems are interconnected. The interests of participants are frequently in conflict; thus we need common workplaces to sort through management objectives and priorities. Resulting plans need the legitimacy that comes from the hard work and agreement of numerous disciplines attacking a problem. These reasons all funnel into the larger justification for integrating management approaches. Our current political climate calls for an improved understanding of the scope of sustainable forest management and its relationship to different geographic, temporal, and normative contexts in which it will be applied.

Third, we need more experimentation with institutions outside the realm of traditional governance structures. Certification is but one example. The rise in prominence of grass-roots groups and the different approaches they represent is another. This movement seems to just be getting started. The AMAs and Model Forests have been a significant departure from traditional government approaches. Although Canada's system appears to be on more solid ground for the future, many lessons emerged from the U.S. experiment that would benefit the next iteration of AMAs regardless of what we might call them. We urge the expansion of these programs in both countries. It is also likely there are other market-based, incentive-oriented institutions that can contribute to sustainable forestry. As we search for new institutions to implement sustainability, we should think about "carrots" first, and "sticks" only if no carrots can be found.

Fourth, our institutions must experiment at different scales. Large federal institutions typically have believed their mandate to be one of serving the needs and interests of all citizens. This often plays out in the need to interpret federal or provincial legislation onto both large and small landscapes, a difficult assignment at best. It is a struggle to account for ecological and socioeconomic differences at numerous scales. On the other hand, ecosystem-based management implies that managers will pay attention to forest conditions at the most suitable scale to address the inherent problems. In chapter 15, Stankey et al. refer to this as scale-matching. In today's forestry, this often means the search for solutions will need to consider multiple scales, with the most frequent scale of planning taking place at the watershed or sub-watershed level. These are relatively small planning units, but in these early days of ecosystem management they are the scope at which we may be able to achieve sustainability goals. Each management unit must be able to conduct experiments and develop strategies that address localized conditions. This different scale approach also encourages experimentation with local partnerships. People are more likely to become involved when the issue is in their backyards. This also seems to be a positive way to harness the NIMBY syndrome. The proliferation of watershed groups is a good example of scale-matched institutions attempting to come to terms with ecosystem-based forestry problems.

Fifth, we need to instill the capacity in our institutions to practice post-normal science. That is, we need institutions that can integrate knowledge from both science and civic discourse into our management policies. Science can tell us what is and what might be, but it does not tell us what *should* be; the latter is society's job. When we engage our communities of interest in civic science,[23] we have increased the likelihood that outcomes will be based on shared community values rather than the preferences of individuals or interest groups. This approach reflects decisions that come through mutual learning, or what "we" working together come to understand.[24] Policies based on shared civic responsibility are more likely to reflect what people think is right—not merely what they prefer—and result in outcomes that benefit the larger community of interest. The end goal of such processes is to create a climate in which achieving sustainable ecosystems becomes everyone's responsibility, not just the responsibility of our government institutions. The Model Forest program, AMAs, and other grass-roots initiatives are good examples of post-normal experimentation. From an institutional standpoint these efforts must be recognized as a significant change in how work is done, and these changes must permeate agency actions not compartmentalized to the Model Forests or isolated community projects.[25] A likely byproduct of building institutions for doing post-normal civic science will be renewed trust among the full range of agency and public participants.

It is important we recognize that institutional change alone is not sufficient to bring us to a sustainable future. While we have focussed extensively on institutional reform and institution building, the fact remains that institutions are comprised of, and created by, people. Changing institutions will mean social change as well. Both Weber and Herzog and Stankey et al. make the point that institutional transformation will come out of the aggregation of thousands of individual transformations. While this is true, it has been our experience that the capacity to change institutions comes from strong, stable leadership. On their own, individual personnel may desire to move in a certain direction, but the enabling influence of a good leader legitimizes the actions and attributes that are necessary to create change.

Among the key attributes the leaders of our institutions must foster is a willingness to work across traditional boundaries. This includes traditional disciplinary boundaries in agencies and academia, but also boundaries between stakeholder groups. We will need to internalize the concept of partnerships—not just strategic partnerships, but partnerships that foster understanding across lines of interest. This will encourage the public to stop thinking like a customer to whom sustainable forestry will be delivered. The intended outcome is to get everyone to take responsibility for forest sustainability.

A second attribute to embed in our post-normal way of thinking is the flexibility to make mistakes. Currently agency personnel work largely in a risk-adverse atmosphere, where actions are by conservative policies in which "no fault" can be identified, and consequently less chance of litigation.[26] The result is that little experimentation occurs and, thus, little learning as well. When we do begin to experiment more boldly with new institutions and new management practices, there are likely to be localized ecological, social, or economic failures. Administrative leadership must demonstrate that the ability to fail is part of the experimental process, and then carry the message to our constituents that we will ultimately need to trust in the long-term resilience of nature and our institutions.

Finally, we also need a genuine willingness to relinquish traditionally held power and authority. This last attribute may well be our greatest challenge. Most sociologists and political scientists will argue that power, once gained, is rarely surrendered willingly. However, the failure of our existing institutions to deliver management programs that are widely supported has resulted in many forestry disputes landing in the judicial system. Few argue that this forum is the best for making responsible and sustainable resource management decisions. The alternative is to open the door to more collaborative, less adversarial decision-making arrangements. From the public standpoint, these will look more like "we" making decisions together, rather than "us" not liking what "they" decided.

As we have pointed out in this final chapter, a number of incremental steps have been taken toward creating institutional change. They may have been slow and tentative, but there is evidence we are lurching ahead in the right direction. It is important to recognize there may be no final destination called "sustainable forestry;" at best, we may only be able to judge our success by the continuous progress and agreements reached. We have used the "two paths" metaphor extensively throughout our introductory and concluding chapters. There is a Chinese proverb that reminds us that every journey begins with one step. More and more people are making that first step. Our hope is that forward progress continues, and that our two nations can learn from one another's experiences in navigating the path toward sustainability.

Notes and References

Chapter 1

1. W. Dietrich, *The Final Forest* (New York: Simon and Schuster, 1992), 20.
2. See R. P. Gale and S. M. Cordray. 1994. "Making sense of sustainability: Nine answers to 'What should be sustained?'" *Rural Sociology*. 59(2):311-32.
3. G. H. Stankey and R. N. Clark, *Social Aspects of New Perspectives in Forestry: A Problem Analysis* (Milford, PA: Grey Towers Press, 1992).
4. M. Howlett (ed), *Canadian Forest Policy: Adapting to Change*. Toronto: University of Toronto Press, 2001; B. Cashore, G. Hoberg, M. Howlett, and J. Rayner, *In Search of Sustainability: British Columbia Forest Policy in the 1990s*. Vancouver, BC. University of British Columbia Press 2001.
5. R. H. Nelson, *A Burning Issue: The Case for Abolishing the U.S. Forest Service*. New York. Rowan and Littlefield, 2000; R. A. Sedjo (ed). *A Vision for the U.S. Forest Service: Goals for its Next Century in Memory of Marion Clawson*. Washington, DC: Resources for the Future, 2000.
6. Eighth International Symposium on Society & Resource Management "*Transcending Boundaries: Natural Resource Management from Summit to Sea*." http://www.ac.wwu.edu/~issrm8th/.
7. G. H. Stankey et al., *Learning to learn: adaptive management and the Northwest Forest Plan* USDA-PNW Forest Service Research Report (Portland, OR: in press).
8. CFS, *The State of Canada's Forests 2000-2001—Sustainable Forestry: A Reality in Canada* (Ottawa: Canadian Forest Service, 2001).
9. B. A. Shindler, M. Brunson, and G. H. Stankey, *The social acceptability of forest practices and conditions: a problem analysis* Gen. Tech. Rep. PNW-GTR-537 (Portland, OR: U.S. Department of Agriculture, Forest Service, Pacific Northwest Research Station, 2002).
10. Environment Canada website. http://www.ns.ec.gc.ca/index_e.html
11. E. May, *At the Cutting Edge: The Crisis in Canada's Forests*. (Toronto: Key Porter Books, 1998).
12. G. Horowitz, "Conservatism, socialism, and liberalism in Canada: an interpretation," *Canadian Journal of Economic and Political Science* 32 (1966): 143-71; S. M. Lipset, "Canada and the United States: a comparative view," *Canadian Review of Sociology and Anthropology* 1 (1964): 173-85; B. Steel et al., "The role of scientists in the natural resource and environmental policy process: a comparison of Canadian and American publics," *Journal of Environmental Systems* 28, no. 2 (2001): 133-55.
13. S. M. Lipset, *Continental Divide: The Values and Institutions of the United States and Canada* (New York: Routledge Press, 1990).
14. C. F. Doran, *Forgotten Partnership: US-Canadian Relations Today* (Baltimore: John Hopkins University Press, 1984); R. Cook, "Imagining a North American garden: some parallels and differences in Canadian and American culture," *Canadian Literature* 103 (1984): 10-23.
15. e.g., see M. P. Marchak, *Logging the Globe* (Montreal: McGill-Queens University Press, 1995); R. Hayter, "International trade relations and regional industrial adjustments: the implications of the 1982-86 Canadian-U.S. softwood lumber dispute for British Columbia," *Environment and Planning* 24:153-70 (1992).
16. G. Hoberg, *Regulating forestry: a comparison of institutions and policies in British Columbia and the U.S. Pacific Northwest*, University of British Columbia Forest Economics and

Policy Analysis Research Unit Working Paper 185 (1993).
17. BC Treaty Commission, *What's the Deal with Treaties? A Lay Person's Guide to Treaty Making in British Columbia* (Vancouver, BC: British Columbia Treaty Commission, 2000).
18. Joel Garreau, *The Nine Nations of North America* (New York: Avon Books, 1981).
19. S. E. Moffett, *The Americanization of Canada* (Toronto: University of Toronto Press, 1972); A. Smith, *Canada: An American Nation? Essays on Continentalism, Identity, and the Canadian Frame of Mind* (Montreal: McGill Queen's University Press, 1994).
20. M. Howlett (ed), *Canadian Forest Policy: Adapting to Change.* (Toronto: University of Toronto Press, 2001).
21. J. W. Thomas, "Are there lessons for Canadian foresters lurking south of the border?" *The Forestry Chronicle* 78(3) (2002):382-87.
22. J. Helms (ed.), *The Dictionary of Forestry* (Bethesda, MD: Society of American Foresters, 1998).
23. J. Rayner "Implementing sustainability in west coast forests: CORE AND FEMAT as experiments in process," *Journal of Canadian Studies* 3, no.1 (1996).
24. B. A. Shindler, M. Brunson, and G. H. Stankey, *Social acceptability.*
25. P. Sabatier, " Policy change over a decade or more," in *Policy Change and Learning: An Advocacy Coalition Approach* (Boulder, CO: Westview Press, 1993); K. Lertzman, J. Rayner, and J. Wilson, "Learning and change in the British Columbia forest policy sector: a consideration of Sabatier's advocacy coalition framework," *Canadian Journal of Political Science* 29, no. 1 (1996): 111-34.
26. e.g. J. Rayner "Implementing sustainability;" Shindler, Brunson, and Stankey, *Social acceptability*; Stankey et al., *Learning to learn.*
27. B. A. Shindler, M. Brunson, and G. H. Stankey, *Social acceptability.*
28. M. Howlett, "The politics of long-term policy stability: tenure reform in British Columbia forest policy," in *In Search of Sustainability: British Columbia Forest Policy in the 1990s* ed. B. Cashore et al. (Vancouver: University of British Columbia Press, 2001), 94-119.
29. J. N. Clarke and D. McCool, *Staking Out the Terrain* (Albany: State University of New York Press, 1985).
30. J. H. Cushman, "Forest Service is rethinking its mission," *New York Times,* sec.1 (24 April 1994):22 as cited in T. M. Beckley and D. Korber, "Sociology's potential to improve forest management and inform forest policy," *The Forestry Chronicle* 71, no. 6 (1995): 712-19.
31. M. Brunson and J. Kennedy, "Redefining "multiple-use": agency responses to changing social values." In: R. Knight, S. Bates (eds.), *A New Century for Natural Resource Management* (Washington, DC: Island Press, 1995).
32. Natural Resources Canada, *Evaluation of the Model Forest Program. National Advisory Committee for the Model Forest Program Evaluation* (Ottawa (Ontario): Canada, 1996); Stankey et al. *Learning to learn.*
33. "Bush will overhaul Northwest Forest Plan," *Oregonian* sec. A (Portland, OR: 8 April 2002): 1, 5.

Chapter 2

1. M. Poffenberger, *Communities and forest management in Canada and the United States.* Working Group on Community Involvement in Forest Management. Gland, Switzerland: The World Conservation Union, 1998.
2. J. S. Rowe, *Forest regions of Canada.* Ottawa: Information Canada, Canadian Forest Service, No. 1300 (1972).
3. L. C. Walker, *The North American Forests: Geography, Ecology, and Silviculture.* Boca Raton, FL: CRC Press, 1999.

4. E. Ricciuti, *The Natural History of North America.* Wayne, NJ: BHB International, Inc., 1997.
5. *Ibid.*
6. L. C. Irland, *The Northeast's Changing Forest.* Petersham, MA: Harvard University Press, 1999; C. Merchant, *Ecological Revolutions: Nature, Gender and Science in New England.* Chapel Hill, NC: University of North Carolina Press, 1989.
7. Poffenberger, 1998.
8. S. Foster, *Forest Pharmacy: Medicinal Plants in American Forests.* Durham, NC: Forest History Society, 1995.
9. Merchant, 1989; W. Cronon, *Changes in the Land: Indians, Colonists and the Ecology of New England.* New York: Hill and Wang, 1983.
10. Merchant, 1989.
11. H. A. Innis, *The Fur Trade in Canada: An Introduction to Canadian Economic History.* Toronto: University of Toronto Press, 1956.
12. D. MacCleery, *American Forests: A History of Resilience and Recovery.* Durham, NC: Forest History Society, 1992.
13. C. F. Carroll, *The Timber Economy of Puritan New England.* Providence, RI: Brown University Press, 1973.
14. *Ibid.*
15. G. G. Whitney, *From Coastal Wilderness to Fruited Plain: A History of Environmental Change in Temperate North America.* Cambridge, England: Cambridge University Press, 1994.
16. MacCleery, 1992.
17. *Ibid.*
18. Poffenberger, 1998.
19. A. R. M. Lower, *Great Britain's Woodyard: British America and the Timber Trade, 1763-1867.* Montreal: McGill-Queen's University Press, 1973.
20. MacCleery, 1992.
21. *Ibid.*
22. D. Mackay, *Heritage Lost: The Crisis in Canada's Forests.* Toronto: Macmillan of Canada, 1985; N. Ohanian, *The American Pulp and Paper Industry, 1900-1940.* Westport, CT: Greenwood Press, 1993.
23. Beckley, T. M., "The nestedness of forest-dependence." *Society and Natural Resources* 11, no. 2 (1998): 101-20.
24. MacCleery, 1992.
25. C. R. Humphrey, T. Lewis, and F. H. Buttel, 2002. *Environment, Energy, and Society,* second edition. Belmont, CA: Wadsworth, 2002.
26. Poffenberger, 1998.
27. M. Apsey, D. Laishley, V. Nordin, and G. Paille, "The perpetual forest: Using lessons from the past to sustain Canada's forests in the future." *The Forestry Chronicle* 76, no. 1 (2000): 29-53.
28. H. Steen, *The U.S. Forest Service: A History.* Seattle: University of Washington Press, 1976.
29. Apsey et. al., 2000.
30. MacCleery, 1992.
31. M. Clawson, *Forests For Whom and For What?* Baltimore, MD: John Hopkins University Press, 1975.
32. K. Drushka, *Stumped: The Forest Industry in Transition.* Vancouver, B.C: Douglas & McIntyre Ltd., 1985.
33. L. Pratt and I. Urquhart, *The Last Great Forest: Japanese Multinationals and Alberta's Northern Forests.* Edmonton, AB. NeWest Press, 1994.
34. F. W. Thomas, "Are there lessons for Canadian foresters lurking south of the border?" *The Forestry Chronicle.* 78, no. 3 (2002):382-87; B. A. Shindler, "Does the public have a role in forest management? Canadian and U.S. perspectives." *The Forestry Chronicle.* 74, no. 5 (1998):700-702.
35. Mackay, 1985.
36. R. P. Gillis and T.R. Roach, *Lost Initiatives: Canada's Forest Industries, Forest Policy and Forest Conservation.* New York: Greenwood Press, 1986.
37. D. Haley and M. K. Luckert, "Forest tenures in Canada. A framework for policy analysis." Information Report E-X-43. Ottawa: Forestry Canada, 1990.
38. Natural Resources Canada, *The State of Canada's Forests: Reflections of a Decade, 2001-2002.* Ottawa: Natural Resources Canada, 2002.

Chapter 3

1. e.g., W. C. Clark and R. E. Munn, eds., *Sustainable Development of the Biosphere* (Cambridge, England: Cambridge University Press, 1986).
2. World Commission on Environment and Development, *Our Common Future* (Oxford, England: Oxford University Press, 1987).
3. e.g., Natural Resources Canada, *Sustainable Development Strategy: Now and for the Future* (Ottawa: Natural Resources Canada, 2001).
4. G. F. Weetman, "Recent developments in Western Canadian forest management," *International Forestry Review* 3 (2001): 124-218.
5. H. Hammond, *Seeing the Forest Among the Trees: The Case for Wholistic Forest Use* (Vancouver, B.C.: Polestar Press Ltd., 1991); H. Kimmins, *Balancing Act: Environmental Issues in Forestry* (Vancouver, B.C.: UBC Press, 1992).
6. P. H. Pearse, *Ready for Change: Crisis and Opportunity in the Coast Forest Industry* (Vancouver B.C.: Ministry of Forests, 2001).
7. R. Hayter and T. Barners, eds., *Troubles in the Rainforests: British Columbia's Forest Economy in Transition*, Vol. 33 of *Canadian Western Geographical Series* (Victoria, B.C.: Western Geographical Press, 1997); P. Marchak, S. Aycock, and D. Herbert, *Falldown: Forest Policy in British Columbia* (Vancouver, B.C.: David Suzuki Foundation and Ecotrust Canada, 1999).
8. D. Floyd, *Forest Sustainability: the History, the Challenge, the Promise.* (Durham, NC: the Forest History Society, 2002).
9. *Ibid.*
10. University of Montana Select Committee, *A University View of the Forest Service* (Washington, DC: GPO, 1970).
11. Congressional Research Service, *Ecosystem Management: Federal Agency Activities* (Washington, DC: Congressional Research Service, Library of Congress, 1994).
12. T. Hennessy and D. Soden, "Ecosystem management: the governance approach," in: D. Soden and B. Steel (eds.) *Handbook of Global Environmental Policy and Administration* (New York: Marcel Dekker, Inc., 1999).
13. Jack W. Thomas, "Are there lessons for Canadian foresters lurking south of the border?" *Forestry Chronicle* 78, no. 3 (2002):382-87.
14. CCFM, *Sustainable Development and Forest Management: National Forum Proceedings, Canadian Council of Forest Ministers* (Hull, QC, 1990).
15. Forestry Canada, *The State of Forestry in Canada: 1990 Report to Parliament* (Ottawa: Forestry Canada, 1991).
16. CFS, *The State of Canada's Forests 2000-2001—Sustainable Forestry: A Reality in Canada* (Ottawa: Canadian Forest Service, 2001).
17. W. B. Smith et al., *Canada's Forests at a Crossroads: An Assessment in the Year 2000. A Global Forest Watch Canada Report* (Washington, DC: Worldwatch Institute, 2000).
18. CFS, *State of Canada's Forests.*
19. CFS, *Canada's Model Forest Program: Achieving Sustainable Forest Management Through Partnership* (Ottawa: Canadian Forest Service, 1999).
20. M. von Mirbach, *A User's Guide to Local Level Indicators of Sustainable Forest Management: Experiences form the Canadian Model Forest Network* (Ottawa: Canadian Forest Service, 2000).
21. K. N. Lee, *Compass and Gyroscope: Integrating Science and Politics for the Environment* (Washington, DC: Island Press, 1993).
22. J. L. Young and P. N. Duinker, "Canada's national forest strategies: a comparative analysis," *Forestry Chronicle* 74 (1998): 683-93.
23. CCFM, *Sustainable Forests: A Canadian Commitment* (Hull, QC: National Forestry Strategy, Canadian Council of Forest Ministers, 1992).

24. Anonymous, *Canada Forest Accord* (Hull, QC: Canadian Council of Forest Ministers, 1992).
25. CCFM, *Sustainable Forests* (1992); CCFM, *Sustainable Forests: A Canadian Commitment* (Hull, QC: National Forestry Strategy, Canadian Council of Forest Ministers, 1998).
26. Blue Ribbon Panel, *Final Evaluation Report: National Forest Strategy "Sustainable Forests: A Canadian Commitment"* (Ottawa: National Forest Strategy Coalition, 1997); Independent Expert Evaluation Panel, *A Mid-term Evaluation of the National Forest Strategy (1998-2003), Sustainable Forests: A Canadian Commitment* (Ottawa: National Forest Strategy Coalition, 2001).
27. CCFM, *Sustainable Forests* (1998).
28. CCFM, *Sustainable Forests* (1992).
29. CCFM, *Defining Sustainable Forest Management: A Canadian Approach to Criteria and Indicators* (Ottawa: Canadian Council of Forest Ministers, 1995).
30. CCFM, *Criteria and Indicators of Sustainable Forest Management in Canada: Technical Report* (Ottawa: Canadian Council of Forest Ministers, 1997).
31. Duinker, P. N. "Criteria and indicators of sustainable forest management in Canada: progress and problems in integrating science and politics at the local level," in *Criteria and Indicators for Sustainable Forest Management at the Forest Management Unit Level,* ed. A. Franc, O Laroussinie, and T. Karjalainen (Joensuu, Finland: Proceedings 38, European Forest Institute, 2001), 7-27.
32. e.g., C. Wedeles et al., *Screening Analysis of Sustainable Forest Management Indicators* (Toronto: ArborVitae Environmental Services Ltd., 1998); J. Williams, P. N. Duinker, and C. Wedeles, *Assessing Progress in Sustainable Forest Management: Proposed Criteria and Indicators for the Upper Great Lakes Region: Final Report to the Great Lakes Forest Alliance* (Toronto: ArborVitae Environmental Services Ltd., 1998).
33. CSA, *A Sustainable Forest Management System: Guidance Document* CAN/CSA-Z808-96, *Environmental Technology: A National Standard of Canada* (Etobicoke, ON: Canadian Standards Association, 1996).
34. W. L. Adamowicz et al., "The sustainable forest management network: maintaining scientific excellence and relevance in a changing world," *The Forestry Chronicle* (2002) 78 (1): 112-14.
35. K. N. Johnson and M. Herring, "Understanding bioregional assessments," in: *Bioregional Assessments: Science at the Crossroads of Management and Policy,* K. Johnson, F. Swanson, M. Herring, and S. Greene (eds.), (Washington, DC: Island Press, 1999).
36. Southern Appalachian Man and the Biosphere Cooperative (SAMAB), Southern Appalachian Assessment: social/cultural/economic technical report. (Washington, DC: U.S. Department of Agriculture Office of Communications, 1996).
37. Interior Columbia Basin Ecosystem Management Plan. 1999. Available at http://www.icbemp.gov/.
38. Record of Decision (ROD), Amendments to Forest Service and Bureau of Land Management planning documents within the range of the northern spotted owl. (Washington, DC: U.S. Department of Agriculture, 1994).
39. G. Stankey, B. Bormann, C. Ryan, B. Shindler,. V. Sturtevant, R. Clark, and C. Philpot, "Adaptive management and the Northwest Forest Plan: rhetoric and reality," *Journal of Forestry* 101:1 (2003).
40. Floyd, *Sustainable Forestry.*
41. National Report on Sustainable Forests, USDA Forest Service. See http://www2.srs.fs.fed.us/2003.
42. Floyd, *Sustainable Forestry.*
43. Hennessy and Soden, "Ecosystem management."

44. R. Grumbine, "What is Ecosystem Management?" *Conservation Biology* 8 (1994):27-33.

45. K. Danter, D. Griest, G, Mullins, and E. Norland. "Organization change as a component of ecosystem management," *Society and Natural Resources* 13, no. 6:537-47.

46. e.g., OMNR *Timber Management Guidelines for the Provision of Moose Habitat* (Toronto: Ontario Ministry of Natural Resources, 1988).

47. e.g., L. Greig et al., *Habitat Supply Analysis and Modelling: State of the Art and Feasibility of Implementation in Ontario* Report prepared for the Wildlife Branch, Ontario Ministry of Natural Resources (Richmond Hill, ON: ESSA Ltd., 1991); OMNR, *Forest Management Planning Manual for Crown Forests* (Toronto: Ontario Ministry of Natural Resources, 1996).

48. e.g., P. N Duinker and F. D. Doyon, "Biodiversity assessment in forest-management planning: what are reasonable expectations?" in *Ecosystem Management of Forested Landscapes: Directions and Implementations*, ed. R.G. d'Eon, J. Johnson, and E. A. Ferguson (Vancouver, B.C.: UBC Press, 2000), 135-47; MWFP, *Detailed Forest Management Plan, 1997-2006* Four Volumes & Appendices (Edmonton, AB: Millar Western Forest Products Ltd., 2000).

49. NSDNR, "Nova Scotia's Wildlife Habitat and Watercourses Protection Regulations," Information Circular WDL-7 (Halifax: Nova Scotia Department of Natural Resources, 2001).

50. CSA, *Sustainable Forest Management*, CAN/CSA-Z808-96; CSA, *Sustainable Forest Management*, CAN/CSA-Z809-96.

51. G. L. Baskerville, "Adaptive management: wood availability and habitat availability," *Forestry Chronicle* 61 (1985): 171-75; P. N. Duinker and G. L. Baskerville, "A systematic approach to forecasting in environmental impact assessment," *Journal of Environmental Management* 23 (1986): 271-90.

52. See C. Kayonnis, B. Shindler, and G. Stankey, *Understanding the Social Acceptability of Natural Resource Decision-making Processes Using a Knowledge Base Modeling Approach*, Gen. Tech. Rep. PNW-GTR-518. (Portland, OR: USDA Forest Service, Pacific Northwest Research Station, 2001); A. Thompson and A. Mitchell, Collaborative knowledge management for long-term research, *Forestry Chronicle* 75, no. 3 (1999):491-96.

53. T. Hennessy and D. Soden, "Ecosystem management."

54. P. N. Duinker, "Public participation's promising progress: advances in forest decision-making in Canada," *Commonwealth Forestry Review* 77, no. 2 (1998): 107-12.

55. J.W. Thomas, "Are there lessons for Canadian foresters lurking south of the border?" (*Forestry Chronicle* 78, no. 3 (2002):382-87.

56. S. Yaffee and J. Wondolleck, "Building bridges across agency boundaries," in: J. Franklin and K. Kohm (eds.) *Creating a Forestry for the 21st Century*(Washington, DC: Island Press, 1997, 381-96; B. Shindler and J. Neburka, "Public participation in forest planning: eight attributes of success," *Journal of Forestry* 91, no. 7 (1997):17-19; P. Duinker, "Public participation's promising progress."

57. OMNR, *Timber Management Planning Manual for Crown Lands in Ontario* (Toronto: Ontario Ministry of Natural Resources, 1986).

58. OMNR, *Forest Management Planning.*

59. P. N. Duinker and M. A. Wanlin, "Public participation in forest policy making: experience of Ontario's Forest Policy Panel," in *Forest Forum, Special Issue #8—Public Participation,* ed. G. Blouin and R. Comeau (Ottawa: Canadian Forestry Association, 1995), 30-32.

60. H. Cortner, M. Wallace, S. Burke, and M. Moote, "Institutions matter: the

need to address the institutional challenges of ecosystem management," *Landscape and Urban Planning,* 40 (1998):159-66.

61. OMNR, *Forest Management Planning.*
62. MWFP, *Detailed Forest Management Plan.*
63. CSA, *Sustainable Forest Management,* CAN/CSA-Z808-96; CSA, *Sustainable Forest Management,* CAN/CSA-Z809-96.
64. W.B. Shepard, "Seeing the forest for the trees: "new perspectives" in the Forest Service" *Renewable Resources Journal,* 8 (1990):8-11.
65. K.N. Johnson, J. Agee, R. Beschta [et al.], "Sustaining the people's lands: recommendations for stewardship of the national forests and grasslands into the next century," *Rangelands* 21, no. 4 (1999):25-28.
66. USDA Forest Service, *The Process Predicament: How Statutory, Regulatory, and Administrative Factors Affect National Forest Management,* Washington, DC (June 2002).
67. e.g., MWFP, *Detailed Forest Management Plan.*
68. P. N. Duinker, "Climate change and forest management, policy and land use," *Land Use Policy* 7 (1990): 124-37.
69. e.g., K. A. Armson, *Forest Management in Ontario* (Toronto: Ontario Ministry of Natural Resources, 1976).
70. e.g., C. S. Binkley, "Preserving nature through intensive plantation forestry: the case for forestland allocation with illustrations from British Columbia," *Forestry Chronicle* 73 (1997): 553-59.
71. BC Ministry of Forests, *New Opportunities for Community Forests* (Victoria, BC: BC Ministry of Forests, 2000) http://www.for.gov.bc.ca/pab/jobs/community/factsheet02.htm (4 January 2002).
72. e.g., M. M'Gonigle, B. Egan, and L. Ambus, *The Community Ecosystem Trust: A New Model for Developing Sustainability* (Victoria, BC: University of Victoria, The Polis Project on Ecological Governance, 2001).
73. OMNR, *Ontario's Living Legacy: How Ontario's Living Legacy Began* (Toronto: Ontario Ministry of Natural Resources, 2000). http://www.ontarioslivinglegacy.com/overview.htm (3 January 2002).
74. See Ontario Forest Accord Advisory Board, *State of Ontario Forest Accord* (Toronto: Ontario Ministry of Natural Resources, 2001).
75. B. Cashore, G. Hoberg, M. Howlett, J. Rayner, and J. Wilson, *In Search of Sustainability: British Columbia Forest Policy in the 1990s* (Vancouver: University of British Columbia Press, 2001).

Chapter 4

1. W. R. Freudenburg, L. J. Wilson, and D. J. O'Leary, "Forty years of spotted owls? A longitudinal analysis of logging-industry job losses," *Sociological Perspectives* 41, no. 1 (1998): 1-26; T.J. Barnes, R. Hayter, and W. Hay, "Too young to retire, too bloody old to work: forest restructuring and community response in Port Alberni, British Columbia," *The Forestry Chronicle* 75, no. 5 (1999): 781-87.
2. H. J. Cortner and M. A. Moote, *The Politics of Ecosystem Management* (Covelo, CA: Island Press, 1999).
3. D. Robinson, A. Hawley, and M. Robson, *Social Valuation of the McGregor Model Forest: Assessing Canadian Public Opinion on Forest & Forest Management: Results of the 1996 Canadian Forest Survey* (British Columbia: McGregor Model Forest Association, 1997); B. Shindler, P. List, and B. S. Steel, "Managing federal forests: public attitudes in Oregon and nationwide," *Journal of Forestry* 91, no. 7 (1993): 36-42.
4. S. C. Doak and J. Kusel, "Well-being in forest-dependent communities, part II: a social assessment focus," in *Sierra Nevada Ecosystem Project: Final Report to Congress, vol. II, Assessments and Scientific Basis for Management Options* (University of California, Davis: Centers for Water and

Wildland Resources, 1996): 375-401; C.C. Harris, W. J. McLaughlin, and G. Brown, "Rural communities in the Interior Columbia Basin: how resilient are they?" *Journal of Forestry* 96, no. 3 (1998): 11-15.
5. Robert J. Chaskin, "Building community capacity: a definitional framework and case studies form a comprehensive community initiative," *Urban Affairs Review* 36, no. 3 (2001): 291-323.
6. T. M. Beckley, "Pluralism by default: community power in a mill town," *Forest Science* 42, no. 1 (1996): 35-45; Freudenburg, Wilson, and O'Leary, "Forty years."
7. J. Swift, *Cut and Run: The Assault on Canada's Forests* (Toronto: Between the Lines, 1983): 51-65.
8. R. P. Gillis and T. R. Roach, *Lost Initiatives: Canada's Forest Industries, Forest Policies and Forest Conservation* (Westport, CT.: Greenwood Press, 1986): 31-49.
9. Donald MacKay, *Un patrimoine en péril: la crise des forêts Canadiennes* (Québec: Les publications du Québec, 1986): 65-86.
10. C. H. Schallau and R. M. Alston, "The commitment to community stability: a policy or shibboleth," *Environmental Law* 17 (1987): 429-81.
11. C. H. Schallau, "Sustained yield versus community stability: an unfortunate wedding," *Journal of Forestry* 87, no. 9 (1989): 16-23.
12. R. G. Lee, "Sustained yield and social order," in *Community & Forestry: Continuities in the Sociology of Natural Resources*, ed. Robert G. Lee, D. R. Field, and W. R. Burch, Jr. (Boulder, CO: Westview Press, 1990), 83-94.
13. Schallau, "Sustained yield;" J. Kusel and L. P. Fortmann, "What is community well-being," in *Well-Being in Forest Dependent Communities, (Vol. I)*, ed. J. Kusel and L. Fortmann (Berkeley, CA: Forest and Rangeland Resources Assessment Program and California Department of Forestry and Fire Protection, 1991), 1-45.
14. H. F. Kaufman and L. C. Kaufman, "Toward the stabilization and enrichment of a forest community," in *Community & Forestry: Continuities in the Sociology of Natural Resources*, ed. Robert G. Lee, D. R. Field, and W. R. Burch, Jr. (Boulder, CO: Westview Press, 1990), 27-39.
15. T. R. Waggener, "Community stability as a forest management objective," *Journal of Forestry* 75, no. 11 (1977): 710-14.
16. G. E. Machlis and J. E. Force, "Community stability and timber-dependent communities," *Rural Sociology* 53, no. 2 (1988): 220-34.
17. S. E. Daniels, W. F. Hyde, and D. Wear, "Distributive effects of Forest Service attempts to maintain community stability," *Forest Science* 37, no. 1 (1991): 245-60.
18. D.E. Kromm, "Limitations on the role of forestry in regional economic development," *Journal of Forestry* 70, no. 10 (1972): 630-33; Waggener, "Community stability."
19. R. T. Bowles, "Single-industry resource communities in Canada's north," in *Rural Sociology in Canada* (Oxford University Press, 1992), 63-83; R. N. Byron, "Community stability and forest policy in British-Columbia," *Canadian Journal of Forest Research* 8 (1978): 61-66.; W. R. Freudenburg, "Addictive economies: extractive industries and vulnerable localities in a changing world economy," *Rural Sociology* 57, no. 3 (1992): 305-32.
20. USDA, 1982 as cited in: T. M. Power, *Lost Landscapes and Failed Economies: The Search for a Value of Place* (Covelo, CA: Island Press, 1996); SAF (Society of American Foresters), "Report of the Society of American Foresters National Task Force on Community Stability," *SAF Resource Policy Series* (Bethesda, MD, 1989).
21. SAF, "Report," 6.
22. e.g., J. H. Drielsma, "The influence of forest-based industries on rural communities" (Ph.D. dissertation, Yale University, School of

Environmental Studies and Natural Resources, 1984); G. E. Machlis, J. E. Force, and R. G. Balice, "Timber, minerals, and social change: an exploratory test of two resource dependent communities," *Rural Sociology* 55, no. 3 (1990): 441-24; C. R. Humphrey, "Timber-dependent communities," in *American Rural Communities*, ed. A. E. Lullof and L. E. Swanson (Boulder, CO: Westview Press, 1990): 34-61.

23. J. E. Force et al., "The relationship between timber production, local historical events, and community social change: a quantitative case study," *Forest Science* 39, no. 4 (1993): 722-42.
24. Drielsma, "Influence of forest based industries."
25. P. Marchak, "Forest industry towns in British Columbia," in *Community & Forestry: Continuities in the Sociology of Natural Resources* ed. Robert G. Lee, D. R. Field, and W. R. Burch, Jr. (Boulder, CO: Westview Press, 1990): 95-106; C. Overdest and G. P. Green, "Forest dependence and community well-being: a segmented market approach," *Society and Natural Resources* 8 (1995): 11-131.
26. Drielsma, "Influence of forest based industries."
27. C. H. Schallau, "Can regulation contribute to economic stability," *Journal of Forestry* 72, no. 4 (1974): 214-16; R. N. Byron, "Community stability and forest policy in British-Columbia."
28. S. T. Dana, *Forestry and Community Development* (Washington, DC: United States Department of Agriculture, 1918); E. Foster, "A plan to help stabilize rural economy by the wise use of forest resources," *Journal of Forestry* 39 (1941): 793-99.
29. See W. G. Robbins *Hard Times in Paradise: Coos Bay Oregon, 1850-1986.* (Seattle: University of Washington Press, 1988); M. S. Carroll, "Taming the lumberjack revisited," *Society and Natural Resources* 2 (1989): 91-106
30. G. A. Hillery, "Definitions of community: areas of agreement," *Rural Sociology* 20, no. 2 (1955): 111-23.
31. T. M. Beckley, "The nestedness of forest dependence: a conceptual framework and empirical exploration," *Society and Natural Resources* 11, no. 2 (1998): 101-20.
32. FEMAT (Federal Ecosystem Management Assessment Team), *Forest Ecosystem Management: An Ecological, Economic, and Social Assessment* (Washington, DC: USDA Forest Service, 1993).
33. Kusel and Fortmann, "What is community well-being."
34. T. M. Quigley, R. W. Haynes, and R.T.T. E. Graham, *Integrated Scientific Assessment for Ecosystem Management in the Interior Columbia Basin and Portions of the Klamath and Great Basins* Gen. Tech. Rep. PNW-GTR-382 (Portland, OR: U.S. Department of Agriculture, Forest Service, Pacific Northwest Research Station, 1996).
35. M. Herring, "Introduction," in *Bioregional Assessment: Science at the Crossroads of Management and Policy*, ed. K. N. Johnson, F. Swanson, M. Herring, and S. Greene (Covelo, CA: Island Press, 1999): 1-40.
36. SNEP (Sierra Nevada Ecosystem Project), *Status of the Sierra Nevada, vol. I, Assessment summaries and management strategies (Final report to Congress)* (University of California, Davis: Wildland Resources Center, Centers for Water and Wildland Resources, 1996).
37. FEMAT, *Forest Ecosystem Management.*
38. J. Kusel, "Well-being in forest-dependent communities, Part I: A new approach," in *Sierra Nevada Ecosystem Project: Final Report to Congress, vol. II, Assessments and scientific basis for management options* (University of California, Davis: Centers for Water and Wildland Resources, 1996): 361-74.
39. FEMAT, *Forest Ecosystem Management*; C. B. Flora and J. L.

Flora, "Entrepreneurial social infrastructures: a necessary ingredient," *The Annals of the American Academy of Political and Social Science* 529 (1993): 48-58; Kusel, "Well-being in forest-dependent communities."

40. FEMAT, *Forest Ecosystem Management*; Flora and Flora, "Entrepreneurial;" Kusel 1996, "Well-being in forest-dependent communities."
41. FEMAT, *Forest Ecosystem Management*; Flora and Flora, "Entrepreneurial;" Kusel 1996, "Well-being in forest-dependent communities."
42. Flora and Flora, "Entrepreneurial."
43. Cornelia B. Flora, "Social capital and sustainability: agriculture and communities in the Great Plains and Corn Belt," *Research Sociology and Development* 6 (1995): 227-46.
44. Jan L. Flora, "Social capital and communities of place," *Rural Sociology* 63, no. 4 (1998): 481-506.
45. *Ibid.*; and M. W. Mullen and B. E. Allison, "Stakeholder involvement and social capital: keys to watershed management success in Alabama. *Journal of the American Water Resources Association* 35, no. 3 (1999): 655-62.
46. FEMAT, *Forest Ecosystem Management.*
47. See J. C. Bliss, T. L. Walkingstick, and C. Bailey, "Development or dependency? Sustaining Alabama's forest communities," *Journal of Forestry* 96, no. 3 (1998): 24-30; Doak and Kusel, "Well-being in forest-dependent communities."
48. K. P. Wilkinson, *The Community in Rural America,* (Westport, CT: Greenwood Press, 1991).
49. e.g., Annabel Krishner Cook, "Increasing poverty in timber-dependent areas in western Washington," *Society and Natural Resources* 8 (1995): 97-109; Overdest and Green, "Forest dependence."
50. e.g., Kusel and Fortmann, "What is community well-being;" Bliss, Walkingstick, and Bailey, "Development or dependency?"
51. Bliss, Walkingstick, and Bailey, "Development or dependency?"
52. P. Marchak, *Green Gold: the forest industry in British Columbia* (Vancouver, B.C.: University of British Columbia Press, 1983); Kusel and Fortmann, "What is community well-being;" Overdest and Green, "Forest dependence."
53. Kusel and Fortmann, "What is community well-being;" T. M. Beckley and A. Sprenger, "Social, political, and cultural dimensions of forest-dependence: the communities of the lower Winnipeg basin," in *The Economic, Social, Political and Cultural Dimensions of Forest-Dependence in Eastern Manitoba. RDI Report Series 1995-1* (Brandon University, Manitoba: The Rural Development Institute, 1995): 22-61.
54. Doak and Kusel, "Well-being in forest-dependent communities."
55. Quigley, Haynes, and Graham, "Integrated scientific assessment."
56. R. L. Smith, *Ecology and Field Biology*, third edition (New York: Harper & Row Publishers, 1980): 18-20.
57. Harris, McLaughlin, and Brown, "Rural communities."
58. *Ibid.*

Chapter 5

1. This chapter uses several tables that appeared in B. Steel and E. Weber, "Ecosystem Management, Devolution, and Public Opinion," *Global Environmental Change* 11(2001): 119-31.
2. J. Pierce et al., *Citizens, Political Communication, and Interest Groups: Environmental Organizations in Canada and the United States* (New York: Praeger Publishers, 1992); P. M. Wood, *Biodiversity and Democracy* (Vancouver, BC: University of British Columbia Press, 2000).
3. B. Steel and N. Lovrich, "An introduction to natural resource policy and public lands: changing paradigms and values," in *Public*

Lands Management in the West, ed. B. Steel (New York: Praeger, 1997), 3-15; Gary Orren, "Fall from grace: the public's loss of faith in government," in *Why People Don't Trust Government*, ed. J. S. Nye, Jr., P. D. Zelikow, and D. C. King (Cambridge, MA: Harvard University Press, 1997), 77-108.

4. F. Fischer, *Citizens, Experts, and the Environment* (Durham, NC: Duke University Press), 7.
5. J. Kuklinski, D. Metlay, and W. D. Kay, "Citizen knowledge and choices in complex issue of energy policy," *American Journal of Political Science* 26 (1982): 615-16.
6. B. Steel, N. Lovrich and J. Pierce, "Trust in natural resource information sources and postmaterialist values," *Journal of Environmental Systems* 22 (1992-93): 123-36; Steel and Lovrich, "Introduction to natural resource policy."
7. D. Beardsley, T. Davies, and R. Hersh, "Improving environmental management: what works, what doesn't," *Environment* 39 (September 1997): 6-9, 28-35; E. Weber, *Pluralism by the Rules: Conflict and Cooperation in Environmental Regulation* (Washington, DC: Georgetown University Press, 1998).
8. e.g., M. Chertow and D. Esty, *Thinking Ecologically: The Next Generation of Environmental Policy* (New Haven, CT: Yale University Press, 1997); D. John, *Civic Environmentalism: Alternatives to Regulation in States and Communities* (Washington, DC: Congressional Quarterly Press, 1994).
9. D. Osborne and T. Gaebler, *Reinventing Government* (New York: Basic Books, 1993).
10. Z. Lan and D. Rosenbloom, "A Public Administration in Transition?" *Public Administrative Review* 52 (November/December 1992): 535-37.
11. M. E. Kraft and D. Scheberle, "Environmental Federalism at Decade's End: New Approaches and Strategies," *Publius* 28 (Winter 1998): 131-46.
12. E. Weber, *Pluralism by the Rules*, 256.
13. U. S. Environmental Protection Agency, "The common sense initiative: a new generation of environmental protection," *EPA Insight Policy Paper* 4 [EPA 175-N-94-003] (August 1994).
14. K. Johnson, *Beyond Polarization: Emerging Strategies for Reconciling Community and Environment* (Seattle, WA: The Northwest Policy Center at the University of Washington, 1993); Lisa Jones, "Howdy neighbor! As last resort, westerners start talking to each other," *High Country News* 28 (13 May 1996): 5.
15. Applegate Partnership, "An open letter to the environmental community," [letter to the editor] *High Country News* (July 1996): 2.
16. S. Brown and K. Marshall, "Ecosystem management in state governments," *Ecological Applications* 6 (August 1996): 721.
17. Testimony of Thomas Jorling, Commissioner of the New York Department of Environmental Conservation, Taking Stock hearings (U.S. Congress, "Taking stock of environmental problems, part I," *Senate Hearings Before the Committee on Environment and Public Works, 103rd Congress, 1st Session* (24, 31 March and 16 July 1993), 158-59).
18. as cited in Jones, "Howdy neighbor!," 2.
19. B. Rabe, *Beyond NIMBY: Hazardous Waste Siting in Canada and the United States* (Washington, DC: The Brookings Institution, 1994).
20. In a study of government efforts to site hazardous waste facilities, Rabe (*Beyond NIMBY)* demonstrates the effectiveness of locally based collaborative efforts.
21. John, *Civic Environmentalism.*
22. *Ibid.*
23. R. Robert "Environmental truce clears smoke in rice fields," *The New York Times* (12 December 1992): I8.
24. C. Sirianni and L. Friedland, *Participatory Democracy in America* (Cambridge, MA: Cambridge University Press, forthcoming).

25. R. Haueber, "Setting the environmental policy agenda: the case of ecosystem management," *Natural Resources Journal* 36 (Winter 1996): 1-28.
26. S. Yaffee et al., *Ecosystem Management in the United States: An Assessment of Current Experience* (Washington, DC: Island Press, 1996).
27. E. Weber, "A new vanguard for the environment: grassroots ecosystem management as a new environmental movement," *Society and Natural Resources* 12 (2000): 237-39.
28. F. D. Robertson, "The next 100 years of forest management," in *Proceedings of Transitions, North American Wildlife and Natural Resources Conference* (1991): 19.
29. M. Brunson, "A definition of social acceptability in ecosystem management," in *Defining Social Acceptability in Ecosystem Management: A Workshop Proceedings* Gen. Tech. Rep. PNW-GTR-369 (Portland, OR: USDA Forest Service, Pacific Northwest Research Station (1996).
30. R. Costanza and C. Folke, "The structure and function of ecological systems in relation to property-rights regimes," in *Rights to Nature: Ecological, Economic, Cultural, and Political Principles of Institutions for the Environment,* ed. S. Hanna, C. Folke, K. G. Maler (Washington, DC: Island Press, 1997), 17; Haueber, "Setting the Environmental Policy Agenda," 2.
31. e.g., G. Brown and C. Harris, "The U.S. Forest Service: Toward the New Resource Management Paradigm?" *Society and Natural Resources* 5 (1992): 231-45.
32. See B. Greber and K. N. Johnson, "What's all this debate about overcutting?" *Journal of Forestry* 89, no. 11 (1991): 25-30.
33. K. Dake and A. Wildavsky, "Individual differences in risk perception and risk taking preferences," in *The Analysis, Communication, and Perception of Risk*, ed. B. J. Garrick and W. C. Gekler (New York: Plenum Press, 1991), 15-24.
34. e.g., Steel and Lovrich, "Introduction to natural resource policy."
35. L. Milbrath, *Environmentalists: Vanguard for a New Society* (Albany, NY: State University of New York, 1984); L. Siegelman and E. Yanarella, "Public information and public issues: a multivariate analysis," *Social Science Quarterly* 67 (1986): 402-10; Kenneth Van Liere and Riley Dunlap, "The social bases of environmental concern: a review of hypothesis, explanations and empirical evidence," *Public Opinion Quarterly* 44 (1980): 181-97.
36. R. Dalton, *Citizen Politics in Western Democracies* (Chatham, NJ: Chatham House Publishers, 1988); R. Inglehart, *Culture Shift in Advanced Industrial Society* (Princeton, NJ: Princeton University Press, 1991); Pierce et al., *Citizens, Political Communication.*
37. C. Gilligan, *In a Different Voice: Psychological Theory and Women's Development* (Cambridge, MA: Harvard University Press, 1982).
38. See P. Mohai, "Men, women, and the environment: an examination of the gender gap in environmental concern and activism," *Society and Natural Resources* 5 (1992): 1-19; M. Steger and S. Witt, "Gender differences in environmental orientations: a comparison of publics and activists in Canada and the U.S.," *Western Political Quarterly* 42 (1989): 627-50.
39. Milbrath, *Environmentalists*; B. Steel, D. Soden, and R. Warner, "The impact of knowledge and values on perceptions of environmental risk to the Great Lakes," *Society and Natural Resources* 3 (1990): 331-48.
40. S. Howell and S. Laska, "The changing face of the environmental coalition: a research note," *Environment and Behavior* 24 (1992): 141.
41. *Ibid.*; see also M. Brunson, B. Shindler and B. S. Steel, "Consensus and dissension among rural and

urban publics concerning forest management in the Pacific Northwest," in *Public Lands Management in the West: Citizens, Interest, Groups, and Values*, ed. B. Steel (Westport, CT: Praeger, 1997), 85-97.

42. R. Nash, *Wilderness and the American Mind,* Revised ed. (New Haven, CT: Yale University Press, (1973).
43. See Steel and Lovrich, "Introduction to natural resource policy."
44. S. Dennis and E. R. Zube, "Voluntary association membership of outdoor recreationists: an exploratory study," *Leisure Sciences* 10 (1988): 229-45; J. Hendee, R. Gale and J. Harry, "Conservation, politics and democracy," *Journal of Soil and Water Conservation* 24 (1969): 212-15.
45. B. Steel et al., "Consensus and dissension among contemporary environmental activists: preservationists and conservationists in the U.S. and Canadian Context," *Environment and Planning C* 8 (1990): 379-93.
46. Steel, Soden, and Warner, "Impact of knowledge."
47. J. Calvert, *Partisanship and Ideology in State Legislative Action on Environmental Issues* Paper presented at the Annual Meetings of the Western Political Science Association (Anaheim, CA, March 1987), 2.
48. Van Liere and Dunlap, "Social bases."
49. *Ibid.*; see also Brown and Harris, "U.S. Forest Service."
50. These questions were included as part of a larger national survey concerning public orientations toward government and politics. The survey was designed by Washington State University and Oregon State University faculty and was implemented at Project Vote Smart offices in Corvallis, Oregon.
51. M. Brunson et al., "Nonindustrial private forest owners and ecosystem management: can they work together?" *Journal of Forestry* 94 (June 1996): 14-22.
52. The question used was, "What is your highest level of education?" The following response categories were provided: (1) never attended school, (2) some grade school, (3)completed grade school, (4) some high school, (5) completed high school, (6) some college, (7) completed college, (8) some graduate work, and (9) an advanced degree.
53. Respondents were asked: "Which of the following best describes your place of residence?" Response categories were: (1) rural area, (2) city of 2,500 or less, (3) city of 2,501 to 25,000, (4) city of 25,001 to 50,000, (5) city of 50,001 to 100,000, (6) city of 100,001 to 250,000, and (7) city of 250,001 plus.
54. The question was: "Do you or any of your immediate family depend on natural resource extraction or agriculture for your economic livelihood?"
55. Respondents were asked if they were "a member of an environmental organization."
56. The question and response categories were: "how often do you visit public lands and forests during your leisure time?" (1) never; (2) rarely, no more than once or twice a year; (3) occasionally, several times a year; (4) somewhat frequently, at least once a month on average; (5) very frequently, at least once a week on average.
57. Van Liere and Dunlap, "Social bases."
58. The measure of environmental attitudes used to predict environmental behavior and participation is Dunlap and Van Liere's ("Social bases") New Environmental Paradigm (NEP) indicator. The measure of NEP employed contained a subset of six of the twelve items found in the original inventory and has been found to generate results virtually identical to those of the twelve-item version. The items are as follows: (1) The balance of nature is very delicate and easily upset by human activities; (2) The earth is like a spaceship with only

limited room and resources; (3) Plants and animals do not exist primarily for human use; (4) Modifying the environment for human use seldom causes serious problems; (5) There are no limits to growth for nations like the United States; (6) Humankind was created to rule over the rest of nature. A Likert type response format was provided for each item, taking the following format: "strongly agree," "agree," "neutral," "disagree," and "strongly disagree." A pro-NEP position consists of agreement on the first three items and disagreement on the last three items.

59. The question used a seven-point scale ("very liberal" to "very conservative") to ascertain subjective political ideology regarding domestic policy issues.
60. Brunson et al., "Nonindustrial private forest."
61. The EM description was revised from *Ibid.*
62. Thomas Koontz, *Federalism in the Forest: National versus State Natural Resource Policy* (Washington, DC: Georgetown University Press, 2002).
63. B. A. Shindler, B. S. Steel, and P. List, "Public judgments of adaptive management: a response from forest communities," *Journal of Forestry* 94, no. 6 (1996): 4-12; M. Brunson, B. Shindler and B. S. Steel, "Consensus and Dissension Among Rural and Urban Publics Concerning Forest Management in the Pacific Northwest," in *Public Lands Management in the West: Citizens, Interest Groups, and Values* ed. B. Steel (Westport, CT: Praeger, 1997), 85-97.
64. B. S. Steel, B. A. Shindler and M. Brunson, "Social acceptability of ecosystem management in the Pacific Northwest," in *Ecosystem Management: A Social Science Perspective* eds. D. Soden, B. L. Lamb, and J. Tennert (Dubuque, IA: Kendall-Hunt Publishers, 1998), 147-60.
65. B. Shindler, "Citizen survey of public involvement in federal forest management," Oregon State University Research Report (Corvallis, OR: 1998).
66. H. J. Cortner, M. Wallace, S. Burke, and M. A. Moote, "Institutions matter: the need to address the institutional challenges of ecosystem management," *Landscape and Urban Planning* 40 (1998): 159-56.
67. P.J. Harter, "Negotiating Regulations: A Cure for Malaise," *Georgetown Law Journal* 71 (1982), 1-113; L. Susskind, and J. Cruickshank, *Breaking the Impasse: Consensual Approaches to Resolving Public Disputes* (New York: Basic Books, 1987).

Chapter 6

1. J. Pierce et al., *Political Culture and Public Policy in Canada and the United States: Only a Border Apart* (Lewiston, NY: The Edwin Mellen Press, 2000), 3.
2. D. Elazar, *American Federalism: A View from the* States (New York: Thomas Crowell Publishers, 1966); G. A. and S. Verba, *The Civic Culture Revisited* (Boston: Little, Brown and Company, 1980); Ronald Inglehart, *The Silent Revolution: Changing Values and Political Styles Among Western Publics* (Princeton, NJ: Princeton University Press, 1977); R. Inglehart, *Culture Shift in Advanced Industrial Society* (Princeton, NJ: Princeton University Press, 1990).
3. Inglehart, *Silent Revolution*; Inglehart, *Culture Shift.*
4. K. M. Curtis and J. Carroll, *Canadian-American Relations: The Promise and the Challenge* (Lexington, MA and Toronto: D.C. Heath Publishers, 1983).
5. G. Horowitz, "Conservatism, socialism, and liberalism in Canada: an interpretation," *Canadian Journal of Economic and Political Science* 32 (1966): 143-71.
6. e.g., S. M. Lipset, "Canada and the United States: a comparative view," *Canadian Review of Sociology and Anthropology* 1 (1964): 173-85; S. M. Lipset, *Continental Divide: The Values and Institutions of the United States and Canada* (New York and London: Routledge Publishers, 1990).

7. A. Smith, *Canada: An American Nation? Essays on Continentalism, Identity, and the Canadian Frame of Mind* (Montreal: McGill Queen's University Press, 1994), 161-63.
8. S. J. Arnold and D. J. Tigert, "Canadians and Americans: a comparative analysis," *International Journal of Comparative Sociology* 15 (1974): 68-83; R. Presthus, "Aspects of political culture and legislative behavior: United States and Canada," in *Cross-National Perspectives: United States and Canada*, ed. Robert Presthus (Leiden, The Netherlands: E. J. Brill Publishers, 1977); D. Baer, E. Grabb, and W. Johnson, "National character, regional culture, and the values of Canadians and Americans," *Canadian Review of Sociology and Anthropology* 30 (1993): 13-36; J. Pierce et al., *Citizens, Political Communication, and Interest Groups: Environmental Organizations in Canada and the United States* (New York: Praeger Publishers, 1992).
9. T. R. Waggener, "Forest, timber, and trade: emerging Canadian and U.S. relations under the Free Trade Agreement," *Canadian-American Public Policy* 4 (1990): 24-25.
10. *Ibid.*, 25-26.
11. D. J. Salazar and D. Alper, "Politics, policy, and the war in the woods," in *Sustaining the Forests of the Pacific Coast,* ed. D. J. Salazar and D. Alper (Vancouver, B.C.: University of British Columbia Press, 2000), 3.
12. G. Hoberg, "How the way we make policy governs the policy we make," in *Sustaining the Forests of the Pacific Coast,* ed. D. J. Salazar and D. Alper (Vancouver, B.C.: University of British Columbia Press, 2000), 28-29.
13. *Ibid.*, 29.
14. *Ibid.*, 48.
15. Pierce et al., *Political Culture,* 247.
16. T. Tsurutani, *Political Change in Japan* (New York: McKay Publishers, 1977).
17. Inglehart, *Culture Shift.*
18. A.L. Schneider and H. Ingram, *Policy Design for Democracy* (Lawrence: University Press of Kansas, 1997), 150-88.
19. Lipset, *Continental Divide*, 8.
20. P. A. Pross, *Group Politics and Public Policy* (Toronto: Oxford University Press, 1975).
21. e.g., R. E. Dunlap and K. D. VanLiere, "The new environmental paradigm: A proposed measuring instrument and preliminary results," *Journal of Environmental Education* 9 (1977): 10-19; R. E. Dunlap and K. D. VanLiere, "Commitment to the dominant social paradigm and concern for environmental quality," *Social Science Quarterly* 66 (1984): 1013-27.
22. Pierce et al., *Political Culture.*
23. N. P. Lovrich and J. Pierce, "The good guys and the bad guys in natural resource politics: content and structure of perceptions of interest among general and attentive publics," *Social Science Journal* 23 (1986): 309-26.

Chapter 7

1. A. J. Sinclair and D. L. Smith, "The model forest program in Canada: building consensus on sustainable forest management?" *Society and Natural Resources* 12 (1999): 121-38.
2. Natural Resources Canada, *Model Forest Network—About Us* (Ottawa 2001) http://www.modelforest.net (21 May 2001).
3. Foothills Model Forest, *Foothills Model Forest Homepage* (Hinton, AB, 2001) http://www.fmf.ab.ca (3 May 2001).
4. J. Parkins and T. M. Beckley. "Monitoring community sustainbility in the Foothills Model Forest: A social indicators approach." Information Report M-X-211E. Canadian Forest Service, 2001. 148pp.
5. Statistics Canada, *Census: 1996 Statistical Profile of Canadian Communities* (Ottawa: Statistics Canada, 2001) http://www.statcan.ca/start.html (3 May 2001).
6. M. A. Toman and P. M. S. Ashton, "Sustainable forest ecosystems and management: a review article," *Forest Science* 42, no. 3 (1996): 366-77.

7. D. N. Bengston, "Changing forest values and ecosystem management," *Society and Natural Resources* 7, (1994): 515-33.
8. Z. Xu and D. N. Bengston, "Trends in national forest values among forestry professionals, environmentalists, and the news media, 1982-1993," *Society and Natural Resources* 10 (1997): 43-59.
9. Bengston, "Changing forest values."
10. M. Richardson, J. Sherman, and M. Gismondi, *Winning Back the Words: Confronting Experts in an Environmental Public Hearing.* Toronto: Garamond Press, 1993; L. Pratt and I. Urquhart, *The Last Great Forest: Japanese Multinationals and Alberta's Northern Forests.* Edmonton, AB. NeWest Press, 1994.
11. Bengston, "Changing forest values."
12. *Ibid.*; B. S. Steel, P. List, and B. Shindler, "Conflicting values about federal forests: A comparison of national and Oregon publics," *Society and Natural Resources* 7 (1994): 137-53; B. L. McFarlane and P. C. Boxall, *Forest Values and Management Preferences of Two Stakeholder Groups in the Foothills Model Forest* Inf. Rep. NOR-X-364 (Edmonton, AB: Natural Resources Canada, Canadian Forest Service, Northern Forestry Centre and Foothills Model Forest, Hinton, AB, 1999).
13. Bengston, "Changing forest values."
14. McFarlane and Boxall, *Forest Values and Management.*
15. B. L. McFarlane and P. C. Boxall, *Factors influencing forest values and attitudes of two stakeholder groups: the case of the Foothills Model Forest,* Alberta, Canada, *Society and Natural Resources* 13 (2000): 649-61.
16. McFarlane and Boxall, *Forest Values and Management*; B. L. McFarlane and P. C. Boxall, *Forest Values and Attitudes of the Public, Environmentalists, Professional Foresters, and Members of Public Advisory Groups in Alberta* Inf. Rep. NOR-X-374 (Edmonton, AB: Natural Resources Canada, Canadian Forest Service, Northern Forestry Centre and Foothills Model Forest, Hinton, AB, 2000).
17. T. M. Beckley et al., *Forest Stakeholder Attitudes and Values: Selected Social-Science Contributions* Inf. Rep. NOR-X-362 (Edmonton, AB: Natural Resources Canada, Canadian Forest Service, Northern Forestry Centre, 1999).
18. T. Heberlein, "Some observations on alternative mechanisms for public involvement: the hearing, the public opinion poll, the workshops, and the quasi-experiment," *Natural Resources Journal* 16 (1976): 204-12; S. Dennis, "Incorporating public opinion surveys in national forest land and resource planning," *Society and Natural Resources* 1 (1988): 309-16; J. E. Force and K. L. Williams, "A profile of national forest planning participants," *Journal of Forestry* 87, no. 1 (1989): 33-38.
19. McFarlane and Boxall, *Forest Values and Attitudes.*
20. Steel, List, and Shindler, "Conflicting values."
21. McFarlane and Boxall, *Forest Values and Attitudes.*
22. J. Hetherington, T. C. Daniel, and T. C. Brown, "Anything goes means everything stays: the perils of uncritical pluralism in the study of ecosystem values," *Society and Natural Resources* 7 (1994): 535-46.
23. B. L. McFarlane, M. A. Fisher, and P. C. Boxall, *Camper Characteristics and Preferences at Managed and Unmanaged Sites in the Foothills Model Forest* Forest Management Notes 64 (Edmonton, AB: Natural Resources Canada, Canadian Forest Service, Northern Forestry Centre and Foothills Model Forest, Hinton, AB, 1999).
24. B. L. McFarlane and P. C. Boxall, *An Overview and Nonmarket Valuation of Camping in the Foothills Model Forest* Inf. Rep. NOR-X-358 (Edmonton, AB: Natural Resources Canada, Canadian Forest Service, Northern Forestry Centre and Foothills Model Forest, Hinton, AB, 1998).

25. J. R. R. Alavalapati, W. L. Adamowicz, and W. A. White, "A comparison of economic impact assessment methods: the case of forestry developments in Alberta," *Canadian Journal of Forest Resources* 28, no. 5 (1998): 711-19.
26. D. McMenamin and J. Haring, "An appraisal of nonsurvey techniques for estimating regional output models," *Journal of Regional Science* 14, no. 2 (1974): 191-205.
27. See M. Patriquin et al., "A Comparison of Impact Measures for Hybrid and Synthetic Techniques: A Case Study of the Foothills Model Forest" (manuscript submitted for publication, 2001).
28. A. Wellstead, C. Olsen, and W. White, *Measuring the Economic Value of the Visitor Sector of a Regional Economy: A Case Study of the Foothills Model Forest* Inf. Rep. NOR-X-378 (Edmonton, AB: Natural Resources Canada, Canadian Forest Service, Northern Forestry Centre and Foothills Model Forest, Hinton, AB, 2001).
29. A. M. Wellstead, M. N. Patriquin, and W. A. White, *A Study of Visitor-Sector Employment in the Foothills Model Forest* Inf. Rep. NOR-X-377 (Edmonton, AB: Natural Resources Canada, Canadian Forest Service, Northern Forestry Centre and Foothills Model Forest, Hinton, AB, 2000).
30. P. Jagger, A. Wellstead, and W. White, *An Expenditure Based Analysis of Community Dependence: A Case Study of the Foothills Model Forest* Unpublished manuscript (Edmonton, AB: Natural Resources Canada, Canadian Forest Service, Northern Forestry Centre, 1998).
31. J. R. R. Alavalapati, W. White, and M. Patriquin, "Economic impact of changes in the forestry sector: a case study of the Foothills region in Alberta," *Forestry Chronicle* 75, no. 1 (1999): 121-28.
32. E. Golan, I. Adelman, and S. Vogel, *Environmental Distortions and Welfare Consequences in a Social Accounting Framework* Working Paper No. 751 (Berkeley, CA: Department of Agricultural and Resource Economics, University of California at Berkeley, 1995).
33. See J. R. R. Alavalapati and W. L. Adamowicz, "Modeling economic and environmental impacts in tourism and resource extraction dependent regions," *Annual Tourism Resources* 27, no. 1 (2000): 188-202; M. Patriquin et al., *Incorporating Natural Capital into Economy-Wide Impact Analysis: A Case Study from Alberta* (manuscript submitted for publication, 2001).
34. See Patriquin et al., *Incorporating Natural Capital.*

Chapter 8

1. J. Habermas, "The public sphere: an encyclopedia article," *New German Critique* 3 (1974): 49-55; J. Habermas, *The Structural Transformation of the Public Sphere. An Inquiry into a Category of Bourgeois Society* trans. T. Burger and F. Lawrence (Cambridge, MA: MIT Press, 1989 [1962]).
2. L. Pratt and I. Urquhart, *The Last Great Forest. Japanese Multinationals and Alberta's Northern Forests* (Edmonton, AB: Northwest Press, 1994).
3. B. Udell, Proceedings from the forest resource advisory group conference, September 24-26 (Hinton, Alberta: unpublished document, 1999).
4. P. Duinker, "Public participation's promising progress: advances in forest decision-making in Canada," *Commonwealth Forestry Review* 17, no. 2 (1998): 107-12; R. L. Lawrence and S. E. Daniels, "Public involvement in natural resource decision making," *Paper in Forest Policy* 3 (Corvallis, OR: Forest Research Laboratory, College of Forestry, Oregon State University, 1996); T. M. Beckley, *Public Involvement in Natural Resource Management in the Foothills Model Forest* (Edmonton, AB: unpublished manuscript, Northern Forestry Centre, 1999).

5. R. Lawrence, S. E. Daniels, and G. Stankey, "Procedural justice and public involvement in natural resources decision making," *Society and Natural Resources* 10, no. 6 (1997): 578.
6. B. Shindler, K. A. Cheek, and G. H. Stankey, *Monitoring and Evaluating Citizen-Agency Interactions: A Framework Developed for Adaptive Management* Gen. Tech. Rep. PNW-GTR-452 (Portland, OR: U.S. Department of Agriculture, Forest Service, Pacific Northwest Research Station, 1999).
7. See K. Lee, *Compass and Gyroscope* (Washington, DC: Island Press, 1993); K. Lee, "Appraising adaptive management," in *Biological Diversity: Balancing Interests through Adaptive Collaborative Management,* ed. L. Buck, C. G. Geisler, J. Schelhas, and E. Wollenburg (Boca Raton, FL: CRC Press, 2001), 3-26.
8. B. Shindler and K. A. Cheek, "Integrating citizens and adaptive management: a propositional analysis," *Journal of Conservation Ecology* 3, no. 1 (1999): 13.
9. e.g. K. A. McComas, "Public meetings about local waste management problems: comparing participants to nonparticipants," *Environmental Management* 27, no. 1 (2001): 135-47; A. Wellstead, R. Stedman, and J. Parkins, "Understanding the concept of representation in forest decision making," *Forest Policy and Economics* (2003) 5 (1): 1-11; C. Overdevest, "Participatory democracy, representative democracy, and the nature of diffuse and concentrated interests: a case study of public involvement on a National Forest District," *Society and Natural Resources* 13 (2000): 685-96; K. G. Gundry and T. A. Heberlein, "Do public meetings represent the public?" *Journal of the American Planning Association* (Spring, 1984): 175-82.
10. Wellstead et al., 2002.
11. Habermas, "Public sphere."
12. C. Calhoun, "Introduction: Habermas and the public sphere," in *Habermas and the Public Sphere,* ed. Craig Calhoun (Cambridge, MA: MIT Press, 1992).
13. Trust in this context is somewhat distinct from other uses where trustworthy relationships are the defined goal (B. A. Shindler, M. Brunson, and G. H. Stankey, *Social Acceptability of Forest Conditions and Management Practices: A Problem Analysis* Gen. Tech. Rep. PNW-GTR-XXX (Portland, OR: U.S. Department of Agriculture, Forest Service, Pacific Northwest Research Station, forthcoming).
14. Calhoun, "Introduction."
15. J. R. Parkins, "Forest management and advisory groups in Alberta: an empirical critique of an emergent public sphere," *The Canadian Journal of Sociology* 27, no. 2 (2002).
16. J. Forester, "Critical theory and planning practice," in *Critical Theory and Public Life,* ed. John Forester (Cambridge, MA: MIT Press, 1985).
17. J. S. Dryzek, "Critical theory as a research program," in *The Cambridge Companion to Habermas,* ed. Stephen K. White (Cambridge, MA: MIT Press, 1995).
18. E. Skollerhorn, "Habermas and nature: the theory of communicative action for studying environmental policy," *Journal of Environmental Planning and Management* 41, no. 5 (1998): 555-73.
19. R. Kemp, "Planning, public hearings, and the politics of discourse," in *Crticial Theory and Public Life,* ed. John Forester (Cambridge, MA: MIT Press, 1985); T. Webler, "'Right' discourse in citizen participation: an evaluative yardstick," in *Fairness and Competence in Citizen Participation,* ed. O. Renn, T. Webler, and P. Wiedemann (Norwell, MA: Kluwer Academic Publishers, 1995).
20. Kemp, "Planning, public hearings," 186.
21. *Ibid.,* 190.
22. Webler, "'Right discourse," 38.

23. Dryzek, "Critical theory."
24. Skollerhorn, "Habermas and nature."
25. by S. Tuler and T. Webler, "Voices from the forest: what participants expect of a public participation process," *Society and Natural Resources* 12 (1999): 437-53.
26. See J. R. Parkins, R. C. Stedman, and B. L. McFarlane, *Public Involvement and Alberta Forest Management: Do Advisory Groups Represent the Public?* Northern Forest Centre Information Report NOR-X-382 (Edmonton, AB, 2001) for a more detailed methodology.
27. Beckley, *Public Involvement;* Duinker, "Public participation's;" Lawrence and Daniels, "Public involvement."
28. Duinker, "Public participation's."

Chapter 9

1. See Barnett Richling, "'You'd never starve here': Return migration to rural Newfoundland." *Canadian Review of Sociology and Anthropology* 22, no. 2 (1985): 236-49, and John T. Omohundro, John T., *Rough food: The seasons of subsistence in Northern Newfoundland.* (St. John's, NF: Institute of Social and Economic Research, 1994).
2. W. R. Catton, *Overshoot, the Ecological Basis of Revolutionary Change* (Urbana, IL: University of Illinois Press, 1980).
3. Michael Harris, *Lament for an Ocean: The Collapse of the Atlantic Cod Fishery, a True Crime Story* (Toronto: McClellan & Stewart, 1998).
4. See Jeffrey A. Hutchings and Ransom A. Myers, "What can be learned from the collapse of a renewable resource? Atlantic Cod, *Gadus morhua,* of Newfoundland and Labrador." *Canadian Journal of Fisheries and Aquatic Sciences* 51 (1994): 2126-46, and Craig Palmer and Peter Sinclair. 1997. *When the Fish are Gone: Ecological Disaster and Fishers in Northwest Newfoundland.* Halifax, NS: Fernwood Publishing.
5. J. A. Gray. 1981. *The Trees behind the Shore: The Forests and Forest Industries of Newfoundland and Labrador.* (Ottawa: Economic Council of Canada, 1981). Interestingly, a later document lists Newfoundland as having 57 percent productive forest cover. The land has not changed so much as the classification of land to justify harvest levels.
6. J. K. Hiller, "The politics of newsprint: The Newfoundland pulp and paper industry, 1915-1939," *Acadensis.* 19, no 2 (1990): 3-39.
7. *Ibid.*
8. Gray. 1981.
9. SIMFOR, *Socio-economic Indicators for the Model Forest* (2001). http://fms.nofc.cfs.nrcan.gc.ca:8080/Simfor/Main.htm
10. Natural Resources Canada (2001). http://www.nrcan.gc.ca/cfs-scf/national/what-quoi/sof/sof01/profiles_e.html#NF
11. *Ibid.*
12. M. den Otter and T. M. Beckley, *"This is Paradise:" Monitoring Community Sustainability in the Western Newfoundland Model Forest Using Subjective and Objective Approaches* (Fredericton, NB: Atlantic Forestry Centre Information Report M-X-216E. Canadian Forest Service, 2002).
13. D. A. Dillman, *Mail and Telephone Surveys: The Total Design Method* (New York, NY: John Wiley & Sons 1978).
14. M. A. Roy, "Guided change through the community forestry: a case study in Forest Management Unit 17 Newfoundland," *The Forestry Chronicle* 65, no. 5 (1989): 344-47.
15. Elaine DuWors, Pierre Villeneuve, and F. L. Fillion, *The Importance of Nature to Canadians: Survey Highlights* (Ottawa: Environment Canada, 1999).
16. Environics, *1991 National Survey of Canadian Public Opinion on Forestry Issues,* Report to Forestry Canada (1992).
17. D. Robinson, M. Robson and A. Hawley, *Social Valuation of the McGregor Model Forest: Assessing*

Canadian Public Opinion on Forest Values and Forest Management—Results of the Canadian Forest Survey '96. Prepared for the McGregor Model Forest Association, 1997.

18. TAWD Consultants Limited, *Public Attitude Survey*, prepared for the Newfoundland Department of Forestry and Agriculture and the Protected Areas Association, 1991.

Chapter 10

1. E.g., National Research Council, *Water Transfers in the West: Efficiency, Equity, and the Environment* (Washington, DC: National Academy Press, 1992); Daniel A. Mazmanian and Michael Kraft, eds., *Toward Sustainable Communities: Transition and Transformations in Environmental Policy* (Cambridge, MA: MIT Press, 1999); Edward P. Weber, *Pluralism by the Rules: Conflict and Cooperation in Environmental Regulation* (Washington, DC: Georgetown University Press, 1998).
2. Steve Born and Kenneth Genskow, *Exploring the Watershed Approach: Critical Dimensions of State-Local Partnerships* (Portland, OR: River Network, 1999).
3. E. P. Weber, "A new vanguard for the environment: grass-roots ecosystem management as a new environmental movement," *Society and Natural Resources* 13 (2000): 237-59.
4. D. Snow, "Coming home," *Chronicle of Community* 1, no. 1 (1996): 40-43.
5. B. Cestero, *Beyond the Hundredth Meeting: A Field Guide to Collaborative Conservation on the West's Public Lands* (Tucson, AZ: Sonoran Institute, 1999).
6. Born and Genskow, *Exploring the Watershed Approach*; Rieke and Kenney, *Resource Management*; some might be inclined to add the "sustainable communities" label to this long list despite their heavy concentration in urban areas (see Lamont C. Hempel, *Sustainable Communities: From Vision to Action* (Claremont, CA: Claremont Graduate University, 1998, unpublished).
7. See Dan Dagget, *Beyond the Rangeland Conflict: Toward a West that Works* (Layton, UT: Gibbs Smith, 1995); John, *Civic Environmentalism*; K. Johnson, *Beyond Polarization: Emerging Strategies for Reconciling Community and Environment* (Seattle, WA: The Northwest Policy Center at the University of Washington, 1993).
8. Lamont C. Hempel, "Conceptual and analytical challenges in building sustainable communities," in *Toward Sustainable Communities*, ed. Mazmanian and Kraft, 48, 52; Willapa Alliance, *Willapa Indicators for a Sustainable Community: 1996 Summer* (South Bend, WA, 1996), 5; Applegate Partnership "An open letter to the environmental community [letter to the editor]," *High Country News* 2 (1996), 1 http://www.hcn.org/home_page/dir/email_letters.html (20 January 1998).
9. Daniel Mazmanian and Michael Kraft, "The three epochs of the environmental movement," in *Toward Sustainable Communities*, ed. Mazmanian and Kraft, 3-42; Hempel, in *Sustainable Communities.*
10. Willapa Alliance, *Places, Faces and Systems of Home: Connections to Local Learning* (South Bend, WA: 1998), 160.
11. The primary interviews with anonymous respondents were conducted between March of 1998 and October of 1999. The interview sessions were semi-structured but purposefully left open-ended, from forty minutes to two hours in length.
12. C. V. Hollander, "Groups cultivate stake in Willapa Bay's future," in *Willapa: Banking on the Bay* compiled by *The Daily Astorian and the Chinook Observer* (1995), 8-9; vol. 1 of *International Journal of Sustainable Development and World Ecology*, 22-33.
13. See Daniel Kemmis, *This Sovereign Land: A New Vision for Governing the West* (Washington, DC: Island Press, 2001).

14. *Chinook Observer,* "A Good Start: Weekend Conference Made Progress in Uniting Pacific County Citizens"(1996); interviews 3/17/98s; 3/18/98a.
15. Tradeoffs among goals are still inevitable at the level of individual decisions. Yet the philosophy argues that every choice be made with an eye toward the overall health and maintenance of the larger system, defined as it is by the three major goals of environmental, economic, and community health.
16. Hempel, "Conceptual and analytical challenges."
17. See Robert Costanza and Carl Folke, "The structure and function of ecological systems in relation to property-rights regimes," in *Rights to Nature,* ed. Hanna, Folke, and Maler (Washington, DC: Island Press, 1996).
18. Kai N. Lee, *Compass and Gyroscope: Integrating Science and Politics for the Environment* (Washington, DC: Island Press, 1993); interviews 3/17/98a; 5/13/98a; 8/5/98a; 8/11/98b; 5/13/99a; 5/17/99a; 5/18/99b.
19. Moseley, "New Ideas," 63; Kevin Priester and James Kent, "Social ecology: new pathways to ecosystem restoration," in *Watershed Restoration: Principles and Practices,* ed. J. E. Williams, M. P. Dombeck, and C. A. Woods (Bethesda, MD: American Fisheries Society, 1997).
20. Kevin Priester, *Words into Action: A Community Assessment of the Applegate Valley* May (Ashland, OR: The Rogue Institute of Ecology and Economy, 1994), 121; interviews 3/13/98c; 8/6/98b; 8/11/98a; 5/12/99b; 5/14/99b; 5/18/99d; 7/19/99a; 10/8/99b
21. interview 7/19/99c
22. interview 5/18/99b
23. Francis Fukuyama, *Trust: The Social Virtues and the Creation of Prosperity* (New York: The Free Press, 1995); Robert D. Putnam, *Making Democracy Work: Civic Traditions in Modern Italy* (Princeton, NJ: Princeton University Press, 1993).
24. interview 7/19/99a
25. interviews 8/11/98a; 8/11/98b
26. The Emergency Supplemental Appropriations Act (P. L. 104-19, 109 Stat. 240-247).
27. J. Lowe, "Memorandum to Forest Supervisors," August 16 (Portland, OR: U.S. Department of Agriculture, Forest Service, Pacific Northwest Region, 1993). There have always been exceptions to the rule. In a 1992 memorandum, for example, timber planners in the Malheur National Forest were instructed to identify sales as salvage "as long as one board comes off that would qualify as salvage" (B. Boaz, "Message to Staff," December 17 (U.S. Department of Agriculture, Forest Service, Malheur National Forest, Oregon, 1992).
28. Challenges are required to be filed in local U.S. District Courts (where lands in question were located) within fifteen days of the initial advertisement of the sale.
29. R. W. Gorte, "The salvage timber sale rider: overview and policy issues," *Congressional Research Service* June 24 (96-569 ENR, 1996)2.
30. Wilderness Society and National Audubon Society, *Salvage Logging in the National Forests: An Ecological, Economic, and Legal Assessment* July (Washington, DC, 1996); General Accounting Office, *Emergency Salvage Sale Program.*
31. Margaret Shannon, Victoria Sturtevant, and Dave Trask, *Organization for Innovation: A Look at the Agencies and Organizations Responsible for Adaptive Management Areas: The Case of the Applegate AMA,* Report submitted to Interagency Liaison for the U.S. Forest Service and Bureau of Land Management, Applegate Adaptive Management Area, Oregon (February, 1997), 19.
32. interviews 1/23/98; 2/9/98; Marston, "The timber wars evolve;" Rolle, "5 Years and Still Cookin.'"
33. interview 5/17/99a
34. interview 5/18/99b
35. interview 3/17/98a; Hollander "Groups cultivate stake," 8-9; Marston "We're much stronger together," 1.

36. Bell and Morse *Sustainability Indicators*; Richard C. Box, *Citizen Governance: Leading American Communities into the 21st Century* (Thousand Oaks, CA: Sage Publications, 1998); Knopman, *Second Generation*; J. C. Scott, *Seeing Like a State: How Certain Schemes to Improve the Human Condition Have Failed* (New Haven and London: Yale University Press, 1998); Snow, "Coming home."
37. Other topics have included (1) hatcheries and hatchery reform, (2) the Washington State Wild Salmonid Policy, the Governor's Salmon Plan, and the State Legislature's Salmon Recovery Plan, and (3) Water Productivity and Water Quality—Fact, Fiction, and Potential Impacts (Willapa Alliance and Pacific County Economic Development Council, *Willapa Conflict Resolution Forums: Endangered Species, Habitat, Water and Natural Resource Users, and Regulations* (South Bend, WA, 1997)).
38. For example, in *A Home for All of Us*, (Jacksonville, OR: The Applegate Partnership 1996) and Applegate River Watershed Council seek "to provide [citizens] with an understanding of the interconnection of all varieties of life in this valley … [and] to offer simple guidelines to responsible practices for interacting with both the natural resources and your human neighbors.
39. Willapa Alliance, *Cover Letter to Willapa Educators Introducing the* Places, Faces and Systems of Home *School Resource Guide* (South Bend, WA, 1998a); Willapa Alliance, *Places, Faces and Systems of Home: Connections to Local Learning* (South Bend, WA, 1998b).
40. Willapa Alliance, *Willapa—The Nature of Home: A Place Based Community Education Program* (South Bend, WA, 1997a), 1.
41. interview 5/17/99a
42. Landy, "Public policy and citizenship," in *Public Policy for Democracy,* ed. H. Ingram and S. R. Smith (Washington, DC: The Brookings Institution, 1993), 19-44; Michael J. Sandel, *Democracy's Discontent: America in Search of a Public Philosophy* (Cambridge, MA: The Belknap Press of Harvard University Press, 1996), 5-6.
43. interview 5/13/99a
44. interview 5/18/99b
45. Willapa Alliance, *The Willapa Alliance: Completing the Foundation: 1995 Annual Report* (South Bend, WA, 1995), 6; Willapa Alliance, "The Willapa Alliance: Willapa and the Web of Sustainability: Annual Report" (South Bend, WA; 1996b:4); interview 3/17/98a).
46. interview 5/17/99a
47. Dan Daggett, Keynote address to 1995 HFWC Annual State of the Watershed Conference.
48. Snow, "Empire or homelands?" 198.
49. interviews 3/11/98b; 3/17/98a
50. Charles C. Mann and Mark L. Plummer, *Noah's Choice: The Future of Endangered Species* (New York: Alfred Knopf, 1995); interviews 8/12/98a; 5/18/99b; 10/8/99b.
51. interviews 3/13/98d; 5/13/98a; 8/6/98b; 8/7/98a; 8/11/98b; 11/18/98; 5/14/99b; 5/18/99a; 10/8/99b
52. Dagget, *Beyond the Rangeland Conflict.*
53. David H. Getches, "Foreword," in *Beyond the Rangeland Conflict: Toward a West that Works* (Layton, UT: Gibbs Smith, 1995), viii; emphasis added.
54. interviews 3/11/98b; 3/13/98d; 5/19/98c; 8/5/98a; 8/6/98c
55. interview 3/17/98a
56. interviews 8/6/98b; 8/7/98e; 8/12/98a; 5/13/99a; 5/13/99b
57. interview 3/17/98a
58. interviews, 8/7/99a; 8/11/98b; 10/27/99b
59. interview 5/18/99a
60. as quoted in Johnson 1997, *Toward a Sustainable Region,* 32.
61. See also the case of the Feather River Alliance, an example of grassroots ecosystem management in California.

62. interviews 8/10/98d; 11/17/98; 5/13/99b; 5/17/99a
63. This is a key tenet of the emerging field of sustainability science. See Kates et al., "Policy Forum," but also Susan Hanna, Carl Folke, and Karl-Goran Maler, eds. *Rights to Nature: Ecological, Economic, Cultural, and Political Principles of Institutions for the Environment* (Washington, DC: Island Press, 1996).
64. interview 8/13/98a
65. Kates et al., "Policy forum," 641; Scott, *Seeing Like a State.*
66. Jack H. Knott and Gary J. Miller, *Reforming Bureaucracy* (Englewood Cliffs, NJ: Prentice-Hall, 1987), 172-81.
67. Scott, *Seeing Like a State.*
68. *Ibid.*, 6, 7.
69. *Ibid.*, 7.
70. interview 10/27/99a
71. interview 5/14/99b
72. Bell and Morse, *Sustainability Indicators*; J. T. Heinen, "Emerging, Diverging and Converging Paradigms on Sustainable Development," *International Journal of Sustainable Development and World Ecology*, 1 (1994): 22-33; Hempel, "Conceptual and analytical challenges".
73. Hempel, "Conceptual and analytical challenges," 63.
74. Heinen, "Emerging, Diverging and Converging Paradigms on Sustainable Development."
75. as quoted in Bell and Morse, *Sustainability Indicators*, 16.
76. interviews 5/19/98b; 8/10/98b; 5/13/99a; 5/18/99d.
77. Willapa Alliance, *Willapa Indicators for a Sustainable Community.*
78. Rogue Institute for Ecology and Economy, *Proposed Criteria and Indicators for Measuring Sustainable Forest Management in the U.S.* (Draft of Presentation at Applegate Partnership Meeting, September 10).
79. Shipley, "The Applegate Partnership"; Johnson, *Toward a Sustainable Region*; Victoria E. Sturtevant and Jon I. Lange, *Applegate Partnership Case Study: Group Dynamics and Community Context* (Seattle, WA: U.S. Forest Service, Pacific Northwest Research Station, 1995).
80. interviews, 5/14/99a; 5/17/99a; 5/18/99b; 5/18/99c; 10/8/99a; meeting minutes 2/3/99
81. See Weber, *Pluralism by the Rules*, 114.
82. See Weber, *Bringing Society Back In.*
83. See Herbert Kaufman, *The Forest Ranger* (Washington, D.C.: Resources for the Future Press, 1967).
84. interviews 5/24/02; 5/28/02
85. See Chapters 4 through 6 in Weber, *Bringing Society Back In.*

Chapter 11

1. W. J. Allen and O. J. H. Bosch, "Shared experiences: the basis for a cooperative approach to identifying and implementing more sustainable land management practices," in *Proceedings of Symposium Resource Management: Issues, Visions, Practice* (New Zealand: Lincoln University, 5-8 July 1996); C. J. Walters, *Adaptive Management of Renewable Resources* (New York: Macmillan Publishing Company, 1986).
2. E. D. Ford, *Scientific Methods for Ecological Research* (Cambridge, England: Cambridge University Press, 2000).
3. K. N. Lee, *Compass and Gyroscope: Integrating Science and Politics for the Environment* (Washington, DC: Island Press, 1993); Walters, *Adaptive Management.*
4. H. J. Cortner and M. A. Moote, *The Politics of Ecosystem Management* (Washington, DC: Island Press, 1999).
5. J. Pipkin, *The Northwest Forest Plan Revisited* (Washington, DC: U.S. Department of the Interior, Office of Policy Analysis, 1998).
6. H. J. Cortner et al., *Institutional barriers and incentives for ecosystem management: a problem analysis* Gen. Tech. Rep. PNW-GTR-354 (Portland, OR: U.S. Department of Agriculture, Forest Service, Pacific Northwest Research Station, 1996).

7. Pipkin, *Northwest Forest Plan*; Lee, *Compass and Gyroscope*; B. L. Johnson, "Introduction to the special issue: adaptive management—scientifically sound, socially challenged?" *Conservation Ecology* 3, no. 1 (1999): 10.
8. C. S. Holling, *Adaptive Environmental Assessment and Management. International Series on Applied Systems Analysis 3* (Toronto: John Wiley & Sons).
9. G. H. Stankey et al., *Learning to Learn: Adaptive Management and the Northwest Forest Plan*, Draft Report (2000).
10. Walters, *Adaptive Management*.
11. Lee, *Compass and Gyroscope*.
12. L. H. Gunderson, C. S. Holling, and S. S. Light, *Barriers and Bridges to the Renewal of Ecosystems and Institutions* (New York: Columbia University Press, 1995).
13. Stankey et al., *Learning to Learn*.
14. C. L. Halbert, "How adaptive is Adaptive Management? Implementing Adaptive Management in Washington State and British Columbia," *Reviews in Fisheries Science* 1, no. 3 (1993): 261-83.
15. C. J. Walters and C. S. Holling, "Large-scale management experiments and learning by doing," *Ecology* 71, no. 6 (1990): 2060-68.
16. Lee, *Compass and Gyroscope*.
17. Forest Ecosystem Management Assessment Team (FEMAT), *Forest Ecosystem Management: An Ecological, Economic, and Social Assessment* (1993).
18. Record of Decision (ROD) for Amendments to Forest Service and Bureau of Land Management Planning Documents Within the Range of the Northern Spotted Owl, Standards and Guidelines for Management of Habitat for Late-Successional and Old-Growth Forest Related Species Within the Range of the Northern Spotted Owl (1994).
19. M. Shannon, V. Sturtevant, and D. Trask, *An Independent Assessment of the Agencies and Organizations Responsible for the Adaptive Management Areas: the case of the Applegate AMA*. A report to the Interagency Liaison, USDA Forest Service and USDI Bureau of Land Management (Medford, OR: Applegate Adaptive Management Area, 1995).
20. J. F. Franklin, "Adaptive Management Areas," *Journal of Forestry* 92, no. 4 (1994): 50.
21. Stankey et al., *Learning to Learn*.
22. Shannon, Sturtevant, and Trask, *An Independent Assessment*.
23. E. T. Tuchmann et al., *The Northwest Forest Plan:* A Report to the President and Congress (Portland, OR: U.S. Department of Agriculture, Forest Service, Pacific Northwest Research Station, 1996).
24. Shannon, Sturtevant, and Trask, *An Independent Assessment*.
25. Canadian Model Forest Network, *Canadian Model Forest Network* (1999). http://www.modelforest.net (10 June 2001).
26. C. Tollefson (ed.), *The Wealth of Forests: Markets, Regulation, and Sustainable Forestry* (Vancouver: University of British Columbia Press, 1999).
27. G. Hoberg, "How the way we make policy governs the policy we make," in *Sustaining the Forests of the Pacific Coast*, ed. D. J. Salazar and D. K. Alper (Vancouver: University of British Columbia Press, 2000).
28. Tollefson, *Wealth of Forests*; J. Rayner, "Implementing sustainability in West Coast Forests: CORE and FEMAT as experiments in process," *Journal of Canadian Studies* 31, no. 1 (1996): 82-101.
29. J. M. Beyers, "The Forest Unbundled: Canada's National Forest Strategy and Model Forest Program 1992-1997," (Ph.D. dissertation, York University, 1998).
30. Canadian Council of Forest Ministers, *Sustainable Forests: A Canadian Commitment* (Hull, Québec, 1992).
31. Tollefson, *Wealth of Forests*.

32. *Ibid.;* A. John Sinclair and D. L. Smith, "The Model Forest Program in Canada: building consensus on sustainable forest management," *Society and Natural Resources Journal* 12 (1999): 121-38; Rayner, "Implementing sustainability;" K. Lertzman, J. Rayner, and J. Wilson, "Learning and change in the British Columbia forest policy sector: a consideration of Sabatier's Advocacy Coalition Framework," *Canadian Journal of Political Science* 29 (1996): 111-33.
33. Sinclair and Smith, "Model Forest Program."
34. Beyers, "Forest Unbundled."
35. *Ibid.*; D. Hebert, "Managing Alberta's boreal mixed-wood ecosystems," *Ecosystem Management* (October, 1994): 40-42.
36. Sinclair and Smith, "Model Forest Program."
37. *Ibid.*
38. *Ibid.*
39. *Ibid.*
40. Tuchmann et al., *Northwest Forest Plan.*
41. USDA, 1998
42. Canadian Model Forest Network, *Canadian Model Forest Network.*
43. *Ibid.*
44. *Ibid.*
45. *Ibid.*
46. FEMAT, *Forest Ecosystem Management*; Tuchmann et al., *Northwest Forest Plan*; Pipkin, *Northwest Forest Plan.*
47. Sinclair and Smith, "Model Forest Program."
48. *Ibid.*
49. *Ibid.*
50. G.H. Stankey and B. Shindler, *Adaptive Management Areas: Achieving the promise, avoiding the peril* Gen. Tech. Rep. PNW-GTR-394 (Portland, OR: U.S. Department of Agriculture, Forest Service, Pacific Northwest Research Station, 1997).
51. Stankey and Shindler, *Adaptive Management Areas.*
52. Beyers, "Forest Unbundled."

Chapter 12

1. Federal Ecosystem Management Team (FEMAT), *Forest Ecosystem Management: An Ecological, Economic, and Social Assessment* [Portland, OR: U.S. Department of Agriculture, U.S. Department of the Interior (and others)], III-27.
2. B. Barber, *Strong Democracy: Participatory Politics for a New Age* (Berkeley, CA: University of California Press, 1984); W. M. Lunch, *The Nationalization of American Politics* (Berkeley, CA: University of California Press, 1987).
3. G. H. Stankey and B. Shindler, *Adaptive Management Areas: Achieving the Promise, Avoiding the Peril* Gen. Tech. Rep. PNW-GTR-394 (Portland, OR: U.S. Department of Agriculture, Forest Service, Pacific Northwest Research Station, 1997).
4. B. Shindler and J. Neburka, "Public participation in forest planning: eight attributes of success," *Journal of Forestry* 91, no. 7 (1997): 17-19.
5. B. Shindler and A. Wright, *Watershed Management in the Central Cascades: A Study of Citizen Knowledge and the Value of Information Sources* (Corvallis: OR, Oregon State University, 2002); B. Shindler and E. Toman, *A Longitudinal Analysis of Fuel Reduction in the Blue Mountains: Public Perspectives on the Use of Prescribed Fire and Mechanical Thinning* (Corvallis, OR: Oregon State University Research Report, 2002).
6. B. Shindler, K. A. Cheek, and G. H. Stankey, *Monitoring and Evaluating Citizen-Agency Interactions: A Framework Developed for Adaptive Management* Gen. Tech. Rep. PNW-GTR-042 (Portland, OR: U.S. Department of Agriculture, Forest Service, Pacific Northwest Research Station, 1999).
7. W. E. Shands, "Public involvement, forest planning, and leadership in a community of interests," in *Proceedings of American Forestry: An Evolving Tradition. Society of American*

Foresters' National Convention (Richmond, VA, 1990), 360-67; R. Lawrence, S. E. Daniels, and G. Stankey, "Procedural justice and public involvement in natural resources decision making," *Society and Natural Resources* 10, no. 6 (1997): 577-89; B. Shindler and K. Aldred-Cheek, "Integrating citizens in adaptive management: a propositional analysis," *Journal of Conservation Ecology* 3, no. 1 (1999): 13. http://www.consecol.org/vol13/iss1/art13

8. For example, D. J. Blahna and S. Yonts-Shepard, "Public involvement in resource planning: toward bridging the gap between policy and implementation," *Society and Natural Resources* 2 (1989): 209-27; J. Delli Priscoli and P. Homenuck, "Consulting the publics," in *Integrated Approaches to Resource Planning and Management* ed. R. Lang (City unknown, AB: The Banff Centre School of Management, 1990); J. Wondolleck and S. Yaffee, *Building Bridges Across Agency Boundaries: In Search of Excellence in the U.S. Forest Service. University of Michigan Report* (Ann Arbor, MI, 1994); Shindler and Neburka, "Public participation."
9. Shindler, Cheek, and Stankey, *Monitoring and Evaluating.*
10. *Ibid.*
11. H. J. Cortner et al., *Institutional Barriers and Incentives for Ecosystem Management: A Problem Analysis* Gen. Tech. Rep. PNW-GTR-354 (Portland, OR: U.S. Department of Agriculture, Forest Service, Pacific Northwest Research Station, 1996).
12. Shindler and Neburka, "Public participation."
13. Cortner et al., *Institutional Barriers.*
14. Delli Priscoli and Homenuck, "Consulting the publics."
15. Stankey and Shindler, *Adaptive Management Areas.*
16. Shindler and Toman, *Longitudinal Analysis*; R. L. Williams, "Public knowledge, preferences and involvement in adaptive ecosystem management" (Master's Thesis, Oregon State University, 2001).
17. K. N. Lee, *Compass and Gyroscope: Integrating Science and Politics for the Environment* (Covelo, CA: Island Press, 1993).
18. Stankey and Shindler, *Adaptive Management Areas.*
19. Lee, *Compass and Gyroscope*

Chapter 13

1. B. Cashore, G. Auld, J. Lawson, and D. Newsom, "Forest Certification (Eco-labeling) Programs and their Policy-Making Authority: Explaining Divergence Among North American and European Case Studies." Paper presented at the International Seminar in Political Consumerism, City University and Stockholm University, Stockholm 2001; E. Meidinger, "Law Making by Global Civil Society: The Forest Certification Prototype." SUNY University at Buffalo, 2001. Available at http://www.law.buffalo.edu/eemeid. 2001
2. R. Crossley, "A Review of Global Forest Management Certification Initiatives: Political and Institutional Aspects" in *UBC-UPM Conference on the Ecological, Social and Political Issues of the Certification of Forest Management Proceedings*, Putrajay, Selangor, Malaysia, 1996.
3. S. Bass, K. Thornber, M. Markopoulos, S. Roberts, and M. Greig-Gran. *Certification's Impacts on Forests, Stakeholders and Supply Chains.* May 2001. A report to the ILED project: Instruments for Sustainable Private Sector Forestry.
4. C. Elliott, *Forest Certification: Analysis from a Policy Network Perspective.* Rural Engineering. Ph. D., Department de genie rural, Ecole Polytechnique Federale de Lausanne, Lausanne, Switzerland, 1999.
5. C. Maser and W. Smith, *Forest Certification in Sustainable Development: Healing the Landscape* (Boca Raton, FL: CRC Press, 2001); V. M. Viana, J. Ervin, R.Z. Donovan, C.

Elliott and H. Gholz (eds) *Certification of Forest Products: Issues and Perspectives* (Boca Raton: CRC Press. 1996).
6. Maser and Smith 2001; Viana et al 1996.
7. Elliott 1999; B Cashore et al 2001.
8. B. Cabarle, R. Hrubes, C. Elliott, and T. Synott. "Certification Accreditation: The Need for Credible Claims," *Journal of Forestry* 93 no. 4 (1995): 12-16.
9. Forest Stewardship Council (FSC). 2000. FSC Principles and Criteria, Available at http://www/fscoax.org/principal.htm
10. FSC-US. 2001. *The FSC-US National Indicators for Forest Certification.* Available at http://www.fscstandards.org
11. *Ibid.*
12. Rainforest Action Network (RAN), "Home Depot Announces Commitment to Stop Selling Old Growth Wood: Landmark policy change could signal end to old growth logging industry." Press release, Aug. 26, 1999. Available at http://www/ran.org/news/newsitem.php?id=74
13. Bass *et al.* 2001; Home Depot, 2001. Company Information, Environment. Web page. Available at http://www.homedepot.com; Certified Forest Products Council (CFPC), 2001. Certification Resource Center. Web page. Available at http://www.certifiedwood.org
14. CFPC 2001
15. S. Berg, R. Cantrell, "Sustainable Forestry Initiative: Towards a Higher Standard," *Journal of Forestry*, 97 no. 11 (1999): 33-35; G. Lapointe, "Sustainable Forest Management Certification: The Canadian Programme," *The Forestry Chronicle*, 74 no. 2 (1998): 227-30; R. Vlosky and L. Ozanne, "Environmental Certification of Wood Products: The US Manufacturer's Perspective," *Forest Products Journal*, 48 no. 9 (1998): 21-26.
16. World Resource Institute (WRI). *Canada's Forests at a Crossroads: An Assessment in the Year 2000* (Victoria: World Resources, 2000).
17. ISO, 2001. ISO Standards Development. Web site. Available at http://www/iso.ch/iso/en/stdsdevelopment/whowhenhow/proc/proc.html
18. R. Krut and H. Gleckman, *ISO 14001: A Missed Opportunity for Sustainable Global Industrial Development* (London: Earthscan Publications Ltd, 1998).
19. Elliott, 1999; P. Hauselmann, "ISO inside out: ISO and environmental management." A WWF International Discussion Paper, 1997; T. Takahashi, *Why Firms Participate in Environmental Voluntary Initiatives: Case Studies in Japan and Canada.* Ph.D., Resource Management and Environmental Studies, Faculty of Graduate Studies, UBC, Vancouver, BC, 2001.
20. Hauselmann 1997.
21. Canadian Sustainable Forestry Certification Coalition, 2002. *Certification Status in Canada.* Retrieved November 30, 2002 at http:www.sfms.com
22. Quality Times, 2001. "ISO 14001 Registrations Break 2,000 Barrier in North America." Available at http://www/thequality times.com/Site/English/iso14001.html
23. Weyerhaeuser, 2001. "Weyerhaeuser Western U.S. Timberland Archives. Environmental Certification; One of the world's largest single blocks or private forest to be certified." Press Release. November 7, 2001. Available at http://www/weyerhasuser.com
24. G. Auld, *Explaining Certification Legitimacy: An Examination of Forest Sector Support for Forest Certification Programs in the United Kingdom, the United States Pacific Coast, and British Columbia, Canada.* Master's Thesis, Auburn University, Auburn, AL 2001; D. Newsom, *Achieving Legitimacy? Exploring Competing Forest Certification Programs' Efforts to Gain Forest Manager Support in the US Southeast, Germany and British*

Columbia, Canada. Master's Thesis, Auburn University, Auburn, AL 2001.
25. Canadian Standards Association, Forest Marketing Program. CAN/CSA-Z808-96, CSA 1996.
26. *Ibid.*; Meridian Institute, *Comparative Analysis of the Forest Stewardship Council and Sustainable Forestry Initiative Certification Programs*. A Report sponsored by the Forest Stewardship Council-US, The Home Depot, The Sustainable Forestry Initiative of the American Forest and Paper Association, 2001.
27. WRI 2000.
28. Elliott 1999.
29. Canadian Standards Association (CSA). 2002. *Certification and Testing*. Forest Product Marking Program. Web page. Available at http://www.csa-international.org/certification/forestry
30. Auld 2001; Elliott 2001; Takahashi 2001.
31. Auld 2001; Cashore et al. 2001
32. S. Wallinger, "A Commitment to the Future: AF&PA's Sustainable Forestry Initiative," *Journal of Forestry*, 93 no. 1 (1995): 16-19.
33. Newsom 2001.
34. Meridian 2001.
35. American Forest Foundation. *AFT's American Tree Farm System and AF&PA's Sustainable Forest Initiative (SFI) Program Collaborate to Expand the Practice of Sustainable Forestry: Mutual Recognition of Forest Management Standards and Certification Systems*. Press Release. June 27, 2000. Available at http://www/treefarmsystem.org/commcorner/06.27.00.html
36. American Forests and Paper Association. *Sustainable Forestry Initiative*. Web page. Retrieved 30 November 2002. http://www.afandpa.org/forestry/sfi_frame2.html.
37. Meridian 2001.
38. G. Hoberg. "How the Way We Make Policy Governs the Policy We Choose," in *Sustaining the Forests of the Pacific Northwest: Forging Truces in the War in the Woods*. Alper, D. and D. Salazar, eds. (Vancouver, University of British Columbia Press, 2000).
39. Forest Ecosystem Management Assessment Team (FEMAT). Forest *Ecosystem Management and Ecological, Economic, and Social Assessment: Report of the Forest Ecocsystem Management Assessment Team*. Produced jointly by the USDA Forest Service, USDC National Oceanic and Atmospheric Administration, USDS National Marine Fisheries Service, USDI Bureau of Land Management, USDI Fish and Wildlife Service, USDI National Parks Service, and the Environmental Protection Agency. (Washington, D.C: US Government Printing Office, 1994).
40. Sierra Club. 1996. *Zero Cut*. Available at http://illinois.sierraclub.org/piasapalisades/zerocut.htm
41. These three local organizations were the Institute of Sustainable Forestry, in California; the Rogue Institute of Ecology and Economy, in Oregon; and the Northwest Natural Resource Group (formerly the Olympic Peninsula Foundation), in Washington.
42. Forest Stewardship Council (FSC). 2000. "Forest Stewardship Standards for the Maritime Forest Region Commission of Enquiry: Final Report."
43. Sierra Club of Canada, 1998. Available at http://www/sierraclub.ca/national/media/appeal-jdi-se98.htm
44. "First Nations" is a term commonly used in Canada to refer to the country's aboriginal peoples.
45. J. Lawson and B. Cashore. "Firm Choices on Sustainable Forestry Forest Certification: The Case of J. D. Irving, Ltd.," in *Forest Policy for Private Forestry: Global and Regional Challenges*, L. Teeter, B. Cashore and D. Zhang, eds. (Wallingford, England: CABI Publishing, in press).
46. *Ibid.*
47. FSC 2000b.
48. Sierra Club of Canada 1998.

49. S. Hammond. "Silva Forest Foundation Conducts Canada's First Wood Certification," *International Journal of Eco-forestry*, 2 no. 4 (1995).
50. Auld 2001.
51. RAN 1999.
52. D. Hogben, "Western Seeks Certification to Satisfy European Buyers." *Vancouver Sun*, June 5, 1998.
53. R Kunin, Introduction and Summary, *Prospering Together: The Economic Impact of the Aboriginal Title Settlements in B.C.* (Vancouver: The Laurier Institute, 1998).
54. FSC-BC, *Forest Stewardship Council (FSC) Regional Certification Standards for British Columbia*, Draft One, 1998.
55. FSC-BC. 2001. *FSC British Columbia*. Web Page. Available at http://www.fsc-bc.org
56. Forest Stewardship Council (FSC). 2001. List of FSC Members. Web Page. Available at http://www.fscoax.org/html/available_documents.html
57. FSC-BC. *Forest Stewardship Council (FSC) Regional Certification Standards for British Columbia*, Draft 3, 2002.
58. Canadian Sustainable Forestry Certification Coalition 2002.
59. Quality Times 2001.
60. Certified Forest Products Council. Web page. http://www.certifiedwood.org/search-modules/compare-systems/comparison-of-systems/content.asp. Retrieved 30 November 2002.
61. *Ibid.*
62. Cashore et al. 2001
63. K. A. Vogt, B. C. Larson, J. C. Gordon, and D. J. Vogt, *Forest Certification: Roots, issues, challenges and benefits* (Boca Raton, FL: CRC Press, 1996).
64. These figures are based on the perspective that the number of certificates is at least as critical in understanding certification dynamics as is the area certified. The fact that, worldwide, the certification of a few very large landowners has dominated total land areas certified is a reflection of land distribution patterns as much as a commentary on certification.

Chapter 14

1. B. Wilson et al., eds., *Forest Policy: International Case Studies* (New York: CABI Publishing, 1999).
2. R. A. Sedjo, A. Goetzl, and S. O. Moffat, *Sustainability of Temperate Forests* (Washington, DC: Resources for the Future, 1998).
3. R. A. Sedjo, "The Forest Sector: Important Innovations," Discussion Paper 97-42 (Washington, DC: Resources for the Future, 1997).
4. R. J. Moulton, "Forestry in U.S. Climate Change Action Plans: from the Arch to Kyoto," in *Proceedings of the 1998 Forest Economics Workshop, March 25-27, 1998*, ed. K. L. Abt and R. C. Abt (Williamsburg, VA, 1998), 204-7; R. A. Sedjo, "Forest 'sinks' as a tool for climate change policy making: a look at the advantages and challenges," *Resources* 143 (Spring, 2001): 21-23.
5. I. J. Bourke, and J. Leitch, *Trade Restrictions and Their Impact on International Trade in Forest Products* (Rome, FAO: 1998); N. Sizer, D. Downes, and D. Kaimowitz, *Tree Trade: Liberalization of International Commerce in Forest Products: Risks and Opportunities* (Washington, DC: World Resources Institute, 1999).
6. The government of Florida, under Florida Forever program, has committed to spend $3 billion between 2001 and 2010 to acquire rural lands for preservation.
7. R. A. Sedjo and D. Botkin, "Using forest plantations to spare natural forests," *Environment* 39, no. 10 (1997): 14-20, 30; D. L. Dekker-Robertson, "American forestry policy – global ethics tradeoffs," *BioScience* 48, no. 6 (1998): 471-78.
8. G. C. van Kooten and E. H. Bulte, *The Economics of Nature: Managing Biological Assets* (Oxford, England: Blackwell Publishers Inc., 2000).
9. B. Shongen and R. Mendelsohn, "Valuing the impact of large scale ecological change in a market: the effect of climate change on U.S.

timber," *American Economic Review* 88, no. 4 (1998): 686-710.

10. G. C. van Kooten, C. S. Binkley, and G. Delcourt, "Effect of carbon taxes and subsidies on optimal forest rotation age and supply of carbon services," *American Journal of Agricultural Economics* 77 (1995): 365-74.
11. G. A. Stainback and J. R. R. Alavalapati, "Economic analysis of slash pine carbon sequestration: implications for private forest landowners in the U.S. South," *Journal of Forest Economics* (in review, 2002).
12. *Ibid.*
13. M. E. Harmon, W. K. Ferrell, and J. F. Franklin, "Effects on carbon storage of conversion of old-growth forest to young forest," *Science* 247, no. 4943 (1990): 699-701.
14. J. R. R. Alavalapati and G. Wong, *Forest Carbon Sequestration Policies: Implications for Forest Management in Canada and the U.S.* Project report to the USDA Research and Scientific Exchanges Division, under the Scientific Cooperation Program (2002).
15. They chose this scenario to maintain consistency with the Kyoto Protocol, where commitments to reduce CO^2 emissions are required by the OECD countries only.
16. Alavalapati and Wong, *Forest Carbon Sequestration.*
17. *Ibid.*
18. L. Wu et al., *Assessing the Impact of Trade Policy and Technology Change in U.S. Forestry Sectors* Proceedings of the Southern Forest Economics Workers Annual Conference, March 26-27, 2001 (Atlanta, GA, 2001): 84-89.
19. J. Gan and S. Ganguli, *Global Trade Liberalization and Forest Product Trade Patterns* Proceedings of the Global Initiatives and Public Policies: First International Conference on Private Forestry in the 21st Century, March 25-27, 2001 (Atlanta, GA, submitted, 2001).
20. Wu et al., *Assessing the Impact*; Gan and Ganguli, *Global Trade Liberalization.*
21. B. Cashore, *Flights of the Phoenix: Explaining the Durability of the Canada-U.S. Softwood Lumber Dispute,* Canadian-American Public Policy No. 32 (December, 1997).
22. J. R. R. Alavalapati, W. L. Adamowicz, and W. White, "Random variables in forest policy: a systematic sensitivity analysis using CGE models," *Journal of Forest Economics* 5, no. 2 (1999): 321-35.
23. D. Zhang, "Welfare impacts of the 1996 United States – Canada Softwood Lumber (trade) Agreement," *Canadian Journal of Forest Research* 31 (2001): 1958-67.
24. American consumption of wood and wood products is expected to increase from 530 million cubic meters in 1991 to nearly 580 million cubic meters by the year 2000 (Dekker-Robertson, "American forestry policy").
25. Biotechnology is broadly defined as the use of biological organisms, systems, and processes for practical or commercial purposes (P. C. Trotter, "Biotechnology and the economic productivity of commercial forests," *Tappi Journal* 69, no. 7 (1986): 22-28). Another definition is "the commercial application of living organisms or their products, which involves the deliberate manipulation of their DNA molecules" (R. A. Sedjo, *Biotechnology and Planted Forests: Assessment of Potential and Possibilities* Discussion Paper 00-06 (Washington, DC: Resources for the Future, 1999).
26. Sedjo and Botkin, "Using forest plantations;" Dekker-Robertson, "American forestry policy."
27. C. Gaston, S. Globerman, and I. Vertinsky, "Biotechnology in forestry: technological and economic perspectives," *Technological Forecasting and Social Change* 50, no. 1 (1995): 79-92.
28. *Ibid.*
29. *Ibid.*; D. Parker and D. Zilberman, "Biotechnology and future of

agriculture and natural resources—an overview," *Technological Forecasting and Social Change* 50, no. 1 (1995): 1-7.

30. Trotter, "Biotechnology and the economic."
31. R. A. Sedjo, "Land use changes and innovation in U.S. forestry," in *Productivity in Natural Resource Industries: Improvement Through Innovation*, ed. R. David Simpson (Washington, DC: Resources for the Future, 1999).
32. Sedjo, "Biotechnology and Planted Forests."
33. G. G. Das and J. R. R. Alavalapati, "Trade mediated biotechnology transfer and its effective absorption: an application to the U.S. Forestry Sector," *Technological Forecasting and Social Change* (accepted, in press, 2002).
34. See Thomas W. Hertel, ed., *Global Trade Analysis: Modeling and Applications* (Cambridge University Press, 1997) for more details about the model.
35. I. W. H. Parry, *Productivity Trends in the Natural Resources Industries* Discussion Paper 97-39 (Washington, DC: Resources for the Future, July, 1997); Das and Alavalapati, "Trade mediated biotechnology." In the results reported, they consider that technological change occurs only in the United States. However, consideration of such technological change in Canada or Western Europe may generate similar results by differing only in the magnitude. For space limitations, we do not elaborate technological changes in other regions as well.
36. Das and Alavalapati, "Trade mediated biotechnology."
37. J. Kaiser, "Words (and axes) fly over transgenic trees," *Science* 292 (6 April 2001).
38. M. Palo, J. Uusivuori, and G. Mery, "World forests, markets and policies: towards a balance," in *World Forests, Markets and Policies*, ed. M. Palo, J. Uusivuori, and G. Mery (Dordrecht, London, Boston: Kluwer Academic Publishers, 2001).
39. J. Uusivuori and S. Laaksonen-Craig, *Foreign Direct Investment, Exports and Exchange Rates: The Case of the U.S., Finnish and Swedish Forest Industries* (Helsinki: Finnish Forest Research Institute, 2000).
40. K. S. Ho, *Workplace Improvements Program for the Furniture Industry in Malaysia* Paper presented at the World Congress of Forest Research Organizations (IUFRO) (Kuala Lumpur, Malaysia, 7-12 August 2000).
41. C. Barr, *The Political Economy of Fiber, Finance and Debt in Indonesia's Pulp and Paper Industries* (Bogor, Indonesia: Center for International Forestry Research, 2000).
42. International Labor Organization (ILO), *Globalization and Sustainability: The Forestry and Wood Industries on the Move* Report for discussion at the Tripartite Meeting on the Social and Labour Dimensions of the Forestry and Wood Industries on the Move (Geneva, 2001).
43. *Ibid.*
44. *Ibid.*
45. UNCTAD, *World Investment Report 2000* (New York and Geneva: United Nations, 2000).
46. European Union, *Panorama of EU Industry* (1997) (Brussels: European Commission, 1997).

Chapter 15

1. J. W. Thomas, "Ecosystem Management," *Natural Resource News,* Special Edition. (LaGrande, OR: Blue Mountain Natural Resource Institute, 1993), 19.
2. Unidentified federal forest manager, cited in M. Shannon and A. Antypas, "Open institutions: uncertainty and ambiguity in 21st-century forestry," in *Creating a Forestry for the 21st Century,* ed. K. Kohm and J. Franklin (Washington, DC: Island Press, 1997), 438.
3. U.S. Fish and Wildlife Service, *An Ecosystem Approach to Fish and Wildlife Conservation: Concept Document* (Washington, DC, 1995).

4. R.E. Grumbine, "What is ecosystem management," *Conservation Biology* 8, no. 1 (1994): 27-38.
5. e.g., D. Slocombe, "Implementing ecosystem-based management: development of theory, practice, and research for planning and managing a region," *Bioscience* 43, no. 9 (1993): 612-21.
6. V. Ostrom, *The Meaning of Democracy and the Vulnerability of Democracies: A Response to Tocqueville's Challenge* (Ann Arbor: University of Michigan Press, 1997).
7. J. Pierce and N. Lovrich, "Trust in the technical information provided by interest groups: the views of legislators, activists, experts, and the general public. *Policy Studies Journal* 11 (1983): 627.
8. L. Bardwell, "Problem-framing: a perspective on environmental problem-solving," *Environmental Management* 15, no. 5 (1991): 603-12.
9. R. Bellah et al., *The Good Society* (New York: Alfred A. Knopf, 1991): 10-11.
10. H. Cortner et al., *Institutional Barriers and Incentives for Ecosystem Management: A Problem Analysis,* Gen. Tech. Rep. PNW-GTR-354 (Portland, OR: U. S. Department of Agriculture, Forest Service, Pacific Northwest Research Station, 1996), 8.
11. E. Ostrom, "An agenda for the study of institutions," *Public Choice* 48 (1986): 3-25.
12. S. Trask Dana and S. K. Fairfax, *Forest and Range Policy: Its Development in the United States,* second edition (New York: McGraw-Hill Book Company, 1980).
13. H. Cortner and M. Moote, *The Politics of Ecosystem Management* (Washington, DC: Island Press, 1999).
14. J. Nienaber Clark and D. McCool, *Staking Out the Terrain: Power and Performance Among Natural Resource Agencies,* second edition (Albany, NY: State University of New York Press, 1996).
15. C. Wood, "Ecosystem management: achieving the new land ethic," *Renewable Resources Journal* 12, no. 1 (1994): 6-12.
16. R. O'Toole, *Reforming the Forest Service* (Washington, DC: Island Press, 1988).
17. R. Nelson, *Public Lands and Private Rights: The Failure of Scientific Management* (Lanham, MD: Rowman & Littlefield, 1995).
18. Cortner et al., *Institutional Barriers.*
19. J. Forester, *Critical Theory, Public Policy, and Planning Practice: Toward a Critical Pragmatism* (Albany, NY: State University of New York Press, 1993).
20. J. Wondolleck, *Public Lands Conflict and Resolution: Managing National Forest Disputes* (New York: Plenum Press, 1988), 136.
21. R.W. Behan, "The succotash syndrome, or multiple use: a heartfelt approach to forest land management," *Natural Resources Journal* 7, no. 4 (1967); George R. Hall, "The myth and reality of multiple use forestry," *Natural Resources Journal* 3 (1963): 276-90.
22. Dana and Fairfax, *Forest and Range Policy.*
23. Wondolleck, *Public Lands Conflict,* 136
24. University of Montana Select Committee, *A University View of the Forest Service* (Washington, DC: GPO, 1970).
25. cited in R. Wambach, "Forestry in the environmental seventies," in *Forest Land Use and the Environment,* ed. R. Weddle (Missoula, MT: Montana Forest and Conservation Experiment Station, School of Forestry, University of Montana, 1971), 64.
26. Wondolleck, *Public Lands Conflict.*
27. Interagency Scientific Committee (ISC), *A Conservation Strategy for the Northern Spotted Owl* (Portland, OR: USDA and USDI, 1990), 427.
28. K. N. Johnson et al., *Bioregional Assessments: Science at the Crossroads of Management and Policy* (Washington, DC: Island Press, 1999).
29. A. Raedeke, J. Rikoon, and C. Nilon, "Ecosystem management and landowner concern about regulations: a case study in the Missouri Ozarks," *Society and Natural Resources* 14, no. 9 (2001): 741-59.

30. Raedeke et al., "Ecosystem Management," 742.
31. T. O'Keefe, "Holistic (new) forestry: significant difference or just another gimmick?" *Journal of Forestry* 88, no. 4 (1990): 23-24.
32. B. Norton, "A new paradigm for environmental management," in *Ecosystem Health: New Goals for Environmental Management*, ed. R. Costanza, B. Norton, and B. Haskell (Washington, DC: Island Press, 1992), 23-41; W. Kessler et al., "New perspectives for sustainable natural resources management," *Ecological Applications* 2, no. 3 (1992): 221-25.
33. R. Lackey, "Ecosystem management: desperately seeking a paradigm," *Journal of Soil and Water Conservation* 53, no. 2 (1998): 92-94.
34. M. Moote et al., *Principles of Ecosystem Management* (Tucson, AZ: Water Resources Research Center, University of Alabama, 1994).
35. D. Bengston, "Changing forest values and ecosystem management," *Society and Natural Resources* 7, no. 6 (1994): 517.
36. Wondolleck, *Public Lands Conflict.*
37. J. Thompson and A. Tuden, "Strategies, structures, and processes of organizational decisions," in *Comparative Studies in Administration,* ed. J. Thompson et al. (New York: Garland Publishing Company, 1987), 197-216.
38. K. Lee, *Compass and Gyroscope: Integrating Science and Politics for the Environment* (Washington, DC: Island Press, 1993).
39. Thompson and Tuden, "Strategies, structures, and processes," 202.
40. M. Schwarz and M. Thompson, *Divided We Stand: Redefining Politics, Technology, and Social Choice* (Philadelphia: University of Pennsylvania Press, 1990).
41. D. Yankelovich, *Coming to Public Judgment: Making Democracy Work in a Complex World* (Syracuse, NY: Syracuse University Press, 1991).
42. Shannon and Antypas, *Open Institutions.*
43. A. Miller, "The role of analytical science in natural resource decision making," *Environmental Management* 17, no. 5 (1993): 563-74.
44. Thompson and Tuden, "Strategies, structures, and processes."
45. *Ibid.*, 202.
46. E. Roe, *Taking Complexity Seriously: Policy Analysis, Triangulation and Sustainable Development* (Boston, MA: Kluwer Academic Publishers, 1998).
47. *Ibid.*, 3
48. *Ibid.*, 4
49. *Ibid.*, 4
50. Miller, "Role of Analytical Science."
51. H. J. Rittel and M. M. Webber, "Dilemmas in a general theory of planning," *Policy Sciences* 4 (1973): 155-69.
52. R. L. Ackoff, *Redesigning the Future: A Systems Approach to Societal Problems* (New York: Wiley, 1974).
53. G. Allen and E. Gould, Jr., "Complexity, wickedness, and public forests," *Journal of Forestry* 84, no. 4 (1986): 20-23.
54. R. Socolow, "Failures of discourse: obstacles to the integration of environmental values into natural resource policy," in *When Values Conflict: Essays on Environmental Analysis, Discourse, and Decisions,* ed. L. H. Tribe, C. S. Schelling, and J. Voss (Cambridge, MA: Ballinger Publishing Company, 1976), 2.
55. John Friedmann *Planning in the Public Domain: From Knowledge to Action* (Princeton, NJ: Princeton University Press, 1987), citing Lindblom (1979) on page 32.
56. Socolow, "Failures of discourse."
57. C. Perrow, *Normal Accidents: Living with High Risk Technologies* (New York: Basic Books, 1984).
58. D. Botkin, *Discordant Harmonies: A New Ecology for the Twenty-First Century* (New York: Oxford University Press, 1990).
59. J. Dryzek, *Rational Ecology: Environment and Political* Economy (Oxford: Basil Blackwell Ltd., 1987), 181.

60. Roe, *Taking Complexity Seriously.*
61. Lee, *Compass and Gyroscope.*
62. A. Wildavsky, *Searching for Safety* (Bowling Green, OH: Social Philosophy and Policy Center, 1988).
63. Friedmann, *Planning in the Public Domain.*
64. Roe, *Taking Complexity Seriously*, 5
65. L. Caldwell, *Between Two Worlds: Science, the Environmental Movement and Policy Choice* (Cambridge: Cambridge University Press, 1990).
66. Thompson and Tuden, "Strategies, structures and processes," 202.
67. Roe, *Taking Complexity Seriously*, 87
68. D. Michael, "Barriers and bridges to learning in a turbulent human ecology," in *Barriers and Bridges to the Renewal of Ecosystems and Institutions*, ed. Lance H. Gunderson, C. S. Holling, and Stephen S. Light (New York: Columbia University Press, 1995), 468.
69. J. Welles, "The survival advantage of stupidity," *Speculation in Science and Technology* 7, no. 1 (1984): 17-21.
70. The planning continuum depicted in these figures is similar to the distinction drawn by Friedmann (*Planning in the Public Domain*) between social reform planning (steeped in rationality, calculation and control, and objective, value-free analysis), and social learning, (emphasizing the interplay of actions, political strategy and tactics, theories of reality, and values).
71. Friedmann, *Planning in the Public Domain*, 44.
72. Polanyi 1962, cited in *Ibid.*, 43.
73. J. Kloppenburg, Jr., "Social theory and the de-reconstruction of agricultural science: a local knowledge for an alternative agriculture," *Rural Sociology* 56 (1991): 528.
74. G. H. Stankey, P. J. Brown, and R. N. Clark, "Allocating and managing for diverse values of forests: the market place and beyond," in *Proceedings from the IUFRO International Conference, Integrated Sustainable Multiple-Use Forest Management Under the Market System*, compilers N. E. Koch and N. A. Moiseev (Copenhagen, Denmark: The Danish Agricultural and Veterinary Research Council, 1992): 257-71.
75. Friedmann, *Planning in the Public Domain*; Shannon and Antypas, *Open Institutions.*
76. Grumbine, "What is Ecosystem Management;" Cortner and Moote, *Politics of Ecosystem Management.*
77. R.N. Clark et al., "Toward an ecological approach: integrating social, economic, cultural, biological, and physical considerations," in *Ecological Stewardship: A Common Reference for Ecosystem Management*, ed. N. C. Johnson et al. (Oxford: Elsevier Science Ltd., 1999), 3: 297-318.
78. R. N. Clark et al., *Integrating Science and Policy in Natural Resource Management: Lessons and Opportunities from North America* Gen. Tech. Rep. PNW-GTR-411 (Portland, OR: U.S. Department of Agriculture, Forest Service, Pacific Northwest Research Station, 1998).
79. Senge, *The Fifth Discipline: The Art & Practice of Learning Organizations* (New York: Doubleday/Currency, 1990), 289.
80. B. Lee, *The Power Principle: Influence With Honor* (New York: Fireside, 1997); Donald M. Michael, *On Learning to Plan—And Planning to Learn* (San Francisco: Jossey-Bass Publishers, 1973); Senge, *The Fifth Discipline.*
81. L. Gunderson, "Resilience, flexibility and adaptive management—antidotes for spurious certitude?" *Conservation Ecology* 3, no. 1 (1999): 7 [online].
82. R. Costanza et al., "Managing our environmental portfolio," *Bioscience* 50, no. 2 (2000) 149-55.
83. R. Paehlke and D. Torgerson, "Environmental politics and the administrative state," in *Managing Leviathan: Environmental Politics and the Administrative State*, ed. R. Paehlke and D. Torgerson (Peterborough, Ontario: Broadview Press, Ltd., 1990), 285-301.

84. Costanza et al., "Managing our environmental portfolio."
85. e.g., see Wildavsky, *Searching for Safety.*
86. Senge, *The Fifth Discipline*; J. Dentico, "Games leaders play: using process simulations to develop collaborative leadership practices for a knowledgeable based society." http://mcb.co.uk/services/conferen/spet98/lim/paper_b5.html (1998).
87. Dentico "Games leaders play."
88. Lee, *The Power Principle.*
89. Dentico "Games leaders play," 3.
90. H. Kaufman, *The Forest Ranger: A Study in Administrative Behavior* (Baltimore: The John Hopkins University Press, 1960); H. Kaufman, *The Limits of Organizational Change* (University, AL: University of Alabama Press, 1975).
91. This situation is changing. Today, resource agencies are increasingly diverse in terms of ethnicity, gender, and disciplinary training; all seem key to promoting more integrative, ecosystem-based management.
92. W. Bridges, *Managing Transitions: Making the Most of Change* (Reading, MA: Addison-Wesley Publishing Company, 1991) 3.
93. Clarke and McCool, *Staking out the Terrain.*
94. Senge, *The Fifth Discipline.*

Chapter 16

1. Beckley and Bonnell, and McFarlane et al., in this volume, chapters 9 and 7; B. Shindler, P. List and B. S. Steel, "Managing federal forests: public attitudes in Oregon and nationwide," *Journal of Forestry* 91, no. 7 (1992): 36-42.
2. J. P. Kimmins, "Future shock in forestry," *The Forestry Chronicle* (2002); F. L. Bunnell, "Operation criteria for sustainable forestry: Focussing on the essence," *The Forestry Chronicle* 73, no. 6 (1997): 679-684; J. W. Thomas, "Are there lessons for Canada lurking south of the border?" Presentation at the University of Alberta. http://www.apbbc.bc.ca/htdocs/LessonsforForestersinCanada.htm.
3. J. P. Kimmins, "Future shock in forestry."
4. T. M. Beckley, "The nestedness of forest-dependence: A conceptual framework and empirical exploration," *Society and Natural Resources* 11, no. 2 (1998): 101-120.
5. Alavalapati et al. in this volume, chapter 14.
6. A. Gore, *Earth in the Balance: Ecology and the Human Spirit* (Boston, MA: Houghton Mifflin, 1992).
7. M. Wackernagel and W. Rees, *Our Ecological Footprint: Reducing Human Impact on the Earth* (Gabriola Island, BC: New Society Publishers); Editorial Services, Communications New Brunswick, "Forest certification to be implemented on Crown land by 2003," (Fredericton, NB: CANADA, 2002).
8. S. Funtowicz and J. R. Ravetz, *Three Types of Risk Assessment and the Emergence of Post-Normal Science,* in S. Krimsky and D. Golding (eds.), *Social Theories of Risk* (Westport, CT: Greenwood, 1992), 251-73.
9. F. Egler, "Commentary: 'Physics envy' in ecology," *Bulletin of the Ecological Society of America* 67 (1986): 233-35.
10. McDermott and Hoberg in this volume, chapter 13.
11. See *Ibid.*
12. M. Whelan, Communications officer with the Canadian Model Forest Network, Personal communication, 28 May 2002.
13. G. H. Stankey and B. Shindler, *Adaptive Management Areas: Achieving the Promise, Avoiding the Peril* Gen. Tech. Rep. PNW-GTR-394 (Portland, OR: U.S. Department of Agriculture, Forest Service, Pacific Northwest Research Station, 1997).
14. Bruce Shindler, Mark Brunson, and George Stankey, *Social Acceptability of Forest Conditions and Mangement Practices: A Problem Analysis* Gen. Tech. Rep. PNW-GTR-537 (Portland, OR: U.S. Department of Agriculture, Forest Service, Pacific Northwest Research Station, 2002).

15. T. Beckley and D. Korber. "Sociology's potential to improve forest management and inform forest policy." *The Forestry Chronicle* (1995) 71(6):712-19.
16. For more than a decade after Earth Day in the United States there were just a handful of non-economist social scientists studying North American Forestry issues in forestry, but a few, such as Don Field, Rabel Burdge, Bob Lee, Bill Burch, and in Canada Pat Marchak, were instrumental in breaking down the barriers between social science and forestry.
17. The answer, of course, is that sociology can help determine whether a stream crossing should be built.
18. J. P. Kimmins, "Future shock in forestry."
19. T. Erdle and M. Sullivan, "Forest management design for contemporary forestry," *The Forestry Chronicle* 74, no. 1 (1998): 83-90.
20. Alberta Sustainable Resource Management and Alberta Environment, New Brunswick Department of Natural Resources and Energy and New Brunswick Department of Environment, BC Ministry of Water, Air and Land Protection and Ministry of Forests, and the like. Recently, Saskatchewan Environment and Resource Management became simply, Saskatchewan Environment.
21. G. H. Stankey et al., "Adaptive Management and the Northwest Forest Plan: Rhetoric and Reality," *Journal of Forestry* (2003) 101 (1): 40-46.
22. R. Lang, "Achieving integration in resource planning," in *Integrated Approaches to Planning and Management*, ed. R. Lang (Banff, Alberta: Banff School of Management, 1990): 27-50.
23. Kai N. Lee, *Compass and Gyroscope: Integrating Science and Politics for the Environment* (Washington, DC: Island Press, 1993).
24. D. Yankelovich, *Coming to Public Judgment: Making Democracy Work in a Complex World* (Syracuse, NY: Syracuse University Press, 1991).
25. Stankey et al., "Adaptive Management."
26. A. Wildavsky, *Searching for Safety* (Bowling Green, OH: Social Philosophy and Policy Center, 1988).

The Authors

Janaki R. R. Alavalapati, Ph.D., *Assistant Professor, School of Forest Resources and Conservation, University of Florida*

Thomas M. Beckley, Ph.D., *Associate Professor, Faculty of Forestry and Environmental Management, University of New Brunswick*

Brian Bonnell, *Manager, Canada's Model Forest Program, Canadian Forest Service, Ottawa, Ontario*

Gary Q. Bull, Ph.D., *Assistant Professor, Department of Forest Resources Management, University of British Columbia*

Roger N. Clark, Ph.D. *Research Social Scientist, Human Natural Resources Interactions Program, USDA Forest Service Pacific Northwest Research Station, Seattle*

Gouranga G. Das, *Post-Doctoral Fellow, School of Forest Resources and Conservation, University of Florida*

Peter N. Duinker, Ph.D., *Professor and Director, School for Resource and Environmental Studies, Dalhousie University*

Mary Carmel Finley, *Doctoral Candidate, Department of Political Science, University of California at San Diego.*

Christina L. Herzog, *Doctoral Candidate, Department of Political Science, Washington State University*

George Hoberg, Ph.D. *Professor and Head, Department of Forest Resources Management, University of British Columbia*

Christina Kakoyannis, *Doctoral Candidate, Department of Forest Resources, Oregon State University*

Nicholas P. Lovrich, Jr., Ph.D., *Professor and Director, Division of Governmental Studies and Services, Department of Political Science, Washington State University*

Stephen F. McCool, Ph.D. *Professor, School of Forestry, University of Montana*

Connie McDermott, *Doctoral Candidate, Department of Forest Resources Management, University of British Columbia*

Bonita L. McFarlane, Ph.D., *Northern Forestry Centre, Canadian Forest Service, Edmonton*

Solange Nadeau, Ph.D., *Atlantic Forestry Centre, Canadian Forest Service, Fredericton, New Brunswick*

John R. Parkins, *Doctoral Candidate, Department of Sociology, University of Alberta and Northern Forestry Centre, Canadian Forest Service, Edmonton*

John C. Pierce, Ph.D., *Research Professor, Washington State University at Vancouver*

Clare M. Ryan, Ph.D., *Associate Professor, College of Forest Resources, University of Washington*

Bruce A. Shindler, Ph.D., *Associate Professor, Department of Forest Resources, Oregon State University*

George H. Stankey, Ph.D., *Research Social Scientist, Human Natural Resources Interactions Program, USDA Forest Service, Pacific Northwest Research Station, Corvallis*

Richard C. Stedman, Ph.D., *Assistant Professor, Department of Agricultural Economics and Rural Sociology, The Pennsylvania State University*

Brent Steel, Ph.D., *Professor, Department of Political Science, Oregon State University*

David O. Watson, M.S. *Northern Forestry Centre, Canadian Forest Service, Edmonton*

Edward O. Weber, Ph.D., *Associate Professor, Department of Political Science and Director, Thomas S. Foley Institute for Public Policy and Public Service, Washington State University*

Cynthia Wilkerson, *Research Assistant, Wildlife Ecology and Conservation, University of Florida*

Index